FOCUS ON ATHEROSCLEROSIS RESEARCH

Focus on Atherosclerosis Research

Leon V. Clark
Editor

Nova Biomedical Books
New York

Production Coordinator: Donna Dennis
Coordinating Editor: Tatiana Shohov
Senior Production Editors: Susan Boriotti and Donna Dennis
Production Editors: Marius Andronie and Rusudan Razmadze
Office Manager: Annette Hellinger
Graphics: Magdalena Nuñez
Editorial Production: Maya Columbus, Aloysha Klestov, Vladimir Klestov, Matthew Kozlowski and Lorna Loperfido
Circulation: Ave Maria Gonzalez, Vera Popovic, Luis Aviles, Jonathan Boriotti, Alexandra Columbus, Raymond Davis, Cathy DeGregory, Melissa Diaz, Marlene Nuñez, Jeannie Pappas and Frankie Punger
Communications and Acquisitions: Serge P. Shohov
✸

Library of Congress Cataloging-in-Publication Data
Focus on atherosclerosis research / Leon V. Clark, editor.
p. cm.
Includes bibliographical references and index.
1-59454-044-6 (hardcover)
1. Atherosclerosis.
[DNLM: 1. Atherosclerosis--etiology. 2. Atherosclerosis--physiopathology. 3. Infection--complications. 4. Lipids--adverse effects. 5. Oxidative Stress. 6. Risk Factors. WG 550 F652 2004] I. Clark, Leon V.
RC692.F693 2004
616.1'36--dc22 2004008862

An imprint of Nova Science Publishers, Inc.
400 Oser Ave, Suite 1600
Hauppauge, New York 11788-3619
Tele. 631-231-7269 Fax 631-231-8175
e-mail: Novascience@earthlink.net
Web Site: http://www.novapublishers.com

Printed in the United States of America

Contents

Preface

Atherosclerosis is a degenerative condition in which arteries build up deposits called plaques (atheromas) which consist of lipids (mainly cholesterol), connective tissue and smooth muscle cells originating from the arterial wall. Plaques develop quietly over a period of years and are unnoticeable until there is an interruption in the normal flow of blood. Plaques may partially or totally block the blood's flow through an artery. Two things that can happen where plaques occur are: bleeding (hemorrhage) into the plaque; and formation of a blood clot (thrombus) on the plaque's surface. Atherosclerosis affects large and medium-sized arteries. The type of artery and where the plaque develops varies with each person. Atherosclerosis research has witnessed startling progress in recent years, partially due to new drugs as well as to new breakthroughs in molecular medicine.

Great interest has been generated in the past decade in the medical and public community on the relationship of infections and atherosclerotic vascular diseases. A large body of data has been accumulated on *Chlamydia pneumoniae* association with cardiovascular and cerebrovascular diseases. Less but intriguing data on the link with vascular disease exists for periodontitis, cytomegalovirus, *Helicobacter pylori* and even influenzavirus. Chapter I summarizes studies on epidemiological, pathological, in vito, cell biology, animal model, and human antibiotic treatment trials. In general epidemiological case-control and cross sectional studies support the association of *C. pneumoniae*, periodontitis, and influenzavirus with atherosclerotic vascular complications, but prospective studies have been less supportive and more contentious. Cell biology, in vitro and animal models suggest that certain pathogens can modulate development of atherosclerosis, including lipid and inflammatory related mechanisms.

Physical fitness and activity status have been well-documented major risk predictors of cardiovascular and total mortality. The aim of chapter II is to show the associations of cardiorespiratory fitness, as measured during clinical exercise testing, with cardiovascular diseases. Cardiorespiratory fitness is independently related to mortality in clinical and population-based studies. Poor maximal oxygen uptake (VO_{2max}) has been shown to be a comparable with elevated systolic blood pressure, smoking, hypercholesterolemia, obesity and diabetes in importance as a risk factor for mortality as well as a predictor of atherosclerotic cardiovascular diseases and the progression of atherosclerosis. In addition to many previously known mechanisms explaining the protective role of physical exercise, there is growing evidence that aerobic exercise training via an increase in endothelial nitric oxide synthase activity can increase the capacity of endothelial cells to evoke vasodilation and decrease blood pressure, however, there seems to be heterogenity in the hemodynamic response to exercise. Exercise training in coronary artery disease leads to an improved agonist-mediated endothelium-dependent vasodilatory capacity. Furthermore, exercise training and increased aerobic capacity can further improve vascular arterial stiffness, and vasodilatory capabilities, leading to decreased systemic vascular resistance. Direct and indirect measurement of VO_{2max} can provide a good estimate for cardiorespiratory fitness level and prognosis. Cardiorespiratory fitness represents one of the strongest and valuable predictors for both fatal and non-fatal cardiovascular events emphasizing the importance of exercise testing in everyday life. Exercise testing should be used as a clinical tool for exercise prescriptions on individual basis. More attention should be directed toward the use of exercise testing to measure the therapeutic response to a life-style change and pharmaceutical interventions.

Cardiovascular (C-V) risk factors can be identified beginning in childhood and these are predictive of C-V disease later in life. Observations in the Bogalusa Heart Study, the focus of chapter III, have shown an important correlation of clinical risk factors in early life with anatomic changes at autopsy in the aorta and coronary vessels with atherosclerosis and cardiac and renal changes related to hypertension. These observations have been extended by echo Doppler studies of carotid arteries and intima media thickness (IMT). A close association of risk factors in young adults, 20-38 years of age, occurs with IMT, and a marked increase in IMT is noted as the numbers of risk factors increase. Obesity and low density lipoprotein cholesterol (LDL-C) are strong predictors of subclinical disease. Observations of risk factors in young individuals and noninvasive studies of structure and function changes of the C-V system have strong implications for prevention by cardiologists.

Chlamydia pneumoniae, a human respiratory pathogen, has been repeatedly linked to atherosclerotic disease based on sero-epidemiologic studies, direct detection of the organism in atherosclerotic lesions, animal experiments, in vitro data, and to a lesser extent tissue culture. Later on, meta-analysis showed no differences in antibody titers between cases and healthy controls. In addition, the detection of *C. pneumoniae* DNA within atherosclerotic lesions by PCR varied from O% to 100% between different laboratories, with evidence of contamination being gathered. Furthermore, the correlation between *C. pneumoniae* DNA or antigen detection in the vessel wall and antibody titres in the serum remains controversial. Recent data have also revealed potential crossreaction with nonchlamydial plaque constituents, indicating that immunohistostaining results obtained with antibodies against *C. pneumoniae* have to be interpreted with caution. Besides methodological difficulties, not all investigators have used the same tests, antigens, strains, primers and protocol conditions. Whether patients with atherosclerosis or acute coronary syndromes benefit from antibiotics against *C. pneumoniae* is still an evolving issue. Animal models and in vitro studies, however, continue to accumulate convincing data for an association. In chapter IV, a literature overview is given of the main studies published on *C. pneumoniae* and its potential relation to atherosclerotic disease.

Oxidative stress is thought to be etiologically related to atherosclerosis. Experimental evidence clearly demonstrates the presence of oxidized LDL (oxLDL) in the intima of arteries and that it contributes to the initiation and progression of atherosclerotic lesions. Eiji Matsuura and Luis R. Lopez have recently demonstrated that oxLDL interacts with an endogenous plasma protein, β2-glycoprotein I (β2GPI), to form complexes and that these complexes circulate in the blood stream of patients with systemic inflammatory diseases. β2GPI/oxLDL complexes have been implicated as autoantigens relevant in atherogenesis. A role of antiphospholipid antibodies (i.e. anti-β2GPI) in arterial thrombosis has been proposed, and circulating IgG immune complexes containing β2GPI and oxLDL have been detected in patients with autoimmune diseases such as systemic lupus erythematosus (SLE) and antiphospholipid syndrome (APS). Their *in vitro* experiments showed that oxLDL quickly interacts with β2GPI via specific ligands generated by Cu^{2+}-oxidation in which the negative charge of oxLDL is neutralized by the complex formed. Novel ELISA systems allowed the measurement of circulating β2GPI/oxLDL complexes and autoantibodies to these complexes to study their association with various thrombotic manifestations. Even though many questions still remain, β2GPI/oxLDL complex can be described as a common metabolic form in atherogenesis and a significant participant in autoimmune-mediated atherosclerosis.

The relationships among lipoprotein metabolism, genetic vascular factors, vascular disease and Alzheimer's disease (AD) suggest that the examination of centenarian populations in relation to certain genes or lipoprotein metabolism provide insights into the human longevity. The findings on the higher frequency of the apolipoprotein-E (APOE) ε4 allele in middle-aged subjects than in centenarians were substantially confirmed. On the contrary, Francesco Panza et al.'s findings did not confirm previous data on increased prevalence of the high-risk angiotensin I converting enzyme 1 (ACE1) D allele in French centenarians. They hypothesized that the variability in the strength of association between ACE1 polymorphism and longevity could be related to regional differences in ACE1*D frequency in Europe showed in our study, as recently reported for apoE ε2 and ε4 allele in centenarians. Indeed some studies of cholesterol profiles in centenarians have also had contradictory outcomes. Notable amongst these is evidence that lower serum levels of high-density lipoprotein cholesterol (HDL-C), with higher HDL2-C subfraction, and higher lipoprotein(a) [Lp(a)] concentration in centenarians which is an apparently disadvantageous lipoprotein profile. Indeed, lower HDL-C concentration was linked with decreased serum albumin, elevated C-reactive protein, and interleukin-6 levels, suggesting that HDL-C could be a sensitive marker for frailty and comorbidity in centenarians. Their own findings suggest that elevated Lp(a) serum levels, increasing the risk for cerebrovascular disease, may play a role in determining clinical AD but that Lp(a) elevation in centenarians, in the absence of other coronary artery disease risk factors, appears as a positive survival factor. They have also shown how in different populations, there are significant trends in the reduction of serum APOE levels from APOE ε2- to ε4-carriers and significant differences in serum APOE levels with respect to age in ε4-carriers but only after adjustment for HDL-C. While further studies are needed to confirm the possible role of APOE concentration as putative longevity factor chapter VI provides an overview of many of the investigated vascular factors that have been investigated with respect to longevity.

The most common metabolic disorders in Western countries, obesity and the insulin resistance syndrome (IRS) are accompanied by an increased risk of coronary heart disease (CHD) and show an alarming increase in prevalence. Both disorders are characterized by increased plasma levels of free fatty acids (FFA) which are thought to play a critical role in the pathogenesis. FFA elevations are associated to obesity and a hypercaloric diet. Furthermore, FFA decrease insulin sensitivity in the adipose tissue, in the liver and in the muscle and aggravate pancreatic dysfunction, ultimately leading to insulin resistance. Insulin resistance at the level of the adipose tissue will further increase plasma FFA, since insulin is a key player in lipid metabolism with in general a lipogenic effect. Impaired adipose tissue FFA trapping, in particular in the postprandial phase, due to insulin resistance or dysfunction of the C3/ASP system (complement component 3/acylation stimulating protein) maintains this vicious circle. Under these conditions there is enhanced hepatic FFA flux, leading to the characteristic IRS lipid phenotype with high triglycerides (TG), low levels of high density lipoprotein (HDL) cholesterol and high levels of small-dense highly atherogenic low density lipoprotein (LDL) cholesterol. Furthermore as studied by Antonie J.H.H.M. van Oostrom and Manuel Castro Cabezas, FFA may impair endothelial function via activation of proinflammatory pathways and production of reactive oxygen species by leukocytes and endothelial cells, processes that eventually may contribute to atherogenesis. Certain fatty acids, the polyunsaturated omega-3 fatty acids in particular, exert beneficial effects on lipid metabolism, endothelial function and inflammation. These effects are most certainly PPAR (peroxisome proliferator activating receptor) mediated since FFA are the natural ligands of these receptors. Treatment of FFA-related metabolic disorders first of all should include life-style intervention consisting of weight reduction and diet modification in favor of polyunsaturated fatty acids. In addition, drugs with PPAR activating properties, like fibrates and the novel class of thiazolidinediones, may be helpful. Furthermore, statins may reduce FFA as well, either via TG lowering or possibly also via PPAR activation.

Healthy endothelium plays a central role in cardiovascular control. Therefore, endothelial dysfunction (ED) may have a particularly significant role in the pathogenesis of atherosclerosis. ED is a consequence of the harmful effects of risk factors of atherosclerosis on the vessel wall and is closely related to the number of risk factors, to their intensity and their duration. ED has been demonstrated in subjects with hypercholesterolemia, diabetes, hypertension, smoking and in patients with atherosclerotic disease (coronary, peripheral arterial). It was also shown that ED is an early event in type I and II diabetes and that it is related to the development and progression of diabetic vascular complications. Chapter VIII has shown that ED (demonstrated by flow mediated endothelium dependent dilation) is inversely related to the extent of microalbuminuria. The involvement of risk factors in ED is also supported by results of intervention studies that showed regression of ED with treatment of risk factors. Improvement of ED may be achieved by elimination of risk factors, by substitution of natural protective endothelial substances (e. g, L-arginine), inhibitors of endothelium-derived contracting factors (e.g., ACE inhibitors, angiotensin II receptor antagonists), cytoprotective agents (e.g., free-radical scavengers such as superoxide dismutase), lipid-lowering drugs or diets and by physical exercise. All these data show that ED is reversible and by treatment of risk factors it is possible to restore vascular function. ED promotes progression of atherosclerosis and probably plays an important role in the development of thrombotic complications in the late stages of the disease. As ED is a key underlying factor in the atherosclerotic process, markers of endothelial abnormalities have been sought, particularly those involving disturbed endothelium-dependent vasomotion or related cellular products – circulating markers. Using these tests it is possible to follow the dose - response of harmful effects of risk factors, and the effects of preventive procedures on vessel wall function. Determination of ED also has important clinical implications. It was shown that ED is significantly and directly correlated with the occurrence of cardiac events and that cardiac events increased as ED worsens.

Atherosclerosis is a complex, irreversible disease of the artery tree causing an excess of morbility and most of the deaths due to myocardial infarction and stroke. The disease process initiates as early in childhood and progresses for decades showing a spectrum of lesions in the arteries ranging from the fatty streak to the advanced plaque. Two elements are present in all the stages of the disease: macrophages and lipids. Striking evidences suggest that the very early event of the atherosclerotic process is the accumulation of circulating LDL into macrophages in the intima that leads to foam cell formation. LDL entry in macrophages is mediated by the scavenger receptor whose ligand is not a native LDL but a modified form of LDL believed to occur in vivo by free radical-mediated oxidation. Furthermore, it has been demonstrated by Fausta Micheletta, Silvia Natoli and Luigi Iuliano that the oxidation of LDL produces an array of oxidized bioactive lipids with a potential contribution in cellular events of atherogenesis, including endothelial dysfunction, platelet aggregation, and apoptosis. The role of oxidative stress in atherosclerosis has also been supported by the inihibitory effect of antioxidants in experimental atherosclerosis.

Mildly elevated plasma homocysteine levels are an independent risk factor for atherothrombotic vascular disease in the coronary, cerebrovascular, and peripheral arterial circulation. Endothelial dysfunction contributes to the vascular disorders linked to hyperhomocysteinemia. Dysfunctional endothelium becomes manifested by impaired endothelium-dependent regulation of vascular tone and blood flow, by increased recruitment and adhesion of circulating inflammatory cells to the endothelium, and by a loss of endothelial cell antithrombotic function. There is increasing evidence that implicates oxidative stress as being contributory to homocysteine's deleterious effects on the vasculature. Elevated levels of homocysteine may lead to increased generation of superoxide by a biochemical mechanism involving nitric oxide synthase, and, to less extent, by an increase in the chemical oxidation of homocysteine and other aminothiols in the circulation. The resultant increase in superoxide levels is further amplified by homocysteine-dependent alterations in the function of cellular antioxidant enzymes such as cellular glutathione peroxidase or extracellular superoxide dismutase. The increased vascular oxidant stress finally results in decreased bioavailability of the endothelium-derived antiatherogenic signaling molecular nitric oxide via oxidative inactivation. This plays a central role in the molecular mechanisms underlying the effects of homocysteine on endothelial function. Scavenging of superoxide anion by either superoxide dismutase or 4,5-dihydroxybenzene 1,3-disulfonate (Tiron) reverses endothelial dysfunction in hyperhomocysteinemic animal models and in isolated aortic rings incubated with homocysteine. Similarly, homocysteine-induced endothelial dysfunction is also reversed by increasing the concentrations of the endogenous antioxidant glutathione or by overexpressing cellular glutathione peroxidase in animal models of mild hyperhomocysteinemia. Taken together, these findings strongly suggest that the adverse vascular effects of homocysteine are at least partly mediated by oxidative inactivation of nitric oxide. Clinical studies have shown that lowering of plasma homocysteine levels by supplementation of folic acid reverses homocysteine-induced endothelial dysfunction. Whether or not this translates into a reduction of clinical events associated with hyperhomocysteinemia is currently under investigation by Norbert Weiss.

In: Focus on Atherosclerosis Research
Editor: Leon V.Clark, pp. 1-36

ISBN 1-59454-044-6

Chapter I

The Role of Infection in Atherosclerosis

I. W. Fong
Professor of Medicine, Chief of Infectious Diseases,
University of Toronto, St. Michael's Hospital

Abstract

Great interest has been generated in the past decade in the medical and public community on the relationship of infections and atherosclerotic vascular diseases. A large body of data has been accumulated on *Chlamydia pneumoniae* association with cardiovascular and cerebrovascular diseases. Less but intriguing data on the link with vascular disease exists for periodontitis, cytomegalovirus, *Helicobacter pylori* and even influenzavirus. This review summarizes studies on epidemiological, pathological, in vito, cell biology, animal model, and human antibiotic treatment trials. In general epidemiological case-control and cross sectional studies support the association of *C. pneumoniae*, periodontitis, and influenzavirus with atherosclerotic vascular complications, but prospective studies have been less supportive and more contentious. Cell biology, in vitro and animal models suggest that certain pathogens can modulate development of atherosclerosis, including lipid and inflammatory related mechanisms.

Introduction

Great interest has been generated in the past decade in the medical and public community on the relationship of infections and atherosclerotic vascular diseases. The concept of infectious agents playing a role in development of atherosclerosis is not new and existed in Europe in the latter part of the nineteenth century [1]. Proponents of this theory in North America in the early part of the twentieth century included such notables as Osler and Billings [2,3].

Interest in this area waned after the elegant studies of Antischkow on cholesterol induced atherosclerosis in the rabbit model in 1913 [4]. Renewed interest of this concept resurfaced in the latter part of the nineteen eighties after a seroepidemiological study by Finnish investigators showed an association between *Chlamydia pneumoniae* antibodies and recent myocardial infarction [5]. Subsequently numerous reports have been published on various infections associated with cardiovascular disease, stroke, aortic aneurysms and peripheral vascular disease. These studies have investigated epidemiological and pathological associations with atherosclerotic vascular disease, as well as biological mechanisms, animal models and interventional human trials to a lesser extent.

Inflammation and Atherosclerosis

The basic principle and understanding of atherosclerosis as a low grade inflammatory disease was recognized by European pathologists in the early part of the nineteenth century [1]. The concept that atherosclerosis is a response to injury (hence a possible response to an infectious agent) was more recently reviewed and popularized by Ross [6].

The accepted paradigm for development of atherosclerosis starts with endothelial injury or dysfunction from one or more inciting factors such as hypertension, free radicals caused by cigarette smoke, elevated cholesterol or low-density lipoprotein (LDL), and possible infections microorganisms among others [7]. This insult to the endothelium leads to increased adhesiveness to leucocytes and platelets, and also increases the permeability. Whatever the nature of the initial insult that leads to endothelial injury, there follows a cascade of events – increased expression of adhesion molecules that stimulate adhesion, aggregation, and migration of leucocytes; upregulation of proinflammatory cytokines that causes further accumulation of leucocytes (T-lymphocytes and macrophages), foam cells, and lipids and at the same time stimulate hepatocytes to increase acute phase reactants; continued stimulation of cytokines and various tissue growth factors influence migration and proliferation of smooth-muscle cells, leucocytes and fibroblasts. The initial lesion that occurs from accumulation of T-lymphocytes and lipid filled macrophages (foam cells) is the fatty streak. Further stimulation of migration and proliferation of smooth muscle cells that become integrated with inflammation in the intima of the artery to form an intermediate lesion. Continuation of this process leads to thickening and remodeling of the arterial wall to form mature atheroma with a lipid core (rich in foam cells, and cholesterol ester) and with a fibrous cap. Hence inflammatory reaction has been demonstrated in various stages of the atheromatous lesions in human and animal models.

Furthermore, systemic markers of inflammation can predict future cardiovascular events in healthy populations [7]. The process of initiation of atherosclerosis is not unlike the clotting cascade and similar to the inflammatory cascade in chronic infection. Thus infections maybe the link between inflammatory markers and atherosclerotic diseases.

It has been postulated that interleukin (IL)-6, an end product of tumor necrosis factor alpha (TNF-α) and IL-1β upregulation, is the central mediator of cardiovascular risk associated with chronic inflammation, smoking, diabetes and visceral obesity [8,9]. IL-6 is a powerful inducer of the hepatic acute phase response, such as fibrinogen, C-reactive protein (CRP), and serum amyloid A (SAA), which are strong risk factors for coronary heart disease (CHD), and is associated with increased blood viscosity, platelet number, and activity.

Infections Associated with Atherosclerosis

It has been recognized each year for decades that there is excess cardiovascular mortality associated with influenza virus outbreaks during the winter. Recent studies have shown that influenza vaccination reduces these cardiovascular events in the elderly [10]. There are several possible explanations for these observations, including nonspecific stressful effect of influenza precipitating acute ischaemic events in subjects with critical narrowing of the coronary arteries; or infection destabilizing a vulnerable plaque, or leading to thrombosis on a plaque by creating a procoagulant environment.

In the past decade numerous studies have assessed the relationship between cardiovascular or cerebrovascular disease, and to a much lesser extent abdominal aneurysm and peripheral vascular disease, with several infections. The data is most extensive for *C. pneumoniae*, less expansive but intriguing for periodontitis, cytomegalovirus, herpes simplex virus, and *Helicobacter pylori*. A few studies have assessed the relationship of atherosclerotic vascular disease with hepatitis A, coxsackie virus, *Mycoplasma pneumoniae*, and human immunodeficiency virus (HIV), but these will not be addressed in this review. To a large extent the increased cardiovascular morbidity seen in HIV subjects is related to dyslipidemia which may be drug induced by highly active antiretroviral agents (HAART), in subjects with other risk factors i.e. cigarette smoking, obesity, insulin resistance etcetera.

Possible Mechanisms of Infection in Atherogenesis

Infections potentially could modify development of atherosclerosis and vascular complications by several mechanisms (see Figure 1). Infectious agents can injure vascular endothelium directly by invasion or by stimulating release of high circulating cytokines via lipopolysaccharide. Endothelial injury (the initiating event in atherogenesis), is associated with activation of leucocytes and expression of endothelial adhesion molecules (e.g. E- and P- selectin and intercellular adhesion molecule -1(ICAM-1). These molecules help localize leucocytes to the area of injury and the inflammatory molecules released from activated leucocytes can produce additional injury. In addition release of endothelial cell molecules, such as IL-8 and platelet activating factor (PAF), attract and activate other cells, amplifying the inflammatory response. Thus leading to migration and accumulation of T-lymphocytes and macrophages in the intima (early inflammatory changes in atherosclerosis). ICAM-1 mediates the adhesion and transmigration of monocytes to the vessel wall [11].

Microbes could potentially accelerate pre-existent atherosclerotic lesions by increasing recruitment of inflammatory cells in these areas. Microbes could be carried to early atherosclerotic lesions by monocyte-macrophages or lymphocytes, and this could enhance the generation of adhesion molecules and proinflammatory cytokines. Indirectly infection at a remote site may increase systemic cytokine levels to have a similar effect. Eventually leading to progression of lesions by increasing uptake in the intima of macrophages, foam cells and proliferation of smooth muscle cells (critical components in the development of atherosclerosis).

The association of various infections with complications of atherosclerosis (myocardial infarction and stroke) maybe an indirect effect due to changes in lipoprotein patterns. Infection and inflammation can produce a wide range of changes in the metabolism that could produce a proatherogenic environment. The main effect of infection and inflammation on the metabolism and lipoprotein is hypertrylyceridemia [12]. This is mainly due to an increase in very low density lipoprotein (VLDL), an established risk factor for cardiovascular disease, as a result of increased production and decreased clearance. Probably one of the major proatherogenic lipid effects of infection or inflammation is a reduction in serum levels of high density lipoprotein (HDL).

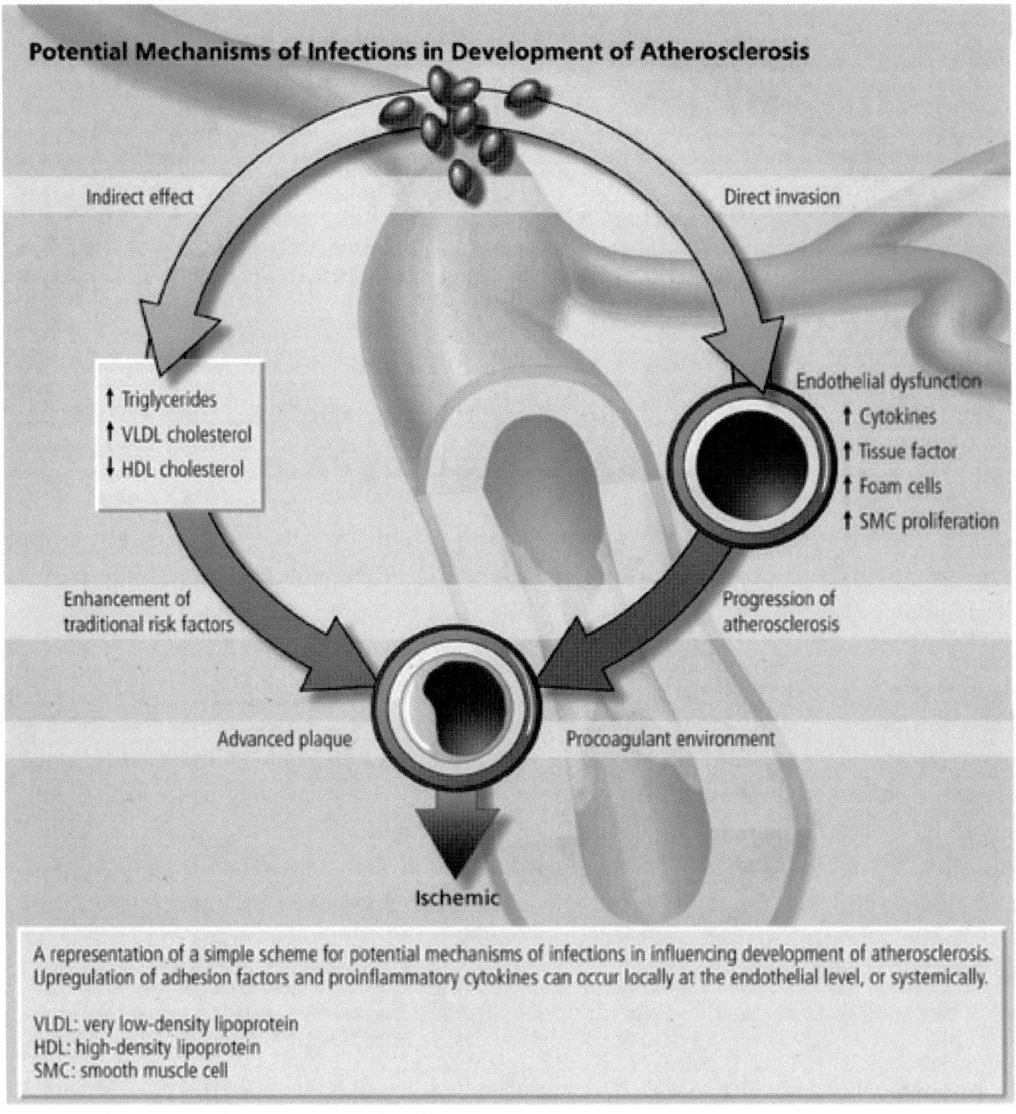

Figure 1. A representation of a simple scheme for potential mechanisms of infections influencing development of atherosclerosis. Upregulation of adhesion factors and proinflammatory cytokines can occur locally at the endothelial level, or systemically. Adapted from Fong, I.W. (Infection & Atherosclerosis: Evidence for possible association. Geriatrics & Aging, 2003; 6: 35-39), with permission of the publisher.

The mechanism for a low HDL in infection and inflammation is not well understood. Chronic infection may also affect the composition and function of HDL, through the effect of apolipoproteins, enzymes, and transfer proteins that regulate the metabolism and function of HDL [13]. HDL has antiatherogenic properties putatively including protection of LDL oxidation and removal of cholesterol from cells to the liver for elimination (reverse cholesterol transport). Clinical and epidemiological studies have established that low serum HDL is a risk factor for cardiovascular disease. In addition bacterial Lps can induce LDL oxidation in vivo in hamsters [14], and in humans with chronic infections such as in patients with the acquired immunodeficiency syndrome (AIDS), a decrease in LDL cholesterol is associated with an increase in small dense LDL [15], which appears to be more proatherogenic than the large buoyant LDL. Thus infection could potentially accelerate atherosclerosis through direct or indirect mechanisms.

Infection and Precipitation of Vascular Events

The current paradigm of sudden death or acute myocardial infarction from atherosclerosis is the occurrence of rupture or erosion of the fibrous cap of an atheroma, exposing thrombogenic material leading to activation of the clotting cascade and development of a thrombus that compromise blood flow. Theoretically infections may precipitate an acute vascular event by predisposing to plaque rupture or creating a vulnerable plaque. Infection induced inflammatory cytokines mediate expression of proteins, especially metalloproteases that may enhance proteolytic digestion of the fibrous cap (capsule) with inability to generate new collagen and smooth-muscle cells through apoptosis. Macrophages and endothelial cells through upregulation of certain proinflammatory cytokines (TNF-α, IL-1 and interferon γ) may enhance vascular smooth-muscle cell apoptosis by a FAS/FAS L-dependent mechanism (death signaling pathway) [16].

Infection or inflammation can potentially precipitate an acute vascular event, as well, by producing a pro-coagulant environment that leads to thrombosis on an eroded, fissured or obstructing plaque. Infection and microbial products such as Lps interact with a variety of lipophilic proteins and inflammatory mediators by which the coagulation system is activated. Lps and TNF-α can induce synthesis and expression of tissue factor on monocytes and endothelium in vitro [17] to activate the extrinsic pathway. TNF- α and IL-1 besides augmentation of tissue factor expression can increase plasminogen activating inhibitor I (PAI-1) by endothelial cells, down regulate thrombomodulin leading to decreased generation of activated protein C and decreased thrombin inactivation [18], thus inhibiting the anticoagulation system. Moreover, infection and inflammation can stimulate platelet activation and assembly of platelet prothrombinase complex and thus further predisposing to thrombosis [19]. Hence infection or inflammation facilitates mechanisms that may trigger initiation and amplification of the coagulation response.

Chlamydia pneumoniae and Atherosclerosis

C. pneumoniae is a common respiratory, intracellular bacterial pathogen capable of causing pharyngitis, sinusitis, bronchitis and pneumonia in the community [20]. Its geographical distribution is world-wide and most of the population becomes infected starting after 5 years of age. By age 20 years 50% of the population would have been exposed and by 65 years up to 80% have antibodies with predominance in males [21]. Most infections have never been recognized and are probably subclinical or undifferentiated from other common respiratory infections. A feature of *C. pneumoniae* that makes it a suitable candidate for playing a role in atherogenesis (besides its widespread distribution) is the ability to persist intracellularly as a chronic latent infection. It is suggested that, in the presence of INF-γ, which depletes available tryptophan, there exist unique host-parasite interactions that may contribute to persistent chlamydial infections [22]. The evidence for association of *C. pneumoniae* with atherosclerotic vascular diseases is derived from seroepidemiological, pathological, in vitro, clinical and animal studies.

Epidemiological Association with *C. pneumoniae*

Since the initial report by Saikku et al [5] in 1988 of the serological association of *C. pneumoniae* with CHD, there have been numerous seroepidemiological studies on this topic in subjects with cardiovascular disease and stroke. The first major review of this topic by Danesh et al [23] in 1997, listed 18 epidemiological studies (2 involved cerebrovascular disease and 3 were prospective), and found an average odds ratio (OR) of 2.0 to support an association of *C. pneumoniae* with vascular disease. Problems noted with these studies include different criteria for cases between studies, adjustment for potential confounders to different degrees and, thus prone to different biases.

Since then over 60 case-controls or cross-sectional studies on the association of *C. pneumoniae* and atherosclerotic vascular diseases and more than 20 prospective or longitudinal studies have been reported. There have been improvements in case definitions and correction for various confounding factors in most of the recent studies, but there are still variability in assays used, end-point definition of seropositivity and type of antibodies (immunoglobulin) associated with CHD or stroke reported. Overall most of the cross-sectional studies (>60%) using different techniques found a positive association with *C. pneumoniae* infection and atherosclerotic vascular diseases [24].

However, in two recent meta-analyses of prospective studies there was not a strong association between *C. pneumoniae* seropositivity and cardiovascular disease (OR 1.1-1.5) [25,26]. In reviewing 26 longitudinal studies only 6 showed an overall positive association with *C. pneumoniae* antibodies and another 5 showed correlation with atherosclerotic disease in subgroups of patients [24]. Different diagnostic serological techniques with lack of standardization and varying cut-off points for positivity hamper the interpretation of the meta-analyses. Moreover, present serological methods do not differentiate between past infection and chronic persistent infection, the latter state being most likely contributing to atherogenesis if it does at all.

Pathological Association of *C. pneumoniae* and Atherosclerosis

Detection of *C. pneumoniae* in atheromatous plaques, with its absence in normal arteries, is another method of establishing association with atherosclerotic vascular disease and its complications. The strongest body of evidence in favor of *C. pneumoniae* as a risk factor for atherosclerosis comes from over 50 pathological studies [24]. However, this does not prove causality as the organism could be entrapped in the plaques as an "innocent bystander" without causing disease. *C. pneumoniae* has been detected on an average of 40% or more from atheromatous plaques from a variety of arteries (aorta, coronary, carotid, iliac, femoral, popliteal) and aorta-coronary vein graft and rarely from normal arteries, with an overall odds ratio for association with atherosclerosis of 24.5 [24]. A variety of techniques have been used including immunohistochemical stain, polymerase chain reaction (PCR) for DNA or mRNA, in situ hybridization, electron microscopy and culture. The studies that failed to detect, or rarely found *C. pneumoniae* in atheroma, primarily used PCR techniques; and the negative findings may be related to small tissue samples (i.e. atherectomy tissue), tissue inhibitors, calcification and insensitive techniques.

However, *C. pneumoniae* DNA has been detected as high as 94.3% in plaques of smokers versus 36% of non-smokers [27]. Jackson et al [28] have reported on the widespread distribution of *C. pneumoniae* in various tissues from autopsy to test the "innocent bystander" hypothesis, and found *C. pneumoniae* mainly located to vascular tissue and rarely in others. Furthermore, in studies assessing the relationship of *C. pneumoniae* detection and severity of disease (or symptoms) or presence of fresh thrombus five of nine reports showed a positive correlation [24].

C. pneumoniae is difficult to culture from tissues but a few centers have successfully recovered viable organisms from atheromatous specimens. It is the only microorganisms cultured from atheromas, and this establishes the association of viable organisms in plaques and not the mere presence of antigenic particles previously phagocytosed by migrating macrophages. *C. pneumoniae* has been cultured from coronary and carotid arteries, coronary bypass vein grafts, and abdominal aneurysms and from a femoral artery in seven reports from six centres [24]. The most successful recovery rates were in 10 of 26 (38%) abdominal aortic aneurysms [29] and in 11 of 70 (16%) atherectomy specimens from coronary arteries and venous bypass grafts [30]. The inability to culture *C. pneumoniae* from vasculature tissues and low recovery rate by several centers maybe due to small amount of tissue cultured, low concentration of organisms, or presence of the organism in a latent, metabolically inactive and unculturable form.

Chlamydial heat shock protein 60kd (CHSP-60), an inflammatory antigen expressed by persistent chlamydia have also been localized to macrophages in human atherosclerotic lesions [31]. Chronic infection of lesions with persistent forms may result in local CHSP-60 production, a potent stimulus to inflammation and progression of lesion. Although the presence of microbes in plaques does not necessarily mean an active role in atherogenesis, the finding that *C. pneumoniae* triggers specific cell-mediated immunity within plaques (detection of chlamydia-specific T-lymphocytes in atherosclerotic lesions), suggest an active role [32,33].

Biological Mechanisms of *C. pneumoniae*

There are substantial in vitro data to support biological plausible mechanisms of initiation or acceleration of atherosclerosis by *C. pneumoniae. C. pneumoniae* is capable of replicating in human macrophages, endothelial and smooth muscle cells derived from human aorta in vitro [34], and can also infect and multiply in lymphocytes [35]. Infection of human endothelial cells can induce proliferation of smooth cells via an endothelial derived soluble factor [36], an integral process in the pathogenesis of atherosclerosis. Furthermore, infection of human endothelial cells in vitro is associated with upregulation of genes encoding cytokines (IL-1, IL-8) chemokines (monocyte chemotactic protein 1 [MCP-1]) and cellular growth factors (heparin-binding epidermal growth factor, basic fibroblast growth factor, and platelet derived growth factor (PDGF) β chain) [37]. Similarly *C. pneumoniae* infection of human smooth muscle cells also result in induction of IL-6 and basic fibroblast growth factor [38]. Inflammation is important in the development of atherosclerosis, and cytokines are believed to play a key role in the initiation and progression of inflammation. Proliferation of *C. pneumoniae* intracellularly in human endothelial cells stimulate adhesion molecules (E-selectin, ICAM-1, VCAM-1), and inflammatory cytokines (IL-1, IL-6 and TNF-α) [39], and acellular components of the bacteria are potent stimuli of the proinflammatory cytokines production in human blood mononuclear cells [40], and these may play an important role in atherogenesis.

These augmented expressions of adhesion molecules and cytokines may promote leucocyte adhesion, migration, intimal inflammation, smooth-muscle cell proliferation and enhance lesion progression.

Moreover, intact *C. pneumoniae* or its Lps can increase the uptake of LDL in human macrophages to become cholesterol laden foam cells in vitro [41], a key component in early and late atherogenesis. In addition chlamydial HSP-60 or viable bacteria can enhance intracellular oxidation of LDL [42], the toxic form of LDL that is putatively the primary mediator of atherogenesis. Chlamydial HSP-60 also localizes in human atheroma in 47% of endarterectomy specimens and co-localizes with human HSP-60 in plaque macrophages, and regulates TNF-α and matrix metalloproteinase expression in macrophages [31], as well can also activate vascular endothelial and smooth muscle cells and macrophages [43]. Humoral immune reaction to chlamydial HSP-60 may also cross react with human HSP-60/65 and may play an important role in the process of vascular endothelial injury, a key element in initiation of atherogenesis, as antibodies to chlamydial HSP-60 can mediate endothelial cyotoxicity [44]. It has been recently shown that *C. pneumoniae* stimulates proliferation of vascular smooth muscle cells through induction of endogenous HSP-60 [45]. There is also recent evidence that polymorphic membrane proteins 20 and 21 of *C. pneumoniae* induce proinflammatory cytokines in human endothelial cells in vitro by activation of the nuclear factor – kappa β (NF-kβ) pathway [46].

Destabilization of plaques can potentially be induced by *C. pneumoniae* through upregulation of matrix metalloproteinase expression [31], and induction of secretion of the 92 k Da gelatinase [47], that could weaken plaques and predispose to rupture. Furthermore, *C. pneumoniae* could predispose to acute thrombosis on plaques as human vascular endothelial cells and smooth-muscle cells infected with this bacteria can induce pro-coagulant protein expression such as tissue factor and plasminogen activator inhibitor- 1 (PAI-1) [48].

Animal models of *C. pneumoniae* and Atherosclerosis

Animal models for assessing the interaction of *C. pneumoniae* on development of atherosclerosis has been performed primarily in rabbits (New Zealand White) and murine models (gene knockout mice), but a few studies have used the pig model. *C. pneumoniae* has been demonstrated to varying degrees to modulate development of atherosclerosis in animal models at various stages of atherogenesis with and without hypercholesterolemia.

The accepted paradigm in the genesis of atherosclerosis is the initial occurrence of endothelial dysfunction. In hyperlidipidemic apolipoprotein E (ApoE) deficient mice repeated infection with *C. pneumoniae* causes endothelial dysfunction and impairs arterial relaxation [49], and the nitric oxide pathway is principally involved. More recently the same group has demonstrated that acute *C. pneumoniae* respiratory infection (after 3 days) causes endothelial dysfunction of both resistance and epicardial coronary arteries, and favours a procoagulant state in the pig model [50]. The subsequent steps in atherogenesis include migration of macrophages and lymphocytes to the intima, and accumulation of cholesterol laden macrophages (foam cells) to form fatty streaks. We have previously demonstrated that respiratory infection with *C. pneumoniae* in normocholesterolemic rabbits can result in patchy areas of foam cell lesions equivalent to fatty streaks, as well as more advanced fibromuscular and calcified lesions of the aorta [51]. However, in this model the mature atheroma typically found in humans with a lipid core and fibrous cap is not seen in rabbits unless fed a lipid rich diet. Inflammatory changes in the aortic intima of normacholesterolemic rabbits and wild type mice (C57BL/6J) chronically infected with *C. pneumoniae* have also been reported [52,53]. In a recent study *C. pneumoniae* injected directly into the wall of coronary arteries (via balloon catheter) or intrabronchially induces neointima formation (moderate intimal proliferation) of the coronary arteries in pigs without a lipid-rich diet [54].

It is generally acknowledged that human subjects with more than one risk factor for cardiovascular disease have a greater incidence of atherosclerotic disease than subjects with only one factor. Thus infection combined with hyperlidemia may produce more rapid and more extensive development of atherosclerosis than from hyperlipidemia alone. This issue has been addressed in animals in both murine and rabbit models. Although rodents are typically resistant to atherosclerosis formation even with lipid rich diet, gene knockout mice with hyperlipidemia readily develop plaques of the aorta.

Wild type mice when fed a high fat and high cholesterol diet develop early mild atherosclerotic changes of the aortic sinus, and *C. pneumoniae* infection increases the lesions by 2.5 to 3.3 fold [55]. In ApoE-deficient mice, which develop mature atheromatous lesions of the aortic arch with spontaneous hypercholesterolemia, increased area of atherosclerosis can be measured following *C. pneumoniae* infections varying from 70% to 2.4 fold increases depending on the age at sacrifice, number of inoculations and sex [56-58]. Another hypercholesterolemic murine model (LDL-receptor deficient) develops atherosclerosis of the aorta with a high cholesterol diet, and *C. pneumoniae* increase the area of involvement from

20% in control to 30% in the infected group [59]. Furthermore, the lesions were more advanced and severe in the *C. pneumoniae* infected group, but not with the *Chlamydia trachomatis* infection.

Other investigators, however, have failed to demonstrate accelerated atherosclerosis with *C. pneumoniae* infection in ApoE-deficient mice [60,61]. This maybe explained by differences in schedule and frequency of inoculations and the use of female instead of male mice, which are less susceptible to atherosclerosis.

Ordinarily NZW rabbits do not develop atherosclerosis on a standard diet but high cholesterol chow induces atherosclerotic changes of the aorta similar to that of human. It was first shown by Muhlestein et al [62] that hypercholesterolemic rabbits had enhanced atherosclerosis of the aorta with *C. pneumoniae* infection, with increase in maximal intimal thickness by 3.4 fold and plaque area index by 8 fold. Moreover, early treatment with a macrolide (azithromycin) could partially reverse these changes. We have also demonstrated that *C. pneumoniae* infection increases the atherosclerotic area of the aorta by 57% in cholesterol fed rabbits, and these changes were not due to increase in total cholesterol, LDL, VLDL, or lowering of HDL [63]. There is no good animal model of unstable plaque to demonstrate that *C. pneumoniae* could precipitate rupture or acute thrombosis. However, a recent study in LDL-receptor/apoE-double knockout mice with advanced atheromas, *C. pneumoniae* infection reduces atheromatous fibrous cap area with increase matrix metalloproteinase-2 and -9 immunoreactivity [64]. Thus, suggesting that *C. pneumoniae* may induce unstability of plaques in vivo.

Clinical Trails on Infection and Atherosclerosis

Interventional trails are often used to prove causality of an observation or association of a risk factor with a disease. However, studies published or ongoing on the value of antimicrobials in atherosclerotic (cardiovascular) disease cannot prove or disprove the role of infection hypothesis for several reasons. A major pitfall of all these studies is the selection of subjects, with advanced atherosclerotic disease, who have experienced previous myocardial infarction or unstable angina. Thus the value of treatment in preventing secondary events is mainly to assess the acute precipitating effect of infections (although theoretically possible this has not been shown nor addressed by the animal models).

A few small, pilot, prospectives clinical trials had stimulated world wide interest in the field of infections and CHD, by showing early decrease in secondary cardiac events in patients treated with short term macrolides after recent myocardial infarction or unstable angina [65,66]. However, in one of these trials the initial beneficial effect was not sustained after 6 months [67]. In another recent small trial of 325 patients with acute myocardial infarction or unstable angina, azithromycin (effective against chlamydia) or combination of amoxicillin, metroinidozole and omeprozole for a week (regimen used for *H. pylori*) were effective in reducing secondary cardiac events after a year compared to placebo [68]. However, a larger trial of 1439 patients with acute coronary syndrome did not show any benefit of 5 days azithromycin in preventing recurrent ischaemic events [69]. Other clinical trails have examined the value of antibiotics in patients with stable CHD. In a small trial of 302 patients (not powered to detect even a 30% difference) no reduction in cardiovascular events was observed after 2 years following 3 months course of azithromycin [70]. However, there was a reduction of global rank score of four serum inflammatory markers (CRP, IL-1, IL-6, and TNF-α) in the treated group. The largest randomized study to date on this topic (The WIZARD Study) has just been published and enrolled over 7700 patients with stable CAD [71]. This was an international study (9 countries), end-point driven with a mean follow up of 14 months (initial design was for 2-3 years follow up). Only patients with *C. pneumoniae* Ig titer of 1:16 or more were enrolled and they were allowed to continue their usual cardiac medication, including aspirin and lipid lowering agents (statins). Previous documented myocardial infarction occurring 6 weeks or more but could be several years was an inclusion criteria. The main outcome measures assessed were the first occurrence of death from any cause, nonfatal re-infarction, coronary revascularization, or hospitalization for angina. The treated group received azithromycin 600 mg/day for three days then 600mg once a week for a total of 12 weeks. There was no significant difference in outcome between azithromycin or placebo.

Several factors might explain the negative results of clinical trials with antibiotic therapy, and these asperities will likely affect current on going trials such as the Azithromycin and Coronary Events Study (ACES) [72], evaluating a year's course of azithromycin, and the Provastatin or Atorvastatin Evaluation and Infection Therapy (PROVE IT) [73] – a gatifloxacin (new quinolone) trail. Although most of the population above 60 years old will have antibodies to *C. pneumoniae* (70-80%), there is no correlation of any serology with the presence of the microorganism in atheromas, [74-76] which is present in only 40-50%of advanced plaques [24]. Thus up to 50-60% of patients enrolled in these trials would not likely be related to *C. pneumoniae* even if it play a role. Furthermore, an effective regimen for eradication of a chronic persistent infection with *C. pneumoniae* has not been established. Although early administration of azithromycin in animal models could ameliorate the accelerated effect of *C. pneumoniae* [62], late treatment has been shown to be ineffective [77,78]. In addition, in experimental mouse pneumonitis with *C. pneumoniae* there was no difference in eradication rates between controls and azithromycin, but combination with rifampin was more effective [79].

Other confounding factors in these trials are the widespread concomitant use of aspirin and hydroxymethylgluteryl coenzyme A reductase inhibitors (statins). Both these agents have been shown to modify or inhibit *C. pneumoniae* induced NF-kβ and cytokine activation in human macrophages and endothelial cells, thus reducing the inflammatory response and chlamydial growth [80-82]. There is compelling evidence that the cardioprotective effects of statins are due not to the lipid lowering but rather to their anti-inflammatory effects [83,84], and potentially their activity or efficacy may partly be related to their suppression of inflammatory response in chlamydial infected cells in the vasculature.

Summary

Table 1 summarizes the evidences of association between *C. pneumoniae* infection and atherosclerosis. Although prospective epidemiological and clinical studies have provided conflicting and controversial results, pathological–based studies have confirmed the association of *C. pneumoniae* with atherosclerotic disease. Moreover, many in vitro studies and experiments in animal models have largely supported a plausible role of *C. pneumoniae* in atherosclerosis. However, large clinical interventional trials, due to problems with current designs, may never prove a benefit of antimicrobials in atherosclerotic vascular diseases. Nevertheless, the current body of evidence establishes this pathogen as a plausible, potentially modifiable risk factor in atherosclerotic diseases [85].

Table 1. Summary of the Evidence for *C. pneumoniae* Association with Atherosclerosis

	Mean Odds Ratio	References
A. Epidemiological Studies		
Case-control/cross-sectional	2.0	23, 24
Prospective/Longitudinal	1.1-1.5	25, 26
B. Pathological Studies	24.5	24
C. Biological Mechanisms	N/A	
• Can infect human vascular endothelial cells, macrophages, SMC		34
• Stimulate adhesion molecules & cytokines in human endothelial cells		37-40
• *C. pneumoniae* Lps stimulates uptake of LDL in macrophages		41
• CHSP-60 enhances intracellular oxidation of LDL		42
• CHSP -60 upregulates TNF-α & matrix metalloproteinase		31, 43
• Induces tissue factor in vascular cells		48
D. Animal Models		
• Causes endothelial dysfunction in ApoE -/- mice & pigs		49, 50
• Causes early atheroscleotic lesion (e.g. fatty streaks) in rabbits		51
• Causes intimal proliferation in coronary arteries of pigs		54
• Accelerates hyperlipidemic lesions in rabbits & mice		56-59,62,63
• Reduces fibrous cap area of plaques in double knockout mice		64
E. Clinical Trials		
• Macrolides ↓ cardiovascular events in small trials		65, 66, 68
• Macrolides ↓ inflammatory markers but not clinical events		70
• Macrolides failed to reduce vascular events in larger trials		69, 71

SMC = smooth muscle cell.
LPS = lipopolysaccharide.
LDL = low density lipoprotein.
CHSP = chlamydial heat shock protein.
TNF = tumor necrosis factor.
ApoE -/- = apolipoprotein E deficient.

Periodontitis and Atherosclerosis

Introduction

Periodontal disease is the most frequent cause of dentition loss in adults and the process begins in early adulthood and progresses in severity and frequency with age. Periodontal disease refers to a diverse group of diseases that affect the periodontium (the supporting structures of the tooth that includes the cementum, periodontal membrane, bones of the jaw, and gingivae (gums)). Periodontal diseases are infectious in nature and are initiated by formation of subgingival plaque that leads to inflammation, tissue swelling and destruction to form periodontal pockets (periodontitis). Experts have recently concluded that three or four bacterial species account for most cases of periodontitis – *Porphyromonas* (formerly bacteriodes) *gingivalis*, *Bacteriodes forsythus*, and *Actinobacillus actinomycetemcomitans* (the latter mostly in juvenile periodontitis), [86] and spirochetes [87]. However, most data on pathogenesis of periodontitis and its relationship with vascular disease is with *P. gingivalis*. Although bacteria are essential for the causation of periodontitis, other factors maybe equally important – such as genetic factors, advancing age, male gender, tobacco smoking, diabetes mellitus, poor oral hygiene and lack of dental care (more common in lower socioeconomic groups and minorities). Several of these factors are also predisposing conditions for atherosclerotic diseases.

Epidemiological Association of Periodontal Disease & Atherosclerosis

There has been considerable interest in the relationship between periodontal disease and cardiovascular disease since the initial report of the association of total dental index and acute myocardial infarction in a case control study in 1989 [88]. Since then there has been at least 12 case control or cross-sectional studies reported, and the large majority of these retrospective studies showed an association with poor dental health and CHD or cerebrovascular disease [89]. Some of these studies used a composite of caries, periodontal disease, and oral tooth loss; others tooth loss or edentulous state alone, or periodontal disease alone with different criteria. Other limitations of these studies included biases from the retrospective nature of the investigation; failure to adequately adjust for confounders such as physical activity, diet, smoking and the amount, and socioeconomic status. Moreover, it also possible that people who are more health conscious are at a lower risk of cardiovascular disease and have better dental health [90]. However, after adjustment for traditional cardiovascular risk factors periodontal disease was associated with increased carotid artery intima-media thickness (a surrogate marker of atherosclerosis) in a large cohort [91].

There have been at least 11 prospective, longitudinal cohort studies on periodontitis and CHD or cerebrovascular disease link and one on the association with peripheral vascular disease (PVD) [89]. The associated increased risk of atherosclerotic vascular disease with oral health varied substantially among the studies. Four of these studies showed a small increase in risk (25-70%), three others a moderate increased risk (111-170%) and four no increased risk [89]. The one study on PVD detected an increased risk of 127% associated with alveolar bone loss [92]. A meta-analysis of these studies would not be reliable due to variability of design and methods of assessment for oral health or periodontal disease. Two other prospective studies assessed special issues in specific groups of subjects. As part of the National Health and Nutrition Examination Survey (NHANE S-1) a cohort of 4027 subjects were enrolled to compare rates of CHD events in edentulous subjects (cured of any periodontal disease) with those with periodontitis over a 17 year period [93]. After adjustment for confounders there was no significant difference in CHD events, indicating that elimination of chronic dental infection may not reduce cardiovascular risk. In a small group of diabetics, with and without severe periodontal disease (39 case-control pairs) a higher prevalence of vascular complications was found in the case group after 6 years [94].

Biological Mechanisms in Periodontal Disease

The biological basis for the link between periodontal disease and atherosclerosis may best be explained by indirect mechanisms. Although a few studies have demonstrated oral bacteria in atheromas by immunostain or PCR the data are not convincing [89]. Cells of the monocytic lineage and the proinflammatory cytokines play critical roles in initiating and propagating both atheroma formation and periodontal disease. LPS is released as extracellular blebs from microorganisms within periodontal pockets and may enter the circulation bound to LPS binding protein (LBP). When LPS is bound to LBP it is then able to bind to CD14 receptors, either soluble or on endothelium or on monocytes/macrophages, resulting in cellular activation. Cellular activation can produce upregulation of adhesion molecules and cytokines and chemokines release. These effects may result in subintimal leucocyte infiltration and proliferation of smooth-muscle cells in blood vessels – initiating or enhancing atherosclerosis. Fibrinogen and white blood cell increases noted in periodontitis patients maybe a secondary effect of the proinflammatory cytokine activation or a constitutive feature of those at risk of both cardiovascular disease and periodontitis [95].

CRP, an acute response protein, is a well-established marker of inflammation and predictor of current and future cardiovascular events [96]. There is also evidence that CRP elevation occurs in patients with periodontitis and the height of the increase is related to the severity of the disease [97,98], and specific treatment can reduce the concentration [98]. In an analysis of 10,000 participants of the Third National Health and Nutrition Survey, CRP and fibrinogen showed significant relation with poor periodontal status, but there was no significant association with total cholesterol or HDL cholesterol [100].

Thrombogenesis is intimately related to atherogenesis, and hypercoagulable or prothrombotic environment can precipitate or predispose to acute ischaemic events. *P. gingivalis*, and *Streptococcus sanguis* (another oral bacteria) have been shown to induce platelet aggregation in vitro [101, 102] and proteinases from *P. gingivalis* can cause activation of the blood coagulation factor X [103]. Periodontal pathogens, especially *P. gingivalis*, are capable of invading human coronary artery endothelial and smooth muscle cells [104] and the organism can temporally modulate chemokines response in endothelial cells [105].

Recently *P. gingivalis* and several other oral bacteria were shown to induce foam cell formation (a key characteristic of early lesions in atherogenesis) in the murine macrophage line [106]. This property appears to be mediated by LPS fraction of the cell, and it was also demonstrated that *P. gingivalis* could degrade fibrous caps of atheromas from autopsy samples.

Animal Models of Periodontal Pathogens Associated Atherosclerosis

There are very few studies on periodontal pathogens or periodontitis in animal models of atherosclerosis to address the issue of cause and effect relationship. Repeated systemic inoculation of *P. gingivalis*, once a week for 24 weeks have been reported to accelerate progression of atherosclerosis in a heterozygous apoE-deficient murine model [107]. After 24 weeks the proximal aortic lesion size was nine-fold greater in infected versus non-infected mice ($p<0.001$) fed regular chow, and two-fold greater when fed high fat diet ($p<0.001$). *P. gingivalis* DNA was also found in aortas and hearts at sacrifice. However, this model is not a "natural" or good model to mimic human periodontitis.

In a recent study using NZW rabbits on a high fat diet and silk ligatures around premolars to induce periodontitis with *P. gingivalis*, there was greater accumulation of lipids in the aorta of the infected group versus controls ($p<0.05$) [108]. Thus these results provide direct evidence that periodontitis may contribute to the pathogenesis of atherosclerosis.

Clinical Trails in Periodontal Disease and Atherosclerosis

There are no reports of randomized interventional trial in patients with periodontal disease to assess subsequent development of vascular events. A relatively large interventional trial is underway to assess surrogate cardiovascular markers such as CRP, fibrinogen, cytokines, and adhesion molecules in subjects with periodontitis randomized to receive standard care versus intensive management (personal communication with Robert Genco).

Summary of Periodontal Disease & Atherosclerosis

The concept that periodontal disease can be associated with atherosclerosis is intriguing and worth further investigations. Although prospective epidemiological studies have been conflicting there is sufficient evidence in support of this relationship. In vitro biological studies and a few animal experiments have largely supported a plausible role in acceleration of atherosclerosis and potentially precipitation of acute vascular events. A summary of the evidences for the relationship with atherosclerosis is shown in Table 2.

Cytomegalovirus and Herpes Simplex Virus and Atherosclerosis

The herpes group of viruses, particularly cytomegalovirus (CMV) and herpes simplex virus (HSV), are very common in the population, are often subclinical and remain latent intracellularly for the subject's life – features that are suitable for potentially playing a role in atherogenesis. Evidences of association between atherosclerotic disease and CMV have been examined seroepidemiologically and pathologically in three groups of subjects: i) patients with native CHD; ii) patients who have undergone percutaneous transluminal coronary angioplasty (PTCA); iii) and recipients of cardiac transplantation.

Table 2. Summary of the Evidence for Periodontitis Association with Atherosclerosis

		No. (No. positive)	References
A.	**Epidemiological Studies**		
•	Case-control/cross-sectional	13 (11)	89
•	Prospective/longitudinal	14 (8)	89
B.	**Biological Mechanisms**	N/A	
•	Periodontal pathogens Lps can ↑ cytokines, fibrinogen, leucocytes		95
•	CRP ↑ with periodontitis and treatment ↓ CRP		97, 98
•	Oral pathogens ↑ platelet aggregation and activate factor X		102, 103
•	*P. gingivalis* can invade human endothelial and SMC's		104
•	*P. gingivalis* can modulate chemokines response in endothelial cells		105
•	Oral pathogens can induce foam cell in murine macrophages, and degrade fibrous cap of atheromas		106
C.	**Animal Models**		
•	*P. gingivalis* inoculated systemically ↑ aortic lesions in mice		107
•	Periodontitis induced in rabbits ↑ aortic lipids		108

CRP = C-reactive protein.
Lps = lipopolysaccharide.
SMC = smooth muscle cells.

Epidemiological Association of Cytomegalovirus/Herpes Simplex Virus

There have been at least 18 cross-sectional or case-control studies on the association between CMV serology and native atherosclerotic vascular diseases (CHD or carotid artery disease), and of these only 7 (≃ 39%) found a positive association [109]. At least 15 prospective, longitudinal studies were performed (mean follow up 5.5 years) and only 3 (20%) showed a significant positive association with atherosclerotic diseases and CMV infection [109].

Eight other studies assessed the relationships between restenosis after PTCA and CMV seropositivity, and only three of those studies (37.5%) found a positive association [109].

The most compelling data supporting an association between CMV infection and arteriosclerosis exists for heart transplant recipients. Of the 22 studies on this issue with 2,681 patients followed for 6 months to 10 years, 12 (54.5%) showed on overall association with CMV infection and accelerated allograft vasculopathy [109].

There are fewer studies examining the relationships of HSV with atherosclerotic vascular disease compared to CMV. Only one of six cross-sectional or case-control studies showed a small positive association with HSV-2 serology and atherosclerotic changes. Of eight prospective studies that assessed this association only two showed a significant association of HSV antibodies and CHD [109].

Pathological Studies of Cytomegalovirus and Herpes Simplex Virus

Arterial samples from aorta, coronary, carotid, and femoral arteries have been examined for herpes viruses; predominantly CMV, by PCR or in situ hybridization and a few studies used immunohistorchemistry (IHC) or election microscopy (EM). Of 17 studies on native atherosclerosis (14 analyzed samples for CMV and 8 for HSV), the mean prevalence of CMV was 47.2% in grossly atheromatous arteries and 54.2% in normal arteries; and for HSV 16.6% and 12.1% respectively [109]. Thus suggesting that CMV and HSV are nonspecific findings with no predilection for diseased vessels.

In two studies analyzing coronary atherectomy specimen's post-PTCA for CMV DNA, both detected CMV at a higher rate in restenosis post-PTCA than from native coronary stenosis, 47.5% versus 4.2%, respectively [109].

At least six studies have assessed atheroectomy or autopsy samples from cardiac graft recipients ad controls for CMV nucleic acid with varying results [109]. The detection rate of CMV averaged 16.8% in graft arteriosclerosis versus 4.5% in native CHD. Rates actually lower than other studies of native atherosclerosis overall. The large variation and discrepancies may due to technique and lack of standardization of methods and collection of samples.

Biological Mechanisms for Herpesviruses

Endothelial injury is important in the pathogenesis of atherosclerosis and there are evidences that HSV and CMV can infect human endothelial cell lines, which may cause arterial injury though increased expression of adhesion molecules with enhanced adherence of leucocytes that may cause cell injury, inflammatory and procoagulant response [110-112]. Both CMV and HSV can also infect vascular smooth muscle cells and may induce proliferation and migration of these cells into the intima (through expression of PDGF and other chemokines) [113-115], important developments in atherogenesis. Viral infection of vascular smooth-muscle cells may also contribute to pathogenesis of atherosclerosis or unstable plaque by alteration of extracellular matrix proteins. In vitro HSV-1 infection of bovine smooth-muscle cells can alter extracellular matrix synthesis [116].

Restenosis vasculopathy post-PTCA (with or without stent) is mainly related to migration of vascular smooth-muscle cells from the media of the vessel wall into the lumen – neointimal proliferation. CMV may contribute to the pathogenesis of restenosis by the ability of the immediate early (IE) gene product (IE 2-84) to bind to p53 (tumor suppressive protein) and inhibits its activity [117]. Since p53 regulates gene products which controls cell cycle progression, the inhibitory effect of CMV IE protein on p53 could increase smooth-muscle cell accumulation and contribute to restenosis [118].

Viral infection may also play a role in atherogenesis through their effect on cholesterol and lipoprotein. Previously avian herpes virus was shown to enhance cholesterol and the ester accumulation in cultured arterial smooth-muscle cells [119]. Similarly, HSV-I infection of arterial smooth-muscle cells also enhanced LDL binding and uptake in vitro [120]. CMV can also increase modified LDL or oxidized LDL uptake in human and rat smooth-muscle cells [121]. Infections of endothelial vascular cells by CMV or HSV can result in rapid procoagulant response induced by inflammatory mediators, and activation of platelets, von Willebrand factor, tissue factor, and factor X/Xa [109]. This may precipitate ischaemic events by acute thrombosis on vulnerable plaques or at critical narrowing of arteries. See figure 2 for a schematic model of viral induced atherosclerosis and vasculopathy.

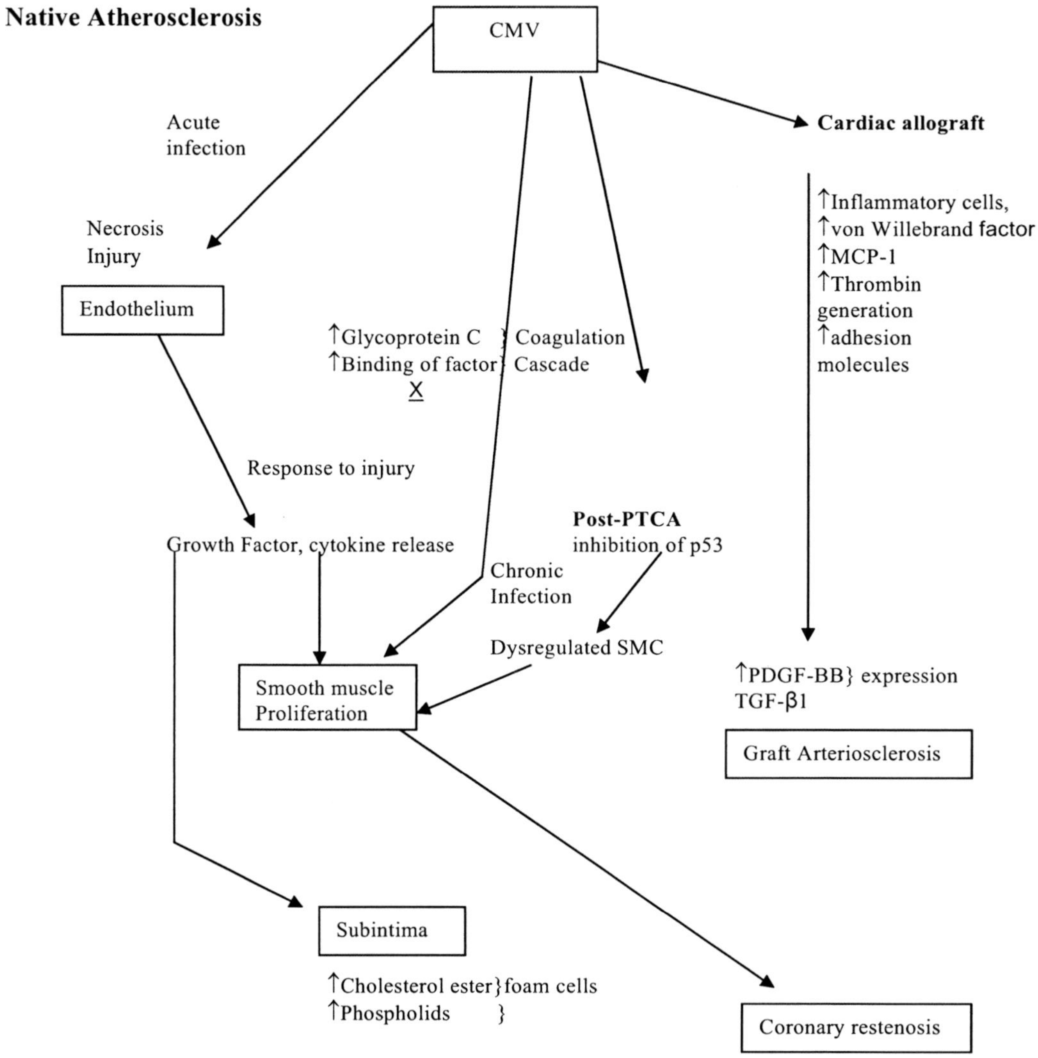

PTCA = percutaneous transluminal coronary angioplasty.
SMC = smooth muscle cell.
MCP-1 = monocyte chemoattractant protein –1.
PDGD = platelet derived growth factor.
TGF-β1 = transforming growth factor beta 1.

Figure 2. A schematic outline for native and cardiac graft atherosclerosis and post-PTCA coronary vasculopathy is shown above for cytomegalovirus. Adapted from Fong, I.W., (ref. 109) with permission of the publisher.

Animal Models of Herpes Viruses and Atherosclerosis

The concept that herpes viruses could play a role in atherogenesis was supported by studies with an avian herpes virus (that causes Marek's disease), with production of lesions in the aorta and coronary arteries of pathogen free chickens, resembling human atherosclerosis [122]. Furthermore, immunization with the related herpes virus of turkeys prevented Marek's disease virus induced atherosclerosis [123].

Studies with CMV in the normocholesterolemic rat model resulted in minimal endothelial cell damage and lipid accumulation in the endothelial cells – early atherosclerotic changes [124]. In addition CMV infection of apoE deficient mice results in enhancement of atherosclerosis over that seen in non-infected hypercholesterolemic mice [125].

In a rodent model to simulate allograft arteriosclerosis, CMV infection has been demonstrated to enhance allograft arteriosclerosis of aortic allograft [126] and cardiac allograft [127]. Moreover, early antiviral therapy with ganciclovir significantly delayed or reduced CMV enhanced allograft arteriosclerosis [128].

Clinical Trial in Transplant Arteriosclerosis

There are no studies to date on intervention with anti-viral agents in native CHD or restenosis post-PTCA. However, a couple of small trials have assessed the effect of anti-viral therapy in transplant arteriosclerosis. In a randomized placebo controlled study ganciclovir for 28 days after cardiac transplantation appear to lower graft coronary arteriosclerosis in patients not receiving calcium channel blockers ($p<0.03$) [129]. In a cohort of heart and heart-lung transplants patients receiving ganciclovir and CMV hyperimmune globulin compared to match controls who received ganciclovir alone, there was lower prevalence and degree of coronary artery thickening in those who received combined therapy [130].

Summary

Herpes viruses, particularly CMV is strongly implicated in cardiac allograft accelerated arteriosclerosis. Although there are biological (in vitro) and animal data to support a plausible role in native atherosclerosis and restenosis of the coronary artery post-PTCA, the evidences in humans are very weak. A summary of the evidences is shown in Table 3.

***Helicobacter pylori* and Atherosclerosis**

Helicobacter pylori is a well established cause of peptic ulcers, and it is strongly associated with gastric carcinoma and gastric lymphomas (mucosa-associated lymphoid tumors) [131]. *H. pylori* has been isolated from persons all over the world and the incidence of colonization is highest in countries or settings of suboptimal sanitary conditions.

Table 3. Summary of the Evidence for CMV/HSV Association with Atherosclerosis

	No (% tve)		References
A. Epidemiological Studies			
Native Atherosclerosis (CMV)			
• Case-control/cross-sectional	18 (39%)		109
• Prospective/longitudinal	15 (20%)		109
Restenosis post-PTCA	8 (37.5%)		109
Cardiac transplantation	22 (54.5%)		109
HSV			
• Case-control/cross-sectional	6 (16.6%)		109
• Prospective/longitudinal	8 (25%)		109
	Mean prevalence		
	Diseased	**controls**	
B. Pathological Studies			
• Native atherosclerosis (CMV)	47.5%	54.2%	109
• Native atherosclerosis (HSV)	16.6%	12.1%	109
• Post PTCA specimens	Restenosis	Native CAD	
CMV tve	47.5%	4.2%	109
• Cardiac transplant	Grafts	Native CAD	
CMV tve	16.8%	4.5%	109
C. Biological Mechanisms			
• CMV/HSV can infect human endothelial & SMC's			110-112
• Can ↑ adhesion molecules, ↑ chemokines and PDGF			113-115
• Cause proliferation and migration of SMC to intima			113-115
• HSV can alter extracellular matrix synthesis			116
• CMV bind and inactivate p53, ↑ SMC accumulation			117, 118
• CMV ↑ oxidized LDL uptake in SMC			121
• CMV/HSV activate platelets, von Willbrand & tissue factor			109
D. Animal Models			References
• Marek's Disease virus ↑ atherosclerosis in chickens			122, 123
• CMV in rats cause damage and ↑ lipids in vascular endothelium			124
• CMV in apoE -/- mice enhance atherosclerosis			125
• CMV accelerate ↑ arteriosclerosis of aortic and cardiac grafts			126, 127
• ganciclovir reduced CMV allograft arteriosclerosis			128
E. Clinical Trials			References
• ganciclovir lower coronary graft arteriosclerosis (small trial)			129
• in case-control cohort ganciclovir & CMV hyperimmune globulin reduce graft arteriosclerosis more than ganciclovir alone			130

CMV = cytomegalovirus; HSV= herpes simplex virus.
PTCA = percutaneous transluminal coronary angioplasty.
SMC = smooth muscle cell; PDGF= platelet derived growth factor.
LDL = low density lipoprotein.
ApoE -/- =apolipoprotein E deficient.

Epidemiological Association of *Helicobacter pylori*

H. pylori has been associated with CHD and cerebrovascular disease in some, generally small, case control or cross-sectional studies, and larger studies with proper adjustment for confounders tended to show no or weaker associations [23]. In a review of 18 prospective, longitudinal studies on this topic only one (5.5%) showed a positive association with IgG antibiotics (OR 2.5) and another showed a trend, OR 2.6 (p=0.08) [132]. Two of these studies measured antibodies to the virulence factor Cag A and found no association with CHD or cardiovascular mortality. Subsequently two recent prospective studies have been reported. In a population based study of 684 subjects common carotid artery intima thickness changes over 5 years were significantly enhanced in subjects seropositive to Cag A but not Cag A-negative *H. pylori* strains [133]. In the other nested case control series over 5.1 years (with 204 controls and 121 first myocardial infarction), although *H. pylori* antibodies was associated with modest increased risk for myocardial infarction (OR=1.90), this was not evident in non-smokers or when adjusted for education [134]. Thus the relationship noted between *H. pylori* and atherosclerotic disease may be due to confounding or co-linearity with socioeconomic status.

Pathological Association of *Helicobacter pylori*

There is no good evidence that *H. pylori* invades the mucosa or submucosal of the stomach or duodenum to even cause bacteremia. However, a few studies (three out of seven) have reported detection of *H. pylori* nuclei acid in atheromatous vessels and not in controls, from 22% to 52.6% [132]. These results are considered unproven as numerous studies have failed to find tissue invasion by *H. pylori*. Furthermore, where as T helper cells in atherosclerotic lesions show *C. pneumoniae* specific reactivity in patients with anti-*C. pneumoniae* antibodies, *H. pylori* specific T cells were found only in the gastric mucosa but not in plaques of the same patient with *H. pylori* antibodies [135]. Indicating that *H. pylori* has no direct role in atherosclerosis in subjects with seropositivity.

Biological Mechanisms of *Helicobacter pylori*

There are few data to support a plausible role of *H. pylori* in the pathogenesis of atherosclerosis. Hypothetical mechanisms include stimulation of an autoimmune reaction to human endogenous heat shock protein (HSP) 60 by *H. pylori* HSP, and an indirect effect of systemic inflammatory mediators by local mucosal inflammation, which may affect homeostasis [136]. Theoretically, *H. pylori* infection could potentiate acute thrombus formation and lead to ischemic events as in experimentally infected mice platelet aggregation and procoagulant activity are enhanced [137].

It has been proposed that *H. pylori* induced chronic gastritis may lead to folate deficiency and hence hyperhomocystinemia, a recently recognized cardiovascular risk factor. However, the data on the relationship between *H. pylori* infection and homocysteine concentrations are conflicting [132,138]. Elevation of soluble endothelial cell adhesion molecules have been found to be a marker of atherosclerotic vascular disease. As discussed earlier in this chapter these markers are early evidence of inflammation that play a role in the adhesion and migration of leucocytes to arterial intima. Thus, stimulation or increased activity of these molecules through infection such as *H. pylori* may contribute indirectly to acceleration of atherosclerosis. However, in a recent study elevated concentration of soluble ICAM-1 and E-selectin were higher in CHD patients seropositive for chlamydia, but not for *H. pylori* antibodies alone [139].

Animal Models for Helicobacter pylori

There has been only one report on the effect of *H. pylori* on atherosclerosis in an animal model, using C57/BL6 and LDL-receptor deficient mice fed a high cholesterol diet. No effect of *H. pylori* infection was found on atherosclerosis in either genotype mouse [140]. The number of animals maybe too small (eight per group) and duration of the experiment maybe too short (12 weeks), to rigorously exclude an enhancing effect.

Table 4. Summary of Evidence of *H. pylori* Association with Atherosclerosis

		References
A.	**Epidemiological Studies**	
•	Case-control/cross-sectional small studies positive/larger studies negative	23
•	Prospective/longitudinal 1/18 (5.5%) showed positive association	132
•	Longitudinal – carotid artery intima thickness associated with Cag A + but not Cag –ve	-133
B.	**Pathological Studies**	
•	A few studies (3/7) detect *H. pylori* nuclei acid in atheromas	132
•	T-cells react to *H. pylori* in gastric mucosa but not in plaques	135
C.	**Biological mechanisms**	
•	*H. pylori* stimulates tissue factor & plasminogen activator inhibition	136
•	Infected mice with *H. pylori* enhance platelet aggregation & procoagulant activity	137
•	*H. pylori* may cause hyperhomocystinemia (conflicting data)	132-139
D.	**Animal Models**	
•	*H. pylori* failed to enhance atherosclerosis in LDL-receptor deficient mice	140
E.	**Clinical Trial**	
•	Regimens for *H. pylori* reduced secondary events in patients with acute myocardial infarction or unstable angina at 1yr but not related to *H. pylori* serology (N=325)	68

LDL = low density lipoprotein.

Clinical Trials in *Helicobacter pylori* Infection

There is limited clinical interventional studies targeting *H. pylori* in cardiovascular subjects. In a recent report of the STAMINA trial, 325 patients with acute myocardial infarction or unstable angina were randomized to receive one week of azithromycin plus metronidzaole and omeprozole, or one week of amoxicillin plus metronidazole and omeprazole, or placebo [68]. Both treatment regimens, which targeted *H. pylori*, reduced cardiac events compared to placebo by 36% after one year [68]. However, *H. pylori* or *C. pneumoniae* antibody status was not significantly related to treatment response, and active infection with *H. pylori* was not investigated by ^{13}C urea breath test.

Summary

The overall epidemiological evidence does not support an association of atherosclerotic disease with *H. pylori*. Although a few pathological studies have reported finding *H. pylori* nuclei acid in atheromas, the majority have not and the data is questionable. There is sparse data on biological mechanism to support a distant role of *H. pylori* in atherogenesis but this is not confirmed by animal experiments. A summary of the evidence is shown in Table 4.

Microbial Burden and Atherosclerosis

Most studies investigating the relationship between infection and atherosclerotic related disease concentrate on the effect of single individual pathogens, a few studies have examined the effect of multiple pathogen burdens. It has been postulated that the aggregate pathogen burden, through molecular mimicry and cytokine stimulation, may have a greater effect on atherogenesis than single or individual pathogen [141].

In a few cross-sectional studies the combined serologies of infectious agents (2 or more of *C. pneumoniae*, *H. pylori*, CMV, HSV, and hepatitis A) have shown stronger association with CHD than antibodies to any single pathogen [142-144]. In a review of five prospective studies addressing the issue of pathogen burden and atherosclerotic disease, four of the five studies found greater association with serologies to multiple pathogens than a single organism [132]. Recently it has been reported that although *C. pneumoniae* antibodies was associated with higher concentrations of soluble endothelial cell adhesion molecules in CHD patients, the concentration of VCAM-1 was even more pronounced when present in combination with *H. pylori* IgA antibodies [139].

Biological Mechanisms and Animal Models for Pathogen Burden

There is little data on biological mechanisms to support the concept of microbial burden as compared to single pathogens, playing a greater role in atherogenesis. In vitro dual infection of He La cells with CMV and *C. pneumoniae* resulted in increased transactivation of the major immediate early promotor of CMV (which is essential for viral gene expression and viral replication) [145]. Thus, *C. pneumoniae* could cause reactivation of latent CMV infection of endothelial or vascular smooth-muscle cells and potentially may contribute to synergistic effect on atherogenesis.

Very few studies have been performed in animal's to investigate the effect of combined pathogens versus single agent. In a recent study the effects of co-infection with *C. pneumoniae* and *H. pylori* on vascular endothelial dysfunction and VCAM-I expression were measured in apoE-knockout mice [146]. Staining for VCAM-I was more intense at the branching sites of aortas from mice with co-infection than in mono-infected or non-infected mice. Relaxation responses to methacholine and the plasma levels of nitrate/nitrite were significantly less in the co-infected group than in other groups ($p<0.05$). These findings suggest an additive interaction of these pathogens in atherogenesis. However, in another study using apoE-knockout mice, murine CMV alone or *C. pneumoniae* alone increased atherosclerotic lesion size by 84% and 70% ($p<0.001$), respectively; but the combination of the two pathogens increased lesion size only by 45% [125]. Thus, this model did not confirm that multiple pathogens increased progression of atherosclerosis to a greater extent than seen with a single pathogen.

Summary

A few prospective seroepidemiological studies have shown a stronger association with antibodies to multiple pathogens than to any single agent. There is sparse data on biological mechanisms and animal models, and the evidence does not strongly support the concept of pathogen burden influencing atherogenesis.

Future Directions

There is clearly need for further investigations in the area of infections and atherosclerosis, but the most promising infections for further research include *C. pneumoniae*, periodontitis and pathogen burden. Further clinical or prospective epidemiological studies assessing *C. pneumoniae* infection and atherosclerotic diseases need to be designed with new investigative diagnostic tools. Currently, there is no established non-invasive technique to determine subjects with chronic persistent chlamydia infection of the vasculature, from those with just past exposure. The initial expectations of PCR in detecting *C. pneumoniae* in peripheral blood mononuclear cells (PBMC), as a reliable predictor of vessel involvement, has not borne fruition [24]. However, further studies with a standardized PCR method in a large number of subjects with histological evidence of *C. pneumoniae* in atheromas is still warranted. We have recently demonstrated that antibodies to CHSP-60 is a good predictor of *C. pneumoniae* antigen in plaques (88.4% positive predictive value) whereas standard serology (using MIF to measure IgG to MOMP) to the organism was not [147]. However, larger studies are needed to confirm this impression. Two recent cross-sectional studies assessing anti-CHSP-60 in patients with CHD have reported promising results. In one study anti-CHSP-60 was found to be a very sensitive and specific marker for acute coronary syndromes, 99% positive in unstable angina and acute myocardial infarction but in only 20%of patients with stable angina and none of the controls [148]. In the other study comparing 250 cases with angiographically significant coronary artery disease ($\geq$ 70% stenosis) with 250 normal controls (< 10% stenosis), elevated anti-CHSP-60, but not anti-human or anti-*Escherichia coli*, was independently associated with CHD after controlling for classical risk factors [149]. Thus, future large prospective studies need to measure anti-CHSP-60 (multiple times) and correlate development of vascular events and progression of atherosclerosis on imaging with this antibody.

Animal studies with *C. pneumoniae* and atherosclerosis has largely been performed on small animals, but future studies need to be extended to larger animals such as pigs and primates to help establish the role in atherogenesis. The preliminary studies on pigs appears promising in inducing endothelial dysfunction and neointimal formation with *C. pneumoniae*, but larger and longer term studies should be executed. Before planning further interventional studies on *C. pneumoniae* and atherosclerotic events, the most effective or optimal regimen in reversing atherosclerotic changes need to be defined in animal models. These trials best involve young adults with no clinical vascular disease, and it would be more cost effective in large, long-term studies to use newer imaging techniques to follow atherosclerotic changes.

With respect to periodontitis, further animal studies are needed in both small and larger animal models to help establish the role in atherogenesis. Large interventional trials in middle-aged subjects (with early periodontal disease) without clinical vascular disease, and properly controlled for confounding factors, and followed for several years for clinical events and atherosclerotic changes on imaging are warranted.

Future investigations on microbial burden and pathogenesis should continue in longitudinal, prospective epidemiological studies utilizing anti-CHSP-60, ^{13}C urea breath test, periodic periodontal assessment, and CMV serology. Animal models should be performed with greater than two microorganisms, with sufficiently large sample size and extended periods (> 20 weeks) to determine the plausibility of pathogen burden on atherosclerosis.

References

[1] Nieto, J.F. Infection and atherosclerosis. New clues from an old hypothesis? *Am. J. Epidemiol.* 1998; 148:937-948.

[2] Osler, W (ed). Diseases of the arteries. In: Modern Medicine: its theory and practice. *Philadelphia: Lea & Febiger*, 1908: p 426-447.

[3] Gibbon, R.V. Germs, Dr. Billings, and the theory of focal infection. *Clin. Infect. Dis.* 1998; 27:627-633.

[4] Finging, G & Hanke, H. Nicolai Nikolajewitsch Anitschkow (1883-1964) established the cholesterol fed rabbit as a model for atherosclerosis research. *Atherosclerosis,* 1997; 135:1-7.

[5] Saikku, P., Leinonen, M., Matilla, K., Ekman, M.R., Neiminen, M.S., Makela, P.H., Huttunen, J.K., & Valtonen, V. Serological evidence of an association of a novel *Chlamydia* TWAR, with chronic heart disease and acute myocardial infarction. *Lancet,* 1988; 2:983-986.

[6] Ross, R. Mechanisms of disease: atherosclerosis – an inflammatory disease (Review). *N. Engl. J. Med.* 1999; 340:115-126.

[7] Rader, D.J. Inflammation markers of coronary risk (Editioral). *N. Eng. J. Med.* 2000; 343:1179-1182.

[8] McCarty, M.F. Interleukin – 6 as a central mediatory of cardiovascular risk associated with chronic inflammation, smoking, diabetes and visceral obesity: down-regulation with essential fatty acids, ethanol and pentoxyphylline. *Med. Hypotheses.* 1998; 52:465-477.

[9] Yudkin, J.S., Kumari, M., Humphries, S.E., Mohamed-Ali, V. Inflammation, obesity, stress and coronary heart disease: is interleukin – 6 the link? (Review). *Atherosclerosis,* 2000; 148:209-214.

[10] Naghavi, M., Barlas, Z., Siadaty, S., Naguib, S, Madjid, M., Casscells, W. Association of influenza vaccination and reduced risk of recurrent myocardial infarction. *Circulation,* 2000; 102:3039-3045.

[11] Bevilacqua, M.P., Nelson, R.M., Mannori, G., Cecconi, O. Endothelial-leucocyte adhesion molecules in human disease. *Annu. Rev. Med.* 1994; 45:361-378.

[12] Gallin, J.I., Kaye, D., O'Leary, W.M. Serum lipids in infection. *N. Engl. J. Med.* 1969; 281:1081-1086.

[13] Khovidhunkit, W., Memon, R.A., Feingold, K.R., Grunfeld, C. Infection and inflammation-induced proatherogenic changes of lipoproteins. *J. Infect. Dis.* 2000; 181 (suppl. 3):S462-472.

[14] Memon, R.A., Staprans. I., Noor, M., Holleran, W.M., Uchinda, Y., Moser, A.H., Feingold, K.R., Grunfeld, C. Infection and inflammation induce LDL oxidation in vivo. *Arterioscler. Thromb. Vasc. Biol.* 2000; 20:1536-1542.

[15] Feingold, K.R., Krauss, R.M., Pang, M., Doerrler, W., Jensen, P., Grunfled, C. The hypertriglyceridemia of acquired immunodeficiency syndrome is associated with an increased prevalence of low density lipoprotein subclass pattern B. *J. Clin. Endocrinol. Metab.* 1993; 76:1423-1427.

[16] Geng, Y.J., Henderson, L.E., Levesque, E.B., Muszynski, M., Libby, P. FAS is expressed in human atherosclerotic intima and promotes apoptosis of cytokine-primed human vascular smooth muscle cells. *Arterioscler. Thromb. Vasc. Biol.* 1997; 17:2200-2208.

[17] Moore, K.L., Andeoli, S.P., Esmon, N.L., Esmon, C.T., Bang, N.U. Endotoxin enhances tissue factor and suppresses thrombomodulin expression of human vascular endothelium in vitro. *J. Clin. Invest.* 1987; 79:124-130.

[18] Colucci, M., Paramo, J.A., Collen, D. Generation in plasma of a fast acting inhibitor of plasminogen activator in response to endotoxin stimulation. *J. Clin. Invest.* 1985; 75:818-824.

[19] Abi-Younes, S., Sauty, A., Mach, F., Sukhova, G.K., Libby, P., Luster, A.D. The stromal cell-derived factor-1 chemokine is a potent platelet agonist highly expressed in atherosclerotic plaques. *Circ. Res.* 2000; 86:131-138.

[20] Grayston, J.T. Infection caused by *Chlamydia pneumoniae* stain TWAR. *Clin. Infect. Dis.* 1992; 15:757-763.

[21] Wang, S.P., Grayston, J.T. Population prevalence of antibody to *Chlamydia pneumoniae* strain TWAR, in: *Chlamydia Infections* (eds Bowie, W.R., Caldwell, H.D., Jones, R.P, et al), Cambridge University Press, Cambridge, England, 1999; p.402-405.

[22] Morrison, R.P. New insights into a persistent problem—chlamydial infections. *J. Clin. Invest.* 2003, 111:1647-1649.

[23] Danesh, J., Collins, R., Peto, R. Chronic infections and coronary heart disease: is there a link? *Lancet* 1997; 350:430-436.

[24] Fong, I.W. (Editor). *Chlamydia pneumoniae and the cardiovascular system. In: Infections and the cardiovascular system: New perspectives.* New York: Kluwer Academic/Plenum Publishers, 2003:121-156.

[25] Danesh, J., Whincup, P., Walker, M., Lennon, L., Thompson, A., Appleby, P., Wong, Y.K., Bernardes-Silva, M., Ward, M. *Chlamydia pneumoniae* IgG titers and coronary heart disease: prospective study and meta-analysis. *BMJ* 2000; 321:208-212.

[26] Bloemenkamp, D.G.M., Mali, W.P.T.M., Visseren, F.L.J., van der Graff, Y. Meta-analysis of sero-epidemiologic studies of the relation between *Chlamydia pneumoniae* and atherosclerosis: Does study design influence results? *Am. Heart J.* 2003; 145:409-417.

[27] Dobrilovic, N., Vudlomari, L., Meyer, M., Wright, C.B. *Chlamydia pneumoniae* in atherosclerotic carotid artery plaques: high prevalence among heavy smokers. *Am. Surgeon*, 2001; 67:589-593.

[28] Jackson, L.A., Campbell, L.A., Schmidt, R.A., Kuo, C.C., Cappuccio, A.L., Lee, M.J., Grayston, J.T. Specificity of detection of *Chlamydia pneumoniae* in cardiovascular atheroma: evaluation of the innocent bystander hypothesis. *Am. J. Pathol.* 1997; 150:1785-1790.

[29] Karlsson, L., Gnarpe, J., Naas, J., Olsson, G., Lindholm, J., Steen, B., Gnarpe, H. Detection of viable *Chlamydia pneumoniae* in abdominal aortic aneurysms. *Eur. J. Vasc. Endovasc. Surg.* 2000; 19:630-635.

[30] Maas, M., Bartels, C., Engel, P.M., Mamat, U., Sievers, H.H. Endovascular presence of viable *Chlamydia pneumoniae* is a common phenomenon in coronary arteries. *J. Am. Coll. Cardiol.* 1998; 31:827-832.

[31] Kol, A., Sukhova, G.K., Lichtman, A.H., Libby, P. Chlamydial heat shock protein localizes in human atheroma and regulates tumor necrosis factor-alpha and matrix metalloproteinase expression. *Circulation* 1998; 98:300-307.

[32] de Boer, O.J., van der wal, A.C., Houtkamp, M.A., Ossewaarde, J.M., Teeling, P., Becker, A.E. Unstable atherosclerotic plaque contain T-cells that respond to *Chlamydia pneumoniae. Cardiovasc. Res.* 2000; 48:402-408.

[33] Mosorin, M., Surcel, H.M., Laurila, Lehtinen, M., Karttunen, R., Juvonen, J., Paavonen, J., Morrison, R.P., Saikku, P., Juvonen, T.A. Detection of *Chlamydia pneumoniae*-reactive T lymphocytes in human atherosclerotic plaques of carotid artery. *Arterioscler. Thromb. Vasc. Biol.* 2000; 20:1061-1067.

[34] Gaydos, C.A., Summergill, J.T., Sahney, N.N., Ramirez, J.A., Quinn, T.C. Replication of *Chlamydia pneumoniae* in vitro in human macrophages, endothelial and aortic artery smooth muscle cells. *Infect. Immun.* 1996; 64:1614-1620.

[35] Haranaga, S., Yamaguchi, H., Friedman, H., Izsumi, S.T., Yamamoto, Y. *Chlamydia pneumoniae* infects and multiplies in lymphocytes in vitro. *Infect. Immun.* 2001; 69:7753-7759.

[36] Coombes, B.K., Mahony, J.B. *Chlamydia pneumoniae* infection of human endothelial cells induces proliferation of smooth muscle cells via an endothelial derived soluble factor(s). *Infect. Immun.* 1999; 67:2909-2915.

[37] Coombes, B.K., Mahony, J.B. cDNA array analysis of altered gene expression in human endothelial cells in response to *Chlamydia pneumoniae* infection. *Infect. Immun.* 2001; 69:1420-1427.

[38] Rodel, J., Woytas, M., Groh, A., Schmidt, K.H., Hartmann, M., Lehmann, M., Straube, E. Production of basic fibrolast growth factor and interleukin-6 by human smooth muscle cells following infection with *Chlamydia pneumoniae. Infect. Immun.* 2000; 68:3635-3644.

[39] Kaukoranta-Tolvanen, S.S., Ronni, T., Leinonen, M., Saikku, P., Laitinen, K. Expression of adhesion molecules on endothelial cells stimulated by *Chlamydia pneumoniae. Microb. Pathog.* 1996; 21:407-411.

[40] Netea, M.G., Selzman, C.H., Kullberg, B.J., Galama, J.M., Weinberg, A., Stulenhoef, A.F., van der Meer, J.W., Dinarello, C.A. Acellular components of *Chlamydia pneumoniae* stimulate cytokine production in human mononuclear cells. *Eur. J. Immunol.* 2000; 30:541-549.

[41] Kalayoglu, M.V., Byrne, G.I. Induction of macrophage foam cell formation by *Chlamydia pneumoniae. J. Infect. Dis.* 1998; 177:725-729.

[42] Kalayoglu, M.V., Hoerrman, B., La Verda, D., Morrison, S.G., Morrison, R.P., Byrne, G.I. Cellular oxidation of low density lipoprotein by *Chlamydia pneumoniae. J. Infect. Dis.* 1999; 180:780-790.

[43] Kol, A., Boucier, T., Lichtman, A.H., Libby, P. Chlamydial and human heat shock protein protein-60's activate human vascular endothelial, smooth muscle cells and macrophages. *J. Clin. Invest.* 1999; 103:571-577.

[44] Mayr, M., Metzler, B., Kiechl, S., Willeit, J., Schett, G., Xu, Q., Wick, G. Endothelial cytotoxicity mediated by serum antibodies to heat shock proteins of *Escherichia coli* and *Chlamydia pneumoniae*: immune reactions to heat shock proteins as possible link between infection and atherosclerosis. *Circulation* 1999; 99:1560-1566.

[45] Hirono, S., Dibrov, E., Hurtado, C., Kostenuk, A., Ducas, R., Peirce, G.N. *Chlamydia pneumoniae* stimulates proliferation of vascualr smooth muscle cells through induction of endogenous heart shock protein 60. *Cir. Res.* 2003; 93:710-716.

[46] Niessner, A., Kaun, C., Zorn, G., Speidl, W., Turel, Z., Christiansen, G., Pedersen, A.S., Birkelund, S., Simon, S., Georgopoulos, A., Graninger, W., de Martin, R., Lipp, J., Binder, B.R., Maurer, G., Huber, K., Wojta, J. Polymorphic membrane protein (PMP) 20 and PMP 21 of *Chlamydia pneumoniae* induce proinflammatory mediators in human endothelial cells in vitro by activation of the nuclear factor kappa β pathway. *J. Infect. Dis.* 2003; 188:108-113.

[47] Vehmaan-Kreula, P., Puolakkainen, M., Sarvas, M., Welgus, H.G., Kovanen, P.T. *Chlamydia pneumoniae* proteins induces secreation of the 92-kDa gelatinase by human monocytes derived macrophages. *Arterioscler. Thromb. Vasc. Biol.* 2001; 21:E1-8.

[48] Dechend, R., Maass, M., Gieffers, J., Dietz, R., Scheidereit, C., Leutz, A., Gulba, D.C. *Chlamydia pneumoniae* infection of vascular smooth and endothelial cells activates NF-kappa β and induces tissue factor and PAI-1 expression: a potential link to accelerated atherosclerosis. *Circulation* 1999; 100:1369-1373.

[49] Luiba, P., Karmani, P., Pesonen, E., Paakkari, I., Forslid, A., Johansson, L., Persson, K., Wadstrom, T., Laurini, R. Endothelial dysfunction after repeated *Chlamydia pneumoniae* infection in apolipoprotein E-knockout mice. *Circulation* 2000; 102:1039-1044.

[50] Luiba, P., Pesonen, E., Paakkari, I., Batra, S., Forslid, A., Kovanen, P., Pentikainen, M., Pesonen, K., Sandström, S. Acute *Chlamydia pneumoniae* infection causes endothelial dysfunction in pigs. *Atherosclerosis*, 2003; 167:215-222.

[51] Fong, I.W., Chiu, B., Viira, E., Jang, D., Mahony, J.B. De novo induction of atherosclerosis by *Chlamydia pneumoniae* in a rabbit model. *Infect. Immun.* 1999; 7:6048-6055.

[52] Laitinen, K., Laurila, A., Pyhala, L., Leinonen, M., Saikku, P. *Chlamydia pneumoniae* infection induces inflammatory changes in the aortas of the rabbits. *Infect. Immun.* 1997; 65:4832-4835.

[53] Blessing, E., Lin, T.M., Campbell, L.A., Rosenfeld, M.E., Lloyd, D., Kuo, C.C. *Chlamydia pneumoniae* induces inflammatory changes in the heart and aorta of normocholesterolemic C57BL/6J mice. *Infect. Immun.* 2000; 68:4765-4768.

[54] Pislaru, S.V., Van Ranst, M., Pislaru, C., Szelid, Z., Theilmeier, G., Ossewaarde, J.M., Holvoet, P., Janssens, S., Verbeken, E., Van de Werf, F.J. *Chlamydia pneumoniae* induces neointimal formation in coronary arteries of normal pigs. *Cardiovasc. Res.* 2003; 57:834-842.

[55] Blessing, E., Campbell, L.A., Rosenfeld, M.E., Chough, N., Kuo, C.C. *Chlamydia pneumoniae* accelerates hyperlipidemia induced atherosclerotic lesion development in C57BL/6J mice. *Atherosclerosis,* 2001; 158:13-17.

[56] Moazed, T.C., Campbell, L.A., Rosenfeld, M.E., Grayston, J.T., Kuo. C.C. *Chlamydia pneumoniae* accelerates the progression of atherosclerosis in Apolipoprotein E-deficient mice. *J. Infect. Dis.* 1999; 180:238-241.

[57] Burnett, M.S., Gaydos, C.A., Madico, G.E., Glad, S.M., Paigen, B., Quinn, T.C., Epstein, S.E. Atherosclerosis in ApoE Knockout mice infected with multiple pathogens. *J. Infect. Dis.* 2001; 183:226-231.

[58] Rothstein, N.M., Quinn, T.C., Madico, G., Gaydos, C.A., Lowenstein, C.J. Effect of azithromycin on murine arteriosclerosis exacerbated by *C. pneumoniae. J. Infect. Dis.* 2001; 183:232-238.

[59] Hu, H., Pierce, G.N., Zhong, G. The atherogenic effects of chlamydia are dependent on serum cholesterol and specific to *Chlamydia pneumoniae. J. Clin. Invest.* 1999; 103:747-753.

[60] Aalto-Setälä, K., Laitinen, K., Erkkilä, L., Leinonen, M., Janhainen, M., Ehnholm, C., Tamminen, M., Puolokkainen, M., Penttilä, I., Saikku, P. *Chlamydia pneumoniae* does not increase atherosclerosis in the aortic root of apoliprotein E-deficient mice. *Arterioscler. Thromb. Vasc. Biol.* 2001; 21:578-584.

[61] Caligiuri, G., Rottenberg, M., Nicoletti, A., Wigzell, H., Hansson, G.K. *Chlamydia pneumoniae* infection does not induce or modify atherosclerosis in mice. *Circulation* 2001; 103:2834-2838.

[62] Muhlestein, J.B., Anderson, J.L., Hammond, E.H., Zhao, L., Trehan, S., Schwobe, E.P., Carlquist, J.F. Infection with *C. pneumoniae* accelerates the development of atherosclerosis and treatment with azithromycin prevents it in a rabbit model. *Circulation* 1998; 97:633-636.

[63] Fong, I.W., Chiu, B., Mahony, J.B., Jang, D., Coombes, B., Dunn, P., Caterini, J., Murdin, A. *C. pneumoniae enhancing effect on atherosclerosis and relationship with lipid profile and fibrinogen in a rabbit model. In: Chlamydial Infections* (Schachter, J., Christiansen, G., Clarke, I.N., Hammerschlag, M.R., Kaltenboeck, B., Kuo, C.C., Rank, R.G., Ridgway, G.L., Saikku, P., Stamm, W.E., Stephens, R.S., Summergill, J.T., Timms, P., Wyrick, P.B., eds), Basim Yeri: Grafmat basin ve Rektam Sanayi Tic, Ltd, Sti, Istanbul, 2002, 321-324.

[64] Ezzahiri, R., Stassen, F.R.M., Kurvers, H.A.J.M., van Pul, M.M.L., Kitsiaar, P.J.E.H.M., Bruggeman, C.A. *Chlamydia pneumoniae* induces unstable atherosclerotic plaque phenotype in LDL-receptor apoE double knockout mice. *Eur. J. Vasc. Endovasc. Surg*. 2003; 26:88-95.

[65] Gupta, S., Leatham, E.W., Carrington, D., Mendall, M., Kaski, J.C., Camm, A.J. Elevated *Chlamydia pneumoniae* antibodies, cardiovascular events and azithromycin, in male survivors of acute myocardial infarction. *Circulation*, 1997; 96:404-407.

[66] Gurfinkel, E., Bozovich, G., Daroca, A., Beck, E., Mautner, B., for the ROXIS study group. Randomized trial of roxithromycin in non-Q wave coronary syndromes: ROXIS pilot study. *Lancet*, 1997; 350:404-407.

[67] Gurfinkel, E., Bozovich, G., Beck, E., Testa, E., Livellara, B., Mautner, B. Treatment with the antibiotic roxithromycin in patients with acute non-Q wave coronary syndrome. Final report of the ROXIS study. *Eur. Heart J.* 1999; 20:121-127.

[68] Stone, A.F., Mendall, M.A., Kaski, J.C., Edger, T.M., Tracey M., Risley, P., Polonieiki, J., Camm, A.J., Northfield, T.C. Effect of treatment for *Chlamydia pneumoniae* and *Helicobacter pylori* on markers of inflammation and cardiac events with acute coronary syndromes. *Circulation*, 2002; 106:1219-1233.

[69] Cercek, B. Azithromycin in Acute Coronary Syndrome Investigators. The effect of short term treatment in patients with acute coronary syndrome. *J. Am. Coll. Cardiol.* 2002, 39 (Suppl. A): 405-413.

[70] Muhlestein, J.B., Anderson, J.L., Carlquist, J.F., Salinkhe, K., Horne, B.D., Pearson, R.R., Burch, T.J., Allen, A., Trehan, S., Nielson, C. Randomized secondary prevention trial of azithromycin in patients with coronary artery disease: primary clinical results of the Academic Study. *Circulation* 2000; 102:1755-1760.

[71] O'Connor, C., Dunne, M.W., Pfeffer, M.A., Muhlestein, J.B., Yao, L., Gupta, S., Benner, R.J., Fisher, M.R., Cook, T.D., for the Investigators, in the WIZARD study. Azithromycin for Secondary Prevention of Coronary Heart Disease events. The WIZARD STUDY: A Randomized Controlled Trail. *JAMA*, 2003; 290:1459-1466.

[72] Jackson, L.A. Description and status of the Azithromycin and Coronary Events study (ACES). *J. Infect. Dis.* 2000; 181 (Suppl. 3): S579-S581.

[73] Cannon, C.P., McCabe, C.H., Belder, R., Breen, J., Braunwald, E. Design of the Provastatin or Atorvastatin Evaluation and Infection Therapy (PROVE IT) – TIMI 22 trail. *Am. J. Cardiol.* 2002; 89:860-861.

[74] Kuo, C.C., Campbell, L.A. Detection of *Chlamydia pneumoniae* in arterial tissues. *J. Infect. Dis.* 2000; 181 (Suppl. 3): S432-S436.

[75] Chiu, B., Viira, E., Tucker, W., Fong, I.W. *Chlamydia pneumoniae*, cytomegalovirus, and herpes simplex virus in atherosclerosis of the carotid artery. *Circulation*, 1997; 96:2144-2148.

[76] Maas, M., Gieffers, J., Krause, E., Engel, P.M., Bartels, C., Solback, W. Poor correlation between microimmunofluorescence serology and polymerase chain reaction for detection of vascular *Chlamydia pneumoniae* infection in coronary artery disease patients. *Med. Microbiol. Immunol.* 1998; 187:103-106.

[77] Fong, I.W., Chiu, B., Viira, E., Jang, D., Fong, M.W., Mahony, J.B. Can an antibiotic (macrolide) prevent *Chlamydia pneumoniae* induced atherosclerosis in a rabbit model? *Clin. Diagn. Lab. Immunol.* 1999; 6:891-894.

[78] Rothstein, N.M., Quinn, T.C., Madico, G., Gaydos, C.A., Lowerstein, C.J. Effect of azithromycin on murine atherosclerosis exacerbated by *C. pneumoniae*. *J. Infect. Dis.* 2001; 183:232-238.

[79] Wolf, K., Malinverni, R. Effect of azithromycin plus rifampin versus that of azithromycin alone on eradication of *Chlamydia pneumoniae* from lung tissue in experimental pneumonitis. *Antimicrob. Agents Chemother.* 1999; 43:1491-1493.

[80] Kothe, H., Dalhoff, K., Rupp, J., Müller, A., Kreuzer, J., Maas, M., Katus, H.A. Hydroxymethylaglutaryl coenzyme A reductose inhibitors modify the inflammatory response of human macrophages and endothelial cells infected with *Chlamydia pneumoniae*. *Circulation*, 2000; 101:1760-1763

[81] Tiran, A., Gruber, H.J., Grain, W.F., Wagner, A.H., Van Leeuven, E.B. Tiran B. Aspirin inhibits *Chlamydia pneumoniae* – induced nuclear factor kappa β activation, cytokine expression, and bacterial development in human endothelial cells. *Arterioscler. Thromb. Vasc. Biol.* 20002; 1075-1080.

[82] Yoneda, E., Miura, K., Matsushima, H., Sugi, K., Murakami, T., Ouchi, K., Yamashita, K., Itoh, H., Nakazawa, T., Suzuki, M., Shirai, M. Aspirin inhibits *Chlamydia pneumoniae* induced NF- (kappa) β activation, cyclooxygenase – 2 expression and prostaglandin E_2 synthesis and attenuates chlamydial growth. *J. Med. Microbiol.* 2003; 52:409-415.

[83] Heart Protection Study Collaborative Group. MRC/BHF heart protection study of cholesterol lowering in 20536 high risk individuals: a randomized placebo controlled trial. *Lancet* 2002; 360:7-22.

[84] Ravnskov, U. Is atherosclerosis caused by high cholesterol? *QJM*, 2002; 95:397-403.

[85] Kalayoglu, M.V., Libby, P., Byrne, G.I. *Chlamydia pneumoniae* as an emerging risk factor in cardiovascular disease. *JAMA,* 2002; 288:2724-2731.

[86] Consensus report on periodontal disease. Pathogenesis and microbial factors. *Ann. Serotypes*, 1996; 1:926-932.

[87] Socransky, S.S., Haffajie, A.D., Cugiri, M.A., Smith, C., Kent, R.L. Mirobial complexes in subgingival plaque. *J. Dent. Res.* 1997; 76 (special issue): 51 [abstract 302].

[88] Matilla, K.J., Nieminen, M.S., Valtonen, V.V., Rasi, V.P., Kesaniemi, Y.A., Syrjala, S.L., Jungell, P.S., Isoluoma, M., Hietaniemi, K., Jokinen, M. Association between dental health and acute myocardial infarction. *BMJ* 1989; 298:779-781.

[89] Fong, I.W., (ed). *Peroidontal disease and the cardiovascular system, in: Infection and the Cardiovascular System: New Perspectives*, Kluwer Academic/Plenum Publishers, New York, 2003, p. 179-200.

[90] Joshipura, K.J., Douglass, C.W., Willett, W.C. Possible explanations for the tooth loss and cardiovascular disease relationship. *Ann. Periodontol.* 1998; 3:175-183.

[91] Beck, J.D., Elter, J.R., Heiss, G., Couper, D., Mauriello, S.M., Offenbacher, S. Relationship of periodontal disease to carotid artery intima-media wall thickness: The atherosclerotic risk in communities (ARIC) study. *Artioscler. Thromb. Vasc. Biol.* 2001; 21:1816-1822.

[92] Mendez, M.V., Scott, T., La Morte, W., Vokonas, P., Menzoian, J.O., Garcia, R. An association between periodontal disease and peripheral vascular disease. *Am. J. Surg.* 1998; 176:153-157.

[93] Hujoel, P.P., Drangsholt, M., Spiekerman, C., De Roven, T.A. Examining the link between coronary heart disease and the elimination of chronic dental infections. *JADA* 2001; 132:883-889.

[94] Thorstensson, H., Kuglensteima, J., Hugoson, A. Medical status and complications in relation to periodontal disease experience in insulin-dependent diabetes. *J. Clin. Peridontol.* 1996: 23:194-202.

[95] Kweider, M., Lowe, E.D.O., Murray, G.D., Kinane, D.F., McGowan, D.A. Dental disease, fibrinogen and white cell count: links with myocardial infarction? *Scot. Med. J.* 1993; 38:73-74.

[96] Hackman, D.G., Anand, S.S. Emerging risk factors for atherosclerotic vascular disease: A critical review of the evidence. *JAMA* 2003; 290:932-940.

[97] Noack, B., Genco, R.J., Trevisan, M., Grossi, S., Zamban, J.S., De Nardin, E. Periodontal infections contributes to elevated systemic C-reactive protein level. *J. Periodontal.* 2001; 72:1221-1227.

[98] Slade, G.D., Offenbacher, S., Beck, J.D., Heiss, G., Parkow, J.S. Acute-phase inflammatory response to periodontal disease in the US population. *J. Dent. Res.* 2000; 79:49-57.

[99] Ebersole, J.L., Machen, R.L., Steffen, M.L., Willmann, D.E. Systemic acute phase reactants, C-reactive protein and haptoglobin, in adult periodontitis. *Clin. Exp. Immunol.* 1997; 107:347-352.

[100] Wu, T., Trevisan, M., Genco, R.J., Falkner, K.L., Dorn, J.P., Sempos, C.T. Examination of the relation between periodontal health status and cardiovascular risk factors: serum total and high density lipoprotein cholesterol, C-reactive protein, and plasma fibrinogen. *Am. J. Epidemiol.* 2000; 151:273-282.

[101] Erikson, P.R., Herzberg, M.C. A collagen-like immunodeterminant on the surface of *streptococcus sanguis* induces platelet aggregation. *J. Immunol.* 1987; 138:3360-3366.

[102] Sharma, A., Novak, E.K., Sojar, H.T., Swank, R.T., Kuramitsu, H.K., Genco, R.J. *Porphyromonas gingivalis* platelet aggregation activity: outer membrane vesicles are potent activators of murine platelets. *Oral Microbial. Immunol.* 2000; 15:393-396.

[103] Imamura, T., Potempa, J., Tanase, S., Travis, J. Activation of blood coagulation factor X by arginine-specific cysteine proteinases (gingipain-RS) from *Porphymonas gingivalis. J. Biol. Chem.* 1997; 272:16062-16067.

[104] Dorn, B.R., Dunn, W.A. Jr., Progulske-Fox, A. Invasion of human coronary artery cells by periodontal pathogens. *Infect. Immun.* 1999; 67:5792-5798.

[105] Nassar, H., Chow, H.H., Khlgation, M., Gibson, F.C. 3rd, Van Dyke, T.E., Genco, C.A. Role for fimbriae and lysine-specific cysteine proteinase gingipain K expression of interleukin-8 and monocyte chemoattractant protein in *Porphyromonas gingivalis* infected endothelial cells. *Infect. Immun.* 2002, 70:268-270.

[106] Kuramitsu, H.K., Qi, M., Kang, I.C., Chen, W. Role for periodontal bacteria in cardiovascular disease. *Ann. Periodontal.* 2001; 6:41-47.

[107] Li, L., Messas, E., Batista, E.L. Jr., Levine, R.A., Amar, S. *Porphymonas gingivalis* infection accelerates the progression of atherosclerosis in a heterozyzous apolipoprotein E-deficient murine model. *Circulation* 2002: 105:861-867.

[108] Jain, A., Batista, E.L. Jr., Serhan, C., Stahl, G.L., Van Dyke, T.E. Role for periodontitis in the progression of lipid deposition in an animal model. *Infect. Immun.* 2003; 71:6012-6018.

[109] Fong, I.W. (ed). *Cytomegalovirus and herpes simplex virus in cardiovascular disease, in: Infection and the Cardiovascular System: New Perspectives.* Kluwer Academic/Plenum Publishers, New York, 2003; p.201-238.

[110] Friedman, H.M., Macarak, E.J., MacGregor, R.R., Wolfe, J., Kefalides, N.A. Virus infection of endothelial cells. *J. Infect. Dis.* 1981; 143:266-273.

[111] Span, A.H., Van Boven, C.P., Bruggeman, C.A. The effect of cytomegalovirus infection on the adherence of polymorphonuclear leucocytes to endothelial cells. *J. Clin. Invest.* 1989; 19:542-548.

[112] van Dam-Mieras, M.C., Muller, A.D., van Hinsberg, V.W., Mullers, W.J., Bomans, P.H., Bruggeman, C.A. The procoagulant response of cytomegalovirus infected endothelial cells. *J. Clin. Invest.* 1989; 19:542-548.

[113] Datta, S.K., Tumilowicz, J.J., Trentin, J.J. Lysis of human arterial smooth muscle cells infected with herpes viridae by peripheral blood mononuclear cells: implications for atherosclerosis. *Viral Immunol.* 1993; 6:153-160.

[114] Zhou, Y.F., Yu, Z.X., Wanishsawad, C., Shou, M., Epstein, S.E. The immediate early gene products of human cytomegalovirus increase vascular smooth muscle migration, proliferation and expression of PDGF β-receptor. *Biochem. Biophys. Res. Commun.* 1999; 256:608-613.

[115] Streblow, D.N., Soderberg-Naucler, C., Viera, J., Smith, P., Wakabayashi, E., Ruchi, F., Mattison, K., Altschuler, Y., Nelson, J.A. The human cytomegalovirus chemokines receptor US28 mediates vascular smooth cell migration. *Cell* 1999; 99:511-520.

[116] Lashgari, M.S., Friedman, H.M., Kefalides, N.A. Suppression of matrix protein synthesis by herpes simplex virus in bovine smooth-muscle cells. *Biochem. Biophys. Res. Commun.* 1987; 143:145-151.

[117] Spier, E., Modali, R., Huang, E.S., Leon, M.B., Shawl, L.F., Finkel, T., Epstein, S.E. Potential role of human cytomegalovirus and p53 interaction in coronary restenosis. *Science* 1994; 265:391-394.

[118] Tanaka, K., Zou, J.P., Takeda, K., Ferrars, V.J., Sandford, G.R., Johnson, T.M., Finkel, T., Epstein, S.E. Effect of human cytomegalovirus immediate early proteins on p53 mediated apoptosis in coronary artery smooth muscle cells. *Circulation*, 1999; 99:1656-1659.

[119] Fabricant, C.G., Hajjar, D.P., Minick, C.R., Fabricant, J. Herpes virus infection enhances cholesterol and cholesteryl ester accumulation in cultured arterial smooth muscle cells. *Am. J. Pathol.* 1981; 105:176-184.

[120] Hsu, H.Y., Nicholson, A.C., Pomerantz, K.B., Kaner, R.J., Hajjar, D.P. Altered cholesterol trafficking in herpes virus infected arterial cell. *J. Biol. Chem.* 1995; 270:19630-19637.

[121] Zhou, Y.F., Guetta, E., Yu, Z.X., Finkel, T., Epstein, S.E. Human cytomegalovirus increases modified low density lipoprotein uptake and scavenger receptor m RNA expression in vascular smooth muscle cells. *J. Clin. Invest.* 1996; 98:2129-2138.

[122] Minick, C.R., Fabricant, C.G., Fabricant, J., Litrenta, M.M. Atherosclerosis induced by infection with a herpes virus. *Am. J. Pathol.* 1979; 96:673-706.

[123] Fabricant, C.G., Fabricant, J., Minick, C.R., Litrenta, M.M. Herpes virus induced atherosclerosis in chickens. *Fed. Proc.* 1983; 42:2476-2479.

[124] Span, A.H.M., Yranuls, G., Bosman, F., van Boven, C.P.A., Bruggeman, C.A. Cytomegalovirus infection induces vascular injury in the rat. *Atherosclerosis*, 1992; 93:41-52.

[125] Burnett, M.S., Gaydos, C.A., Madico, G.E., Glad, S.M., Paigen, B., Quinn, T.C., Epstein, S.E. Atherosclerosis in apoE knockout mice infected with multiple pathogens. *J. Infect. Dis.* 2001; 183:226-231.

[126] Lemström, K.B., Brining, J.H., Bruggeman, C.A., Lautenschlager, I.T., Hayry, P.J. Cytomeglovirus infection enhances smooth muscle cell proliferation and intimal thickening of rat allograft. *J. Clin. Invest.* 1993; 92:549-558.

[127] Lemström, K.B., Koskinen, P.K., Bruning, J.H., Bruggeman, C.A., Hayry, P.J. Cyomegalovirus antigen expression, endothelial cell proliferation, and intimal thickening in rat cardiac allografts after cytomegalovirus infection. *Circulation*, 1995; 92:2594-2604.

[128] Lemström, K.B., Koskinen, P.K., Bruning, J.H., Bruggeman, C.A., Lautenschlager, I.T., Hayry, P.J. Effect of ganciclovir prophylaxis on cytomegalovirus enhanced allograft arterisoclerosis. *Transplant Int.* 7 (Suppl. 1): S383-S384.

[129] Valantine, H.A., Gao, G.Z., Menon, S.G., Renlund, D.G., Hunt, S.A., Oyer, P., Stinson, E.B., Brown, B.W. Jr., Merigan, T.C., Schroeder, J.S. Impact of prophylactic immediate post-transplant ganciclovir on development of transplant atherosclerosis: a post hoc analysis of a randomized, placebo-controlled study. *Circulation*, 1999; 100:61-66.

[130] Valantine, H.A., Luikart, H., Doyle, R., Theodore, J., Hunt, S., Oyer, P., Robbins, R., Berry, G., Reitz, B. Impact of cytomegalovirus hyperimmune globulin on outcome after cardiothorac tansplantation: a comparative study of combined prophylaxis with CMV

hyperimmune globulin plus ganciclovir versus ganciclovir alone. *Transplantation,* 2001; 72:1647-1652.

[131] Blaser, M. *Helicobacter pylori* and related organisms; in: Mandell, G.L., Bennett, J.E., Dolin, R (eds.), *Principles and practice of infectious diseases, 2000*, (5th ed. Vol 2, pp 2285-2293), Churchill Livingstone; Philadelphia.

[132] Fong, I.W. (ed). Miscellaneous infections and atherosclerosis; in: *Infections and the cardiovascular system: New perspectives.* 2003; pp 239-266; Kluwer Academic/Plenum Publishers, New York.

[133] Mayr, M., Kiechl, S., Mendall, M.A., Willeit, J., Wick, G., Xu, Q. Increased risk of atherosclerosis is confined to Cag A-positive *Helicobacter pylori* strains; prospective results from the Brunek study. *Stroke*, 2003; 34:610-615.

[134] Witherell, H.L., Smith, K.L., Friedman, G.D., Ley, C., Thom, D.H., Orentreich, N., Vogelman, J.H., Parsonnet, J. C-reactive protein, *Helicobacter pylori, Chlamydia pneumoniae*, cytomegalovirus and risk for myocardial infarction. *Ann. Epidemiol.* 2003, 13; 170-177.

[135] Benagiano, M., Azzurri, A., Ciervo, H., Amedei, A., Tomburini,C., Ferrari, M., Telford, J.L., Balderi, C.T., Romognario, S., Canone, A., D'Elios, M.M., Del Prete, G. T helper type 1 lymphocytes drive inflammation in human atherosclerotic lesions. *Proc. Natl. Acad. Sci* USA, 2003; 100:6658-6663.

[136] Montemurro, P., Barbuti, G., Dundon, W.G., Del Giudice, G., Rappuoli, R., Colucci, M., De Rinaldis, P., Montecucco, C., Semeraro, N., Papini, E. *Helicobacter pylori* neutrophil activating protein stimulates tissues factor and plasminogen activator inhibition-2 production by human mononuclear cells. *J. Infect. Dis.* 2001; 183:1169-1174.

[137] Elizalde, J.I., Gomez, J., Panes, J., Lozano, M., Casadevall, M., Ramirez, J., Pizcuetta, P., Marco, F., Rojas, F.D., Granger, D.N. Platelet activation in mice and human *Helicobacter pylori* infection. *J. Clin. Invest.* 1997; 100:996-1005.

[138] Yoshino, N., Adachi, K., Takashima, T., Miyaoka, Y., Yuki, T., Ishuhara, S., Kinoshita, Y. *Helicobacter pylori* infection does not affect the serum level of homocysteine. *Am. J. Gastroenterol.* 2002; 97:2927-2928.

[139] Schumacher, A., Seljeflot, I., Lerkerod, A.B., Sommervoll., L., Ottserstad, J.E., Arnesen, H. Does infection with *Chlamydia pneumoniae* and /or *Helicobacter pylori* increase the expression of endothelial cell adhesion molecules in humans? *Clin. Microbiol. Infect.* 2002; 8:654-651.

[140] Mach, F., Sukhova, G.F., Michetti, M., Libby, P., Michetti, P. Influence of *Helicobacter pylori* infection during atherogenesis in vivo in mice. *Circ. Res.* 2002; 90, el (circresaha.org).

[141] Epstein, S.E., Zhu, J., Burnett, M.S., Zhou, Y.F., Vercellotti, G., Hajjar, D. Infection and atherosclerosis: potential roles of pathogen burden and molecular mimicry. *Arterioscler. Thromb. Vasc. Biol.* 2000; 20:1417-1420.

[142] Zhu, J., Quyyumi, A.A., Norman, J.E., Csako, G., Waclawiw, M.A., Shearer, G.M., Epstein, S.E. Effects of total pathogen burden on coronary artery disease risk and C-reactive protein levels. *Am. J. Cardiol.* 2000; 85:140-146.

[143] Anderson, J.L., Carlquist, J.F., Muhlestein, J.B., Horne, B., Elmer, S.P. Evaluation of C-reactive protein, an inflammatory marker, and infectious serology as risk factors for coronary artery disease and myocardial infarction. *J. Am. Coll. Cardiol.* 1998; 32:35-41.

[144] Auer, J., Berent, R., Weber, T., Eber, B. Interleukin-1 receptor antagonist gene polymorphism, infectious burden, and coronary artery disease. *Clin. Infect. Dis.* 2002; 34:35-41.

[145] Wanishsawad, C., Zhou, Y.F., Epstein, S.E. *Chlamydia pneumoniae* induced transactivation of the major immediate early promoter of cytomegalovirus: potential synergy of infectious agents in the pathogenesis of atherosclerosis. *J. Infect. Dis.* 2000; 181: 787-790.

[146] Liuba, P., Pesonen, E., Paakkari, I., Batra, S., Andersen, L., Forslid, M., Yla-Henttuala, S., Persson, K., Wadstrom, T., Wong, X., Laurini, R. Co-infection with *Chlamydia pneumoniae* and *Helicobacter pylori* results in vascular endothelial dysfunction and enhanced VCAM-1 expression in apoE- knockout mice. *J. Vasc. Res.* 2003; 40:115-122.

[147] Fong, I.W, Chiu, B., Virra, E., Tucker, W., Wood, H., Peeling, R. Chlamydial heat-shock protein-60 antibody and correlation with *Chlamydia pneumoniae* in atherosclerotic plaques. *J. Infect. Dis.* 2002; 186:1469-1473.

[148] Biasucci, L.M., Liuzzo, G., Ciervo, A., Petrucca, A., Piro, M., Angiolitto, D.J., Crea, F., Cassone, A., Maseri, A. Antibody response to chlamydial heat shock protein 60 is strongly associated with acute coronary syndromes. *Circulation,* 2003; 107:3015-3017.

[149] Mahdi, O.S., Horne, B.D., Mullen, K., Muhlestein, J.B., Byrne, G.I. Serum immunoglobulin G antibodies to chlamydial heat shock protein 60 but not to human and bacterial homologs are associated with coronary artery disease. *Circulation,* 2002; 106:1659-1663.

In: Focus on Atherosclerosis Research
Editor: Leon V.Clark, pp. 37-62

ISBN 1-59454-044-6

Chapter II

Cardiorespiratory Fitness and Atherosclerotic Cardiovascular Diseases

Jari A. Laukkanen*[*] *and Sudhir Kurl
Research Institute of Public Health, University of Kuopio, Kuopio, Finland

Abstract

Physical fitness and activity status have been well-documented major risk predictors of cardiovascular and total mortality. The aim of article is to show the associations of cardiorespiratory fitness, as measured during clinical exercise testing, with cardiovascular diseases. Cardiorespiratory fitness is independently related to mortality in clinical and population-based studies. Poor maximal oxygen uptake (VO_{2max}) has been shown to be a comparable with elevated systolic blood pressure, smoking, hypercholesterolemia, obesity and diabetes in importance as a risk factor for mortality as well as a predictor of atherosclerotic cardiovascular diseases and the progression of atherosclerosis.

In addition to many previously known mechanisms explaining the protective role of physical exercise, there is growing evidence that aerobic exercise training via an increase in endothelial nitric oxide synthase activity can increase the capacity of endothelial cells to evoke vasodilation and decrease blood pressure, however, there seems to be heterogenity in the hemodynamic response to exercise. Exercise training in coronary artery disease leads to an improved agonist-mediated endothelium-dependent vasodilatory capacity. Furthermore, exercise training and increased aerobic capacity can further improve vascular arterial stiffness, and vasodilatory capabilities, leading to decreased systemic vascular resistance.

[*] Correspondence to: Jari A. Laukkanen, Research Institute of Public Health, University of Kuopio, P.O. 1627, 70211 Kuopio, Finland, Tel: 358-17-162960, FAX: 358-17-162936, e-mail: jariantero.laukkanen@uku.fi

Direct and indirect measurement of VO_{2max} can provide a good estimate for cardiorespiratory fitness level and prognosis. Cardiorespiratory fitness represents one of the strongest and valuable predictors for both fatal and non-fatal cardiovascular events emphasizing the importance of exercise testing in everyday life. Exercise testing should be used as a clinical tool for exercise prescriptions on individual basis. More attention should be directed toward the use of exercise testing to measure the therapeutic response to a life-style change and pharmaceutical interventions.

Introduction

Physical inactivity is currently considered to be one of the most crucial public health problems [1], and physical fitness and activity status have become known risk predictors of various cardiovascular diseases. Clinicians have long been aware that subjects with high levels of physical fitness and functional capacity have a good prognosis. Cardiorespiratory fitness can be seen as a set of attributes that enables an individual to perform physical activity at his or her optimal level. It is determined mainly by physical exercise and, in part, is also influenced by some other factors, including age, gender, heredity, and medical status [2]. Low cardiorespiratory fitness [3-14] has consistently been associated with an increased risk of premature death in many prospective studies. This has been mainly due to an increased cardiovascular mortality in unfit individuals [3,5-6]. Indeed, low cardiorespiratory fitness has been found to be as strong a predictor of mortality as the conventional modifiable risk factors, such as cigarette smoking, diabetes, hypercholesterolemia and hypertension [8, 10-12].

Additional evidence for the antiatherogenic effect is derived from some clinical trials, which have shown that physical activity alone [15] or combined with low-fat diet [16-17] or comprehensive lifestyle modification [18-19] with concomitant improvement in cardiorespiratory fitness, retards the progression of angiographically quantified coronary atherosclerosis in patients with coronary heart disease. Previous population-based studies have reported an association between physical activity or cardiorespiratory fitness and the incidence of coronary heart disease [20-24] and the progression of atheroclerosis [25]. From the public health point of view, it is a highly welcome sign that even a slight amount (and even short bouts] of exercise is sufficient to reduce the risk, and those who already meet the recommended level may benefit by performing more physical activity [20-21,23, 26-27].

The American College of Sports Medicine (ACSM) consensus statement symposium faced issues of dose-response and the challenge of defining the nature of the relationship between regular physical activity and a whole series of health outcomes ranging from premature mortality, cardiovascular risk factors to quality-of-life indicators [28]. These relationships can be described in detail using dose-response curves, but the problem can also be addressed by simplifying in terms of the three main curves (Figure 1). The first pattern (curve B) is a linear relationship, and this may be the most appropriate that represents the relationship between physical fitness and cardiovascular events, whereas the curve A may best describe the dose-response pattern upon which the current physical activity recommendations are based as well as how the some benefits cardiovascular risk factors can be achieved. In contrast, some other benefits such as marked increase in coronary blood flow

[2] can be obtained only when the level of physical exercise is rather high (curve C). Thus, it may be that part of the certain health benefits can conform to this curve (C), especially when fitness level is already high.

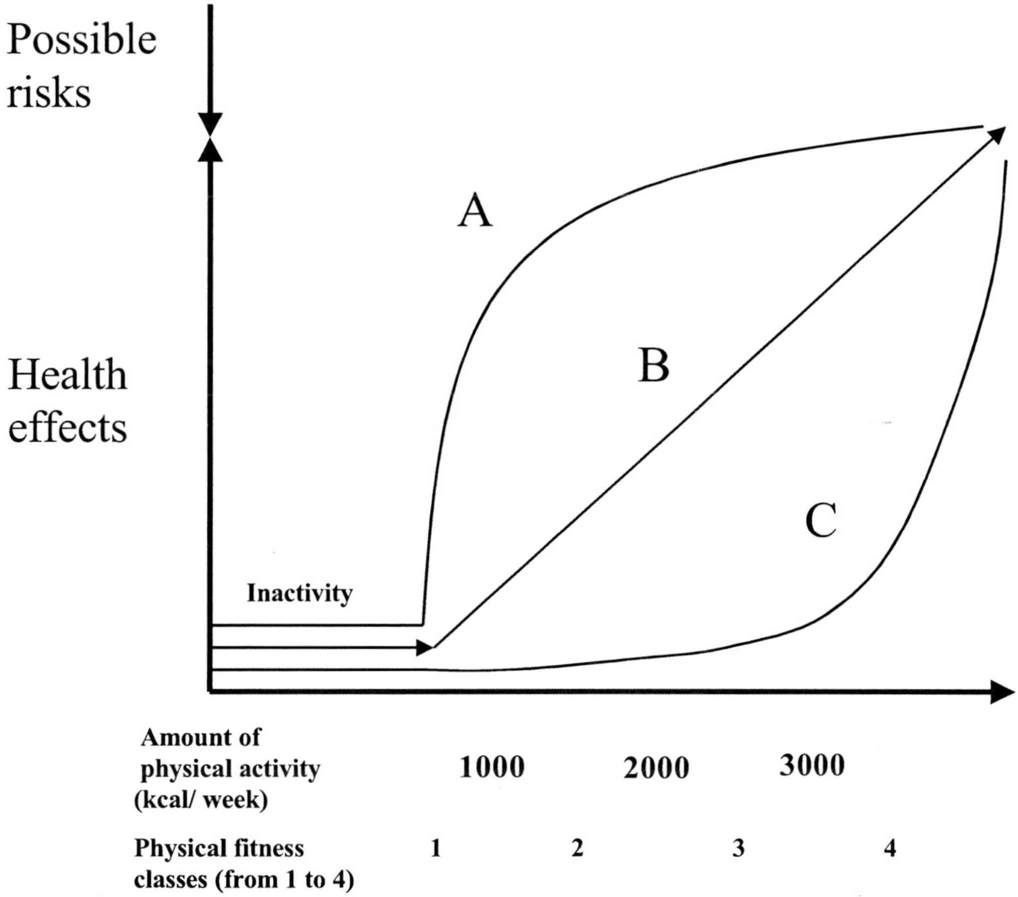

Figure 1. Schematic illustration depicting the relationship between physical activity level and defined in energy expended. See text for explanation.

The Risk Factors for Atherosclerotic Cardiovascular Diseases

Risk factor assessment determines the therapeutic strategy, because the intensity of preventive intervention is tailored to the patient's risk of coronary heart disease [29]. The conventional risk factors for coronary heart disease include elevated serum total cholesterol and LDL cholesterol, low HDL cholesterol, elevated blood pressure, cigarette smoking, diabetes, underlying vascular disease, menopausal status (women only), and age. Aggressive risk factor reduction, formerly used mainly in secondary prevention, may be pivotal to optimal patient management in high-risk primary prevention [29-30]. A number of noninvasive imaging modalities have the potential to measure and to monitor atherosclerosis in asymptomatic individuals and include exercise ECG testing, electron beam computed tomography, magnetic resonance coronary angiography, positron emission tomography, ankle/brachial blood pressure index, and B-mode ultrasound. Novel serum markers, including C-reactive protein and homocysteine, have the ability to gauge risk in the individual patient. Systemic therapy for risk reduction using nonivasive diagnostics tests such as exercise testing in primary prevention may be preferable to local therapy (eg, angioplasty and bypass) and may more effectively prevent acute coronary events than these more invasive approaches.

Global risk assessment including exercise testing is useful for assessing both the relative risk and absolute long-term risk. Even though short-term risk may not be high in subjects who have conventional risk factors, long-term risk can be high. Risk prediction in these subjects will highlight the need for early and prolonged intervention on risk factors. In younger and middle-aged adults, relative risk (RR) ratios help to reveal long-term risk for CVD. Although long-term prevention may not call for the use of risk-reducing drugs, it definitely will require the introduction of lifestyle modification (ie, smoking cessation in smokers, weight control, increased physical activity, and a diet low in cholesterol-raising fats). Besides the assessments of these conventional risk factors many parameters recorded during the exercise testing will help the global risk stratification.

There is a need for easily available noninvasive methods for defining those subjects with an increased risk of atherosclerotic cardiovascular diseases and greatest need for secondary preventive measures. The aim is to identify valuable predictors for cardiovascular diseases and mortality during exercise testing. On the basis of the present knowledge, it would be possible to show the prognostic value of exercise testing.

Exercise Testing, and the Role of Physical Fitness

Exercise testing has been used in everyday clinical practice for a long time, and its use has contributed significantly in the management of patients. The stress test combined with electrocardiogram (ECG) was originally used primarily for the detection of ST changes secondary to myocardial ischemia. Although other clinical findings, such as dyspnea on exertion, resting ECG abnormalities, or multiple risk factors for atherosclerosis may suggest the possibility of coronary artery disease, the most predictive clinical finding is a history of chest pain or discomfort. Clinical exercise testing consists of the continuous monitoring of an ECG with frequent recordings. In current form, the main focus has been on ECG recordings and blood pressure responses to exercise for screening coronary artery disease, although cardiorespiratory fitness may provide additional predictive power to other commonly available test parameters in modern exercise testing. All together, important information is derived from exercise capacity, blood pressure responses, development of arrhythmias, and whether or not symptoms such as chest pain develop during exercise [31].

Some testing scores have been developed to predict accurately the risk by combining the prognostic information of exercise capacity, ischemic findings, heart rate (HR), blood pressure and symptoms [32-34]. This allows for assessment of presence and severity of ischemia, prognosis, overall functional capacity, and efficacy of therapeutic interventions. Although stress testing in cardiological clinical evaluation is sometimes combined with radionuclide or echocardiographic imaging, it would focus primarily on the ECG, blood pressure and exercise capacity [31].

It is a widely recognized fact that exercise capacity can be used as an independent clinical tool for various outcomes supporting the value of the exercise test. It is a noninvasive and easily available test that provides a wealth of clinically relevant diagnostic and prognostic information. Directly measured maximal oxygen uptake (VO_{2max}) is a gold standard for assessing the amount of the oxygen consumption during a maximal effort [2]. VO_{2max}, product of cardiac output and arteriovenous oxygen difference, during exercise represents cardiac, circulatory and respiratory function and muscle oxygen utilization under physiological stress conditions. Many large studies estimate fitness levels by the measurement of the duration of exercise, the peak exercise stage, or estimates of the peak energy expenditure achieved during graded exercise tests. One metabolic equivalent (MET) is defined as the oxygen uptake when a person is at rest, which is equivalent to 3.5 ml of oxygen per kilogram of body weight per minute. The use of directly measured VO_{2max} has the advantage of providing a common assessment of fitness for use with different exercise protocols [2, 35]. The disadvantages of indirect measures may be that relationship between directly measured VO_{2max} and exercise time is not linear due to large increment in work rate per stage and a wide variety of heterogeneous populations with different basic fitness level [36].

In both asymptomatic subjects and patients with CHD, exercise-based measures of functional capacity repeatedly have been shown to provide powerful independent prognostic value. When decisions are made about the value of exercise testing in clinical cardiology, the focus is inevitably on measures believed to reflect the state of the coronary arterial circulation and the potential need for angiography and revascularization. Thus, ischemia in the presence of poor cardiorespiratory fitness means high risk, whereas in the presence of good cardiorespiratory fitness ischemia has little prognostic impact. In asymptomatic subjects, in whom evidence of ischemia is quite infrequent, the utility of exercise capacity has been more of a puzzle.

Cardiorespiratory Fitness and Overall Mortality in General Population

Previous studies concerning the level of cardiorespiratory fitness needed to reduce premature mortality are based on a few population-based studies [3-7, 10-11,Table 1]. The latest findings support U.S. cohort studies suggesting that the risk of death associated with low cardiorespiratory fitness or physical inactivity is comparable with that of conventional risk factors including smoking, hypertension, obesity and diabetes [8,10-12]. Previous findings in normal subjects [5] have shown that the most striking difference in overall mortality rates was observed when fitness increased from low level to next lowest level. This observation concurs with the consensus reflected that the greatest health benefits are achieved by increasing physical activity among the least fit [2,26-28, 37].

In a large study, a total of 10244 men and 3210 women underwent a medical examination at the Cooper Institute, and physical fitness categories were determined according to treadmill time for different age groups and, thus the cut-off values for the cardiorespiratory fitness based on exercise time for unfit men were different in various age groups. Blair and coworkers [5,8] found that at least moderate durations of treadmill test were associated with reduced cardiovascular and all-cause mortality, while higher durations achieved, provided only slight further reduction in the risk of death. An important finding was that the inverse associations were observed within categories of common risk factors including smoking, elevated blood pressure, cholesterol and body mass, emphasizing the protective effect of good cardiovascular fitness among individuals with or without the presence of risk factors [8].

Table 1. Physical Fitness and Mortality: Cohort Studies (Primary Prevention)

Study Population	Fitness Measurement	Categories Compared	Follow-up time	Main Result
U.S. Railroad study (1988)[3](n=3043)	Submaximal heart rate,treadmill	most fit fourth vs.least fit fourth	20 years	1.2-fold risk for unfit men
Lipid Research Clinics Prevalence study (1988)[2](n=4276)	submaximal heart rate,treadmill (modified Bruce protocol)	most fit fourth vsleast fit fourth	8.5 years	2.7-fold adjusted risk for 2 SD increase in heart rate (17 beats/ minute)*
Cooper Clinics study (1989)[4](n=13344)	Exercise duration,treadmill (modified Balke protocol)	all others vs.least fit quintile	8 years	1.6-fold adjusted risk for unfit men and women
Study of Norwegian men (1993)[5] (n=2014)	Maximal workload,electrically braked cycle ergometer	least fit fourth vs. most fit fourth	16 years	0.5-fold adjusted risk for fit men
Kuopio Ischemic Heart Disease Risk Factor Study (2001)[11] (n=2682)	VO_{2max},electrically braked cycle ergometer	most fit fourth vs. least fit fourth	10.7 years	2.8-fold adjusted risk for unfit men
The St. James WTH study (2003)[42] (n=5721)	Treadmill exercise (Bruce protocol)	1 MET increment	8.4 years	17% decrease in adjusted risk for women
Lipid Research Clinics Prevalence study (2003)[44] (n=2994)	Near-maximal treadmill exercise (modified Bruce-protocol)	1 MET decrement	20 years	1.20-fold age-adjusted risk change for women*

* Main outcome measure was cardiovascular mortality.

In the U.S. Railroad Study [4], physical fitness was assessed using a sub-maximal treadmill exercise test in railroad workers without known cardiovascular diseases at baseline. Mortality rates increased with declining physical fitness among the workers even if the most common confounders were taken into account. In a study of Norwegian middle-age men [6] who were free of cardiovascular disease, physical fitness was assessed with a symptom-limited exercise tolerance test on a bicycle ergometer. Among men in the most fit category, the risk of mortality decreases almost to half after adjustment for life style and biological factors. Sandvik and coworkers [6] observed that moderate levels of work capacity in cycle ergometer test were associated with reduced cardiovascular mortality, but there was some further reduction in the risk at the highest levels.

In a prospective population-based study among middle-aged men [11], VO_{2max} and exercise test duration, through their whole ranges, had strong, graded and inverse associations with overall mortality. In fact, VO_{2max} and exercise test duration were one of the strongest predictors for mortality in this unselected male cohort. Although the mortality curves for the quartiles of VO_{2max} continued to diverge over the period of follow-up and the strongest excess mortality was observed at low levels of cardiorespiratory fitness, and VO_{2max} was inversely related to mortality showing a clear dose-response relationship. Furthermore, it is found that good VO_{2max} reduces the risk of premature death among healthy persons and individuals with unfavourable risk profile [11].

Cardiorespiratory Fitness and Mortality among Unhealthy Subjects

Exercise capacity during treadmill exercise has been exhibited the strongest association with all-cause mortality and cardiac events in patients with known or suspected cardiac disease [12-14,38-40] (Table 2). A recent study by Myers et al. [12] provides more insight regarding the relation between fitness and survival in their analysis of 6213 consecutive male patients. These data differs from other cohort data drawn from population-based studies that the subjects were referred due to some symptoms or medical disease. A nearly linear reduction in mortality was observed as fitness levels increased, and each increase of 1 MET in exercise capacity conferred a 12 percent improvement in survival. Notably, the risk of death among those with a peak exercise capacity of less than 5 MET was nearly double the risk among those with a peak exercise capacity of more than 8 MET.

Authors showed that good exercise capacity is a strong predictor of mortality in men with history of hypertension, pulmonary disease, diabetes, smoking, obesity or elevated cholesterol level as well as in subjects with or without underlying cardiovascular disease or the use of β-blockers [12]. It was found that the absolute peak exercise capacity outperformed the percentage of age-predicted exercise capacity. One might expect to find a difference between the risk of death predicted by the absolute level of fitness and the risk predicted by the age-related level among subjects older than 65 years of age because exercise capacity is generally lower in elderly persons [12], but the work capacity has been the only treadmill exercise variable that was strongly associated with outcome, and its prognostic effect was of the same magnitude in elderly and younger persons [14].

Table 2. Physical Fitness and Mortality: Clinical Populations (Secondary Prevention)

Study Population	Fitness Measurement	Categories Compared	Follow-up time	Main Result
Olmsted Country,* Minnesota, 1998 [13] (n=2193)	Workload, treadmill (Bruce, modified Bruce or Naughton protocol)	1 MET change	6.3 years	17 to 23 % decreased adjusted risk for 1 MET increase in exercise capacity
Heart failure study cohort, Cleveland, 1999 [39] (n=470)	VO_{2max}, Ve/ CO_2, HRtreadmill (Naughton protocol)	Highest quartile vs. others	1.5 years	3.2-fold adjusted risk (peak Ve/ $CO_{2\geq}$44.7)
Olmsted Country, elderly population, Mayo Clinic(2000) [14] (n=3107)	Workload, work capacity,treadmill	1 MET change	6 years	18 % decreased adjusted risk for 1 MET increase in exercise capacity
Clinical population, Standford (2002) [12] (n=3679)	Exercise duration,treadmill (graded or ramp protocol)	1 MET change	6.2 years	12 % decreased age adjusted risk for 1 MET increase in exercise capacity

* A retrospective, population-based cohort study.

In the Coronary Artery Surgery Study [41], high exercise capacity (≥10 METS) was associated with excellent prognosis even in patients with 3-vessel disease, and this subgroup showed no significant benefit from surgery. The benefits of surgery seemed to be confined to the coronary artery disease patients with impaired exercise capacity, even among patients with impaired left ventricular function.

The Role Physical Fitness as a Risk Factor among Females

Previous studies showing the importance of physical fitness are based mainly on male populations. However, recent studies are shown that physical fitness is one of the strongest prognostic markers among females. These are supported for many reports showing that habitual physical exercise can decrease the risk of premature death [20-24].

The St James WTH study is a large and provides the most unambiguous evidence to date of its independent prognostic importance [42] (Table 1). These results are, in fact, quite comparable to those previously obtained in both asymptomatic and symptomatic men. The strong prognostic value of exercise capacity is also evident in elderly [43]. Exercise capacity appears to be a remarkably robust and important measure.

In a population-based cohort of asymptomatic North American women, exercise testing clearly discriminated mortality risk using exercise variables other than ST-segment responses [44] (Table 1). Exercise capacity and HR responses during exercise and recovery were strong, graded, and independent predictors of cardiovascular and all-cause mortality, while exercise-induced ST-segment depression was not. Failure to achieve target HR and ventricular arrhythmia were also predictors of poor prognosis. The association of exercise capacity and heart rate recovery HR recovery with cardiovascular and all-cause mortality was strong, consistent, graded, and independent, while exercise-induced ST-segment depression carried no increased risk in this large population-based cohort of women [44]. These findings support the potential use of exercise testing as a risk-stratification tool for primary prevention in asymptomatic women, incremental to the traditional cardiovascular risk factors.

Cardiorespiratory Fitness and Cardiovascular Events

The low to moderate amount and intensity of leisure-time physical activity is associated with decreased mortality from coronary disease [20, 23]. Regular physical exercise is of vital importance, in maintaining or taking up light or moderate physical activity reduces mortality in subjects with and those without cardiovascular disease [37, 45]. In a sample of men without previous cardiovascular disease, the risk of acute myocardial infarction was decreased by performing physical activity for at least two hours a week [22]. In men with normal or abnormal ECG findings, those with low VO_{2max} had a greatly increased risk of myocardial infarction. Secondly, VO_{2max} was inversely related to the risk of death from cardiovascular causes in men without known cardiac disease as well as those with already diagnosed disease. The Lipid Research Clinics Mortality Follow-up Study [3] was a primarily designed to examine the association between lipid profile and health outcomes, but information was also collected on treadmill exercise test. Physical fitness was assessed during submaximal treadmill exercise test by recording heart rate at a standard moderate workload. Ekelund and coworkers [3] showed a marked difference in cardiovascular and coronary disease mortality between high and low levels of cardiorespiratory fitness assessed by heart rate at a speed of 4 km/h (2.5 miles/h) on a treadmill.

Cardiorespiratory Fitness, and the Risk of Stroke

Physical activity may protect against future stroke, although the relation between physical activity and the risk for stroke has been inconsistent [46]. In one study, Herman and coworkers [47] reported a graded dose-response relation, whereas other studies [48] have reported a nonlinear association between physical activity and the risk for stroke.

The authors examined the relationship of cardiorespiratory fitness, as indicated by maximum oxygen consumption VO_2max with subsequent incidence of stroke [49]. They also compared VO_2max with conventional risk factors as a predictor for future strokes. Men with no stroke or pulmonary disease at baseline participated in the study. The VO_2max was measured directly during exercise testing at baseline. The relative risk for any stroke in unfit men was over 3.0-fold and for ischemic stroke, 3.5-fold, compared with fit men, after adjusting for age and examination year. The associations remained significant after further taking into account smoking, alcohol consumption, socioeconomic status, energy expenditure of physical activity, prevalent coronary heart disease, diabetes, systolic blood pressure, and serum low-density lipoprotein cholesterol level for any or ischemic strokes. Low cardiorespiratory fitness was comparable with systolic blood pressure, obesity, alcohol consumption, smoking, and serum low-density lipoprotein cholesterol level as a risk factor for stroke.

This study showed that excessive risk for stroke was observed among men with the lowest level of VO_2max. A threshold was observed between the 2 lowest quartiles of VO_2max, indicating that men with a VO_2max below this level were at high risk for stroke. However, a change in VO_2max of 3.5 mL/kg per minute corresponds to a 17% decrease in the risk for stroke among these men from eastern Finland. Previous studies have found an association between physical activity at work [50] or during leisure time [51] and subsequent risk for stroke. Leisure-time physical activity has been related to a decreased risk for ischemic stroke [51]. Whereas, low levels of occupational physical activity have been shown to increase the risk for stroke. [52].

Some studies have provided an indication of a U-shaped relationship between physical activity and stroke [48, 53]. Although these reports provide some indication that a very high level of intense physical activity could even increase the risk for stroke, there is no consistent evidence of this, and no significant results regarding this kind of U-shaped relationship [48, 53]. Since there are few people doing vigorous activity, and the absolute number of strokes is small in the average population who perform vigorous exercise. High-intensity exercise is more effective than low-intensity exercise for improving VO_2max in healthy persons, whereas lower-intensity physical activity may be sufficient to improve VO_2max in high-risk individuals, but the optimal intensity of exercise in different risk groups for the prevention of stroke remains unclear.

As low cardiorespiratory fitness is a modifiable factor, any physical exercise programs that improve the level of cardiorespiratory fitness should be recommended. On the basis of previous studies, it seems that at least 6 months of regular training with moderate intensity can improve exercise capacity by 1 metabolic unit, correponding to 3.5 mL/kg per minute of directly measured VO_2max. However, training history and the duration and intensity of exercise may have a marked effect on the improvement of VO_2max, and accumulating evidence from exercise intervention trials show that it is possible to achieve a beneficial effect on risk factors without improvements of exercise capacity.

It has been shown that low VO_{2max} was an important predictor for the progression of carotid atherosclerosis [54]. One of the most simple, natural, and cheapest ways of preventing a stroke is to exercise regularly. In spite of this trivial wisdom it is hard to get people to increase their cardiorespiratory fitness, probably because no direct gains are visible and no immediate rewards are attached. Lee and coworkers [55] recently reported in a meta-analysis of all studies including epidemiological as well as case-control studies, showing a clear benefit of physical activity to prevent both stroke incidence and mortality. The magnitude of the effect is vast, homogenous, and significant: highly active individuals had a 21% lower risk of ischemic stroke and a 34% lower risk of hemorrhagic stroke when compared with low-active individuals [55]. This graded effect of reducing the risk of stroke with vigorous levels compared with moderate levels of physical exercise implies more than just a causal relationship. It shows that the moderate physical activity also is effective in a significant, and recommendable way. It shows that people who exercise moderately have some advantage to prevent a stroke.

It is now established beyond doubt that high-level physical activity should be strongly recommended for the prevention of stroke. Primary care physicians and stroke specialists will now be able to recommend this protective measure with a higher level of certainty.

The Role of Cardiorespiratory Fitness in the Presence of Common Risk Factors

Previous studies have shown that physical activity [37] and good cardiorespiratory fitness [8,10,12] reduce the risk of premature death among individuals with unfavourable risk profile. The protective role of higher exercise capacity has been documented in the presence of established cardiovascular risk factors [42]. This study reported that exercise capacity provides additional prognostic information even after well know predictive value of Framingham risk score was taken into account. In eastern Finnish study population [11], VO_{2max} provides additional information on the risk related to common CVD mortality predictors such as smoking, overweight, hypertension and lipid profile. It seems that overweight in the presence of low VO_{2max} was related to an increased risk of CVD, however, overweight was not a significant predictor among fit men. This finding is supported by some previous studies showing that being fit reduces the risks of obesity [10, 56-57]. A partial explanation for this interaction may be that BMI is a more specific measure of obesity in unfit men.

Exercise and Endothelial Function

Endothelium is a physiological barrier between the blood and vascular smooth muscle, which produces relaxing and contracting factors. The endothelium is a primary target for the effect of risk factors as well as the benefits of physical exercise. Human coronary and peripheral arteries show endothelial dysfunction in conditions related to preclinical atherosclerosis. Recent studies have shown that common CVD risk factors including overweight [58] hypertension [59,60] and smoking [59-61] cause endothelial dysfunction in blood vessels. On the other hand, physical exercise could increase the capacity of endothelial cells to evoke vasodilation [62]. This may be a beneficial mechanism in protection against the negative effects of these known risk factors for CVD in early stages of atherosclerosis.

Exercise intervention studies have showed protective physiologic effects on endothelial function in older adults could be even blunted by an active life-style. Exercise training improves endothelium-dependent vasodilatation both in epicardial coronary vessels and in resistance vessels in patients with coronary artery disease [63]. This might represent an important and a proposed principal mechanism by which regular aerobic exercise can decrease the risk of cardiovascular diseases.

These findings showing that exercise training appears to improve endothelial function, which is an integral to many aspects of cardiovascular health. The improvements in endothelial vasodilator function appear to be a systemic effect of exercise. Exercise elevates shear stress on the vessel walls that could be a stimulus for endothelium-derived nitric oxide, leading to smooth muscle relaxation and vasodilation [64]. It is suggested that running but not food restriction or a sedentary condition improves endothelial vasodilator function, presumably because of an increase in nitric oxide [65].

It is shown that patients with type 2 diabetes completed an exercise program consisting of three 1-hour sessions of moderate intensity aerobic and resistance exercise lead to improvements in reactive hyperemic brachial artery vasodilation and forearm blood flow [66]. However, non-diabetic subjects following the same program showed no improvement in endothelial function despite the increase in cardiorespiratory fitness but aerobic exercise training improved endothelial vasodilator function in patients with the metabolic syndrome [67].

The improved vasodilator function was not explained by any known atherosclerosis risk factors, suggesting that chronic exercise hyperemia may up-regulate endothelial release of nitric oxide. In some studies, brisk walking in patients with mild to moderate hypertension improved endothelial vasodilator function through increased nitric oxide release [68]. Although blood pressure was slightly reduced, the improvements in endothelial function did not correlate with blood pressure changes. Thus, enhanced endothelial vasodilator function did not merely reflect the blood pressure reductions.

Exercise and Arterial Stiffness

With aging and hypertension, the arteries become stiff from the progressive degeneration of the arterial bed, increased collagen and calcium content, artery dilation and hypertrophy. The Atherosclerosis Risk in Communities Study, [69] a large sample of middle-aged men and women, observed that several indices of common carotid artery stiffness were higher in persons with type 2 diabetes or impaired glucose tolerance compared with persons with normal glucose tolerance. Glucose intolerance and elevated lipid levels contributed to increased artery stiffness. It has been suggested that glycation-induced cross-linking formation in interstitial collagen adds to increase in arterial stiffness with age and diabetes. Structural changes like arterial medial degeneration reduce arterial compliance, causing more stiffening. These factors lead to the increase in systolic blood pressure and contribute to atherosclerosis and adverse cardiovascular events.

Interaction of physical exercise, genes, life style and environment may influence the endothelial dysfunction with aging. Exercise intervention studies showed protective physiologic effects on endothelial function in older adults and aging could be even blunted by an active life-style [70]. It has been demonstrated that physical exercise could change gene expression leading to increased expression of some proteins and contribute to enhance glucose uptake, glycogen synthesis, and insulin action [71, 72]. Exercise could also induce vascular endothelial growth factor expression. It is suggested that exercise could improve endothelial function through the increase of nitric oxide and calcitonin gene-related peptide release in elderly. Another possible effects of physical exercise on endothelial function with aging would result from cardiovascular ischemic preconditioning which could protect against the progression of atherosclerosis.

It has been postulated that growth factors released during repeated bouts of exercise mediate stiffness or that increases in heart rate and blood pressure during exercise condition artery walls. In the Baltimore Longitudinal Study of Aging, higher VO_{2max} was associated with less arterial stiffness at any age and in both sexes [73]. Moreover, arterial stiffness measurements consisting of pulse-wave velocity were lower by 26%, and carotid arterial pressure pulse augmentation index was lower by 36% in elderly men who had a long history of endurance training compared with their sedentary peers. However, at the moment, there are no large studies on the effects of amount, intensity and frequency of physical exercise and cardiorespiratory fitness on arterial stiffness.

The available data can suggest but do not establish cause and effect between increased fitness and reduced arterial stiffening convincingly. It is not widely known if isolated systolic hypertension, a clinical manifestation of larger artery stiffening, is particularly resistant to exercise compared with essential hypertension. Amount and intensity of exercise needed to reduce arterial stiffness is not known. Thus, exercise-induced mechanisms that reduce arterial stiffness may be of an additional benefit of physical exercise, but the evidence supporting this view is currently limited.

The Effect of Physical Exercise on Glucose Intolerance, Lipids, and Blood Pressure

Essential hypertension is known to be associated with diabetes largely irrespective of age and obesity [74]. Elevated blood pressure is also a part of metabolic syndrome, which is becoming common worldwide, and carries with it high morbidity and mortality. It is estimated that a doubling of cardiovascular events when hypertension and diabetes coexist. Patients with cardiovascular disease have abnormalities in peripheral and central characteristics of cardiovascular structure and function that precede the clinical cardiovascular disease. Elevated blood pressure with glucose abnormalities increase left ventricular mass and arterial wall thickness, left ventricular diastolic filling dysfunction, impairs endothelial function, and increases arterial stiffness. Although glucose control is important for preventing microvascular atherosclerosis, blood pressure reduction is needed to prevent cardiovascular events in diabetic patients with hypertension.

It has been reported that good cardiorespiratory fitness and decreased body weight contribute to improved insulin responses supporting a synergistic effect of physical activity and weight reduction [75]. However, it is possible that physical activity and weight loss may improve insulin sensitivity through separate physiological mechanisms. The present findings confirm these earlier studies by demonstrating that exercise alone can reduce plasma insulin concentration in the presence of unchanged glucose concentrations. The participants in the exercise and weight loss group have shown reduced levels of both glucose and insulin. In addition, physical activity is thought to improve insulin sensitivity by increasing oxidative enzymes and glucose transporters in muscle. Long-term physical activity has been shown to reduce the insulin response to glucose.

Physical activity, including appropriate endurance and resistance training, is a major therapeutic modality for type 2 diabetes as recommended by the American College of Sports Medicine. This view is supported by a previous study that showed the beneficial effect of aerobic and resistance training on glycemic control in type 2 diabetes, although a dose-response relationship is indiscernible [76].

Many intervention trials have shown the net effect of physical exercise on blood pressure. On the basis of previous randomized trials have reported that aerobic exercise was associated with an overall reduction in mean blood pressure of 3.9/2.6 mm Hg across all initial blood pressure levels, independent of body weight and race [77]. Subgroup analysis showed a greater mean reduction of 4.9/3.7 mm Hg in hypertensive persons. Some studies have suggested that the degree of blood pressure reduction did not differ due to frequency or intensity of exercise, suggesting that all forms are effective. However, training from three to five times per week during 30 - 60 min per session at an intensity of about 40 - 50% of net maximal exercise performance appears to be effective with regard to blood pressure reduction. The evidence that higher intensity exercise would be less effective is at present inconsistent.

A previous trial observed a decreases in blood pressure of 6/5 mm Hg in hypertensive persons and decreases of 2/3 mm Hg in normotensive persons [78]. There was regression of left ventricular mass and concentric left ventricular remodeling in older persons who had mild or moderate hypertension and performed endurance exercise. These changes correlated with a reduction in systolic blood pressure.

There are important factors such as type, intensity, and duration of exercise, baseline blood pressure, and concomitant use of antihypertensive medications that may cause a wide variation in the magnitude of blood pressure reduction, although exercise training appears to reduce blood pressure to some degree. Evidence from previous studies including subjects with diabetes, hypertension, prevalent cardiovascular disease, or related conditions suggests that regular physical exercise is also a potentially valuable therapy for improving cardiovascular health in patients with diabetes and hypertension.

Convincing evidence shows that the protective effect of cardiorespiratory fitness may be partly explained through physical exercise that has a favourable effect on lipid profile and fat metabolism, blood pressure [79], incidence of non-insulin-dependent diabetes [80], insulin sensitivity [80] and blood coagulation [81,82] especially among high risk groups. Physical exercise could also decrease the accumulation of collagen in the artery wall [83], pulsatile stress [84] and the atherogenic activity of mononuclear cells as well as inflammation [85].

Physical activity both prevents and helps treat many conventional atherosclerotic risk factors, including hypertension, insulin resistance and glucose intolerance, elevated triglyceride concentrations, low serum HDL concentrations, and obesity. Furthermore, physical exercise in combination with weight reduction can decrease LDL concentrations and limit the reduction in HDL that often occurs with a reduction in dietary saturated fat. The magnitude of the exercise effect is influenced by type of the exercise intervention, age, underlying disease, genetic variations, and whether exercise produces concomitant reductions in body weight. The effect on atherosclerotic risk factors can be large in some individuals whereas it is possible that other individuals do not similarly respond to the exercise intervention. In general, the effect of exercise alone on atherosclerotic risk factors may be less than that achieved by pharmacological therapies. The overall impact of exercise can be observed on the risk factors reduction and well-being of the individual, that cannot achieved by any miracle pill. This can be significantly magnified by other lifestyle changes such as changes in dietary composition and weight loss.

Exercise and Left Ventricular Function

Other direct mechanisms may include that physical exercise improves cardiorespiratory capacity by increasing cardiac output, the formation of collateral vessels, the dimension of coronary arteries, the extraction of oxygen from blood and the threshold for ventricular arrhythmias [2]. Furthermore, individuals with decreased left ventricular function may benefit from these cardiac adaptations including increased oxygen utilization and better cardiac function by performing regular physical exercise [2, 86].

Impaired cardiac function results primarily from reduced left ventricular contractility during systole, diminished left ventricular filling during diastole, or both. The common feature is abnormal early left ventricular diastolic filling with aging, suggesting reduced compliance or prolonged relaxation ventricular muscle. Left ventricular diastolic dysfunction may be an early marker of functional abnormality.

Left ventricular diastolic dysfunction is associated with reduced exercise performance. It has been reported that the left ventricular ejection fraction response to exercise may also be impaired even though left ventricular systolic function is normal at rest. Some causes of left ventricular dysfunction are latent global myocardial ischemia or metabolic myocardial disturbances [87, 88]. Differences in medications, duration of treatment for diabetes and techniques for measuring diastolic function may contribute to the lack of the efficacy of glycemic control and exercise on cardiac function.

The diminution of early diastolic filling is uncommon in healthy older persons with a long history of endurance exercise. Moderate-intensity aerobic and resistance exercise was shown to improve left ventricular diastolic dysfunction in men with hypertension [89]. In healthy normotensive men, aerobic training with improvement in VO_{2max} enhanced early diastolic filling both at rest and during acute exercise by increasing left ventricular mass by 14 %. It is possible that exercise also reduced elevated resting atrial filling rate, which is an indicator of left ventricular diastolic function [90]. The combination of increased left ventricular mass and improved diastolic filling represent a desirable physiological rather than a pathological hypertrophy.

Cardiorespiratory Fitness, Physical Activity and Heredity

Poor exercise capacity may be caused by myocardial ischemia but can also be caused by a host of factors apparently less amenable to "fixing," such as cardiac dysfunction, pulmonary disease, chronotropic incompetence, and physical inactivity. As a modifiable risk factor, poor cardiorespiratory fitness can specifically indicate the need to improve cardiorespiratory endurance with exercise training. If cardiorespiratory fitness can be seen as a nonmodifiable indicator of risk, similar to age, gender, and genetics, it can direct us to higher-risk groups who need intensification of management. In order to assess whether modifying cardiorespiratory fitness does, in fact, alter survival in the expected manner, we need further randomized trial data.

Cardiorespiratory fitness or exercise performance is a physiological measure that has been used as a marker for habitual physical activity. This approach has been criticized on the basis of genetic component of physical fitness. The modest correlation between physical fitness and physical activity measured by questionnaires indicates that the part of the fitness is genetically determined. However, genetic contributions to fitness are important but probably account for less of the variation observed in fitness than is due to environmental factors, principally physical activity. Current estimates place that genetic contribution to aerobic power at approximately 25-40 % [91-93]. Furthermore, it is well established that exercise training can make substantial improvements in aerobic power, typically 15-20 % in adult men and women. Changes in exercise capacity induced by endurance training and adjusted for pretraining values were also characterized by a significant familial resemblance.

The interesting point of effectiveness of exercise is the trainability. Previous findings suggest that trainability of submaximal working capacities is strongly influenced by familial and/or genetic factors [92]. Although the mean increase in cardiorespiratory fitness has been significant, the same exercise dose may be ineffective for some subjects and highly effective for others. For most individuals, increases in physical exercise produces increase in VO_{2max}, but the amount of adaption in VO_{2max} to standard exercise dose varies widely and is under genetic control [93]. In long-term studies, there is evidence that high-intensity training elicits a greater increase in VO_{2max} than low-intensity training for the same total energy expenditure [94]. If fitness is an objectively measurable intermediate factor between physical activity and health benefits, VO_{2max} could be used as an indicator of the beneficial and achieved "dose" of physical activity. Age, sex and underlying diseases are not major determinants of human responses to regular physical activity, whereas the pre-training level of a phenotype and the duration, intensity, frequency and type of physical activity may contribute to the variability in training response in some individuals [93]. Therefore, the level of physical activity sufficient to improve or maintain good VO_{2max} probably depends on the initial health and fitness status, the length of the previous training history and the duration, frequency and intensity of the exercise. However, it is known that VO_{2max} decreases by 5 % to 15 % per decade between the ages of 20 to 80 and the rate of decline in oxygen uptake is directly related to maintenance of physical activity level [2].

Effectiveness of Exercise in the CVD Risk Reduction

The expert panel suggested that every adult should accumulate 30 minutes or more of moderate-intensity physical activity on most, preferably all, days of the week to promote health and to prevent chronic diseases [26,27]. In a recent randomized trial [95], the authors estimated that an increase of 10 % in cardiorespiratory fitness, corresponding to 1 MET increase in exercise capacity, can be achieved by maintaining these exercise recommendations for 2 years, although it is possible that these changes can be achieved by short-time exercise training. More intense structured exercise could increase physical fitness by 1 MET in 6 months [95]. Blair and co-workers [7] reported that an increase of 2 METs in treadmill performance was related to a reduction of 30 % in mortality. In a large study [12], each increase of 1 MET in exercise capacity conferred a 12 percent improvement in survival. Comparable risk reduction in each 1-MET increment in workload was found among younger (less than 65 years old) and older subjects [14]. The real changes over time in fitness or activity have an effect on mortality, because previous studies [7,9,37,45] have suggested that even a small improvement in cardiorespiratory fitness and an increase in physical activity can result in reduced all-cause and cardiovascular mortality. Therefore, the measurement of cardiorespiratory fitness should be repeated in order to investigate the effect over a period of time and recommend exercise programs.

Many studies have concluded that comprehensive, exercise-based cardiac rehabilitation reduces mortality rates in patients surviving from myocardial infarction. These results were based on the cardiac rehabilitation program consisted of exercise only or also included other psychosocial and educational interventions. Total mortality was reduced almost by one-third with the exercise-only intervention and much less with comprehensive rehabilitation, whereas the respective cardiac mortality was reduced by one-third in both rehabilitation programs [96]. Neither the exercise-only program nor the comprehensive intervention significantly reduced the rate of nonfatal myocardial infarction [96].

Interpretation and Implications of the Current Knowledge

The primary clinical implication of previous studies is that there is a need for wider recognitation of the potential value of maintaining at least minimal levels of physical activity, although the recommended dose may vary. An alternative interpretation of primary findings derived from epidemiologic studies is that inactivity and poor fitness is, at least to some extent, a proxy or a marker for other factors contributing to increased mortality (e.g. degree of CAD and functional impairment, severity of underlying chronic disease, or disease progression). This problem is mainly evident in studies showing the association between fitness and outcomes in clinical populations who have some reasons or symptoms for exercise testing in hospital. When the second interpretation is the main reason, the preventive measures and medical treatments may be needed in addition to increased physical exercise knowing the possible risks of vigorous physical activity [97-98]. It is difficult to distinguish an increased risk due to a low level of VO_{2max} from that of increased risk of prevalent asymptomatic or pre-existing disease. For example, an existing but non-diagnosed disease could lead to a decreased exercise capacity because diseases may begin early in life and develop from an asymptomatic phase to a clinical disease over many decades.

The prognostic power of exercise capacity is strong among apparently healthy persons and patients with cardiovascular disease or use of medications. Nevertheless, given the high prognostic value of exercise capacity relative to other markers of risk in recent studies, clinicians who are reviewing exercise-test results should encourage patients to improve their exercise capacity. In terms of reducing mortality from any cause, improving exercise tolerance warrants at least as much attention as other major risk factors from physicians who treat patients with or at high risk for cardiovascular disease [8,11,12]. The present studies demonstrated that VO_{2max} is a strong predictor of hard outcome events [12], as well as soft outcome measures such as acute coronary events [22] and the progression of carotid atherosclerosis [25]. It seems that exercise testing, using either treadmill [3,5] or cycle ergometer [6,9,11], is a useful and reliable way to define cardiorespiratory fitness status and exercise testing methods that are readily available at any exercise laboratory can be used without additional equipment in the prediction of future risk.

The currently available data provides us with an opportunity to reconsider whether exercise capacity should be measured in asymptomatic subjects as part of a cardiovascular risk assessment process. The data are now quite persuasive that this measure provides independent prognostic information in both men and women. It could be argued that there is little harm in a recommendation to screen for the extremely unfit and encourage them into some form of exercise and primary prevention program. Interventions could be applied in a factorial design to modify cardiovascular risk factors or to improve exercise capacity beyond some nominal threshold. An effective intervention applied to the most unfit members of our society before they develop clinical disease has the upside of a huge potential for improving health and even generating long-term cost savings.

Strategies for Further Studies and Clinical Practice

Although clinical exercise testing has been considered lacking of both sensitivity and specificity in order to make it applicable for detecting all individuals at increased risk of cardiovascular disease. This view is based on the reliance on ischemic ST-changes in exercise stress test. Recent reports have shown the usefulness of measuring exercise capacity among clinical and general population and both the genders. On the basis of current knowledge, it could be suggested that those unfit asymptomatic subjects without previously known cardiovascular or pulmonary disease but who have maximal exercise capacity below 7-8 METs and those unfit men with known CVD having exercise capacity below 5-6 METs have an increased risk of premature death although the exact value according to gender and various pre-test probabilities are not well known. The studies correctly emphasize the importance of the promotion of physical activity among healthy subjects and patients with various cardiovascular diseases.

The optimal intensity of exercise and the level of fitness should be defined in different risk groups that exercise will be recommended as a treatment in different stages of atherosclerotic diseases. Thereby, the conclusion that clinicians always should encourage their patients to increase cardiorespiratory fitness may lead to unwanted consequenses such as sudden cardiac death if previous training history is not known [97-98]. Finally, accumulating evidence from controlled trials emphazise the many improvements in health related fitness achieved by even low intensity exercise, which may not increase performance capacity. The definition for the exact cut-off values in different groups of subjects with different underlying health status and pre-test-probability warrants further studies. Sedentary adults should avoid isolated bouts of unaccustomed vigorous physical activity and should follow the standard recommendation to increase physical activity levels gradually over time. The possible value of this strategy is supported by the observation that the risk of exercise-related cardiac events is greatest in the least active individuals

The risk-benefit of exercise is highly favourable for most patients with diabetes or cardiovascular disease, but some precautions are warranted. An exercise stress test should be performed to rule out ischemia, complex arrhythmias, and symptoms. Severe hypertension should be controlled to lower levels before an intensive exercise program is started. Patients with type 2 diabetes not requiring insulin may not need their blood glucose level routinely checked when exercising. There is usually no need for supplementary food except when exercise is exceptionally vigorous and long. Diabetic patients using insulin should be encouraged to exercise and be given instructions about blood glucose monitoring, insulin dosing, and supplementary foods.

Very little is known about prognostic value and cardiovascular benefits derived from very short bouts of exercise performed frequently as well as very long bouts performed infrequently but causing similar energy expenditure. From the risk stratification point of view, respiratory gas analysis and hemodynamic responses may help to describe the exercise characteristics due to limited cardiovascular reasons. In future, VO_{2max} and other specific indexes derived from ventilatory gas analysis, which can provide important information to evaluate functional capacity and distinguish cardiovascular from pulmonary limitations. Some previous studies have been reported that peak oxygen uptake and oxygen pulse, or abnormally high ventilatory response to exercise are powerful predictors of death [14,99-101].

Much has been learned in the past decade about how to use behavioral interventions to help sedentary adults become more active. It would be nice to evaluate the efficacy of disseminating these interventions with modern technology. More research may address the mechanisms by which exercise reduces coronary heart disease risk as well as explore other medical conditions effectively managed by physical exercise. The contribution of physical inactivity to the development of the present obesity epidemic warrants additional investigation, as does the role of physical activity in the treatment of obesity. It is important to develop information on the minimal and optimal amounts of physical activity required to produce specific beneficial outcomes and to understand individual differences that may augment or impede the beneficial effects of physical training. More attention also needs to be given to assessing the health benefits from physical activity when intensity is measured relative to the individual's fitness and age and not as an absolute value. Additional studies are recommended to show the value of physical activity programs superimposed on contemporary therapy for coronary heart disease patients because most data forming our knowledge were obtained many years ago when coronary heart disease mortality rates were substantially higher than at present.

Conclusions

An increasing amount of data supports the measurement of cardiorespiratory fitness in clinical practice. Exercise testing can be recommended for defining cardiorespiratory fitness and future risk in clinical practise among high-risk groups as well as asymptomatic subjects undergoing exercise testing. Direct or indirect measurement of VO_{2max} can provide a good estimate of prognosis and it should be taken into account when estimated the future risk of the patients. When interpreting the exercise test, it is very important to take exercise capacity into account; in addition to abnormal exercise ECG findings, blood pressure and HR responses. Low cardiorespiratory fitness is an important and independent risk factor for CVD, and can be considered to be as important a risk factor as smoking, hypertension, hypercholesterolemia, obesity and diabetes that can provide additional prognostic value. The individualized and population-based approaches to increased physical exercise across the life span are recommended

Acknowledgements

This study was supported by grants from the Finnish Academy, the Finnish Ministry of Education, and the town of Kuopio and the Juho Vainio Foundation, Helsinki, Finland, Finnish Cultural Foundation of Northern Savo, Kuopio, Finland, Aarne Koskelo Foundation, Helsinki, Finland, and Finnish Cardiovascular Foundation, Helsinki, Finland.

References

[1] McGinnis JM, Foege WH: Actual causes of death in the United States. *JAMA* 1993, 270:2207-2212.

[2] Fletcher GF, Balady GJ, Amsterdam EA, et al.: Exercise standards for testing and training: a statement for healthcare professionals from the American Heart Association. *Circulation* 2001, 104:1694-1740.

[3] Ekelund L-G, Haskell WL, Johnson JL, et al.: Physical fitness as a predictor of cardiovascular mortality in asymptomatic North American men: the Lipid Research Clinics Mortality Follow-up Study. *N Engl J Med.* 1988, 319:1379-1384.

[4] Slattery ML, Jacobs DR Jr: Physical fitness and cardiovascular disease mortality: the US Railroad Study. *Am J Epidemiol.* 1988, 127:571-580.

[5] Blair SN, Kohl HW III, Paffenbarger RS Jr, et al.: Physical fitness and all-cause mortality: a prospective study of healthy men and women. *JAMA* 1989, 262:2395-2401.

[6] Sandvik L, Erikssen J, Thaulow E, et al.: Physical fitness as a predictor of mortality among healthy, middle-aged Norwegian men. *N Engl J Med*, 1993;328:533-537.

[7] Blair SN, Kohl HW III, Barlow CE, et al.: Changes in physical fitness and all-cause mortality: a prospective study of healthy and unhealthy men. *JAMA* 1995, 273:1093-1098.

[8] Blair SN, Kampert JB, Kohl HW III, et al.: Influences of cardiorespiratory fitness and other precursors on cardiovascular disease and all-cause mortality in men and women. *JAMA* 1996, 276:205-210.

[9] Erikssen G, Liestol K, Bjornholt J, et al.: Changes in physical fitness and changes in mortality. *Lancet* 1998, 352:759-762.

[10] Wei MW, Kampert JB, Barlow CE, et al.: Relationship between low cardiorespiratory fitness and mortality in normal-weight, overweight, and obese men. *JAMA* 1999, 282:1547-1553.

[11] Laukkanen JA, Lakka TA, Rauramaa R, et al.: Cardiovascular fitness as a predictor of mortality in men. *Arch Intern Med* 2001, 161:825-831.

[12] Myers J, Prakash M, Froelicher V, et al.: Exercise capacity and mortality among men referred for exercise testing. *N Engl J Med* 2002, 346:793-801.

[13] Roger VL, Jacobsen SJ, Pellikka PA, et al.: Prognostic value of treadmill exercise testing: a population-based study in Olmsted County, Minnesota. *Circulation* 1998, 98:2836-2841.

[14] Goraya TY, Jacobsen SJ, Pellikka PA, et al.: Prognostic value of treadmill exercise testing in elderly persons. *Ann Intern Med* 2000, 132:862-870.

[15] Hambrecht R, Niebauer J, Marburger C, et al.: Various intensities of leisure time physical activity in patients with coronary artery disease: effects on cardiorespiratory fitness and progression of coronary atherosclerotic lesions. *J Am Coll Cardiol* 1993, 22:468-477.

[16] Schuler G, Hambrecht R, Schlierf G, et al.: Regular physical exercise and low-fat diet. Effects on progression of coronary disease. *Circulation* 1992, 86:1-11.

[17] Niebauer J, Hambrecht R, Velich T, et al.: Attenuated progression of coronary artery disease after 6 years of multifactorial risk intervention: role of physical exercise. *Circulation* 1997, 96:2534-2541.

[18] Ornish D, Brown SE, Scherwitz LW, et al.: Can lifestyle changes reverse coronary heart disease? The Lifestyle Heart Trial. *Lancet* 1990, 336:129-133.

[19] Haskell WL, Alderman EL, Fair JM, et al.: Effects of intensive multiple risk factor reduction on coronary atherosclerosis and clinical cardiac events in men and women with coronary artery disease. The Stanford Coronary Risk Intervention Project (SCRIP). *Circulation* 1994, 89:975-990.

[20] Leon AS, Connett J, Jacobs DR Jr, Rauramaa R: Leisure-time physical activity levels and risk of coronary heart disease and death. The Multiple Risk Factor Intervention Trial. *JAMA* 1987, 258:2388-2395.

[21] Morris JN, Clayton DG, Everitt MG, et al.: Exercise in leisure time: coronary attack and death rates. *Br Heart J* 1990, 63:325-334.

[22] Lakka TA, Venäläinen JM, Rauramaa R, et al.: Relation of physical activity and cardiorespiratory fitness to the risk of acute myocardial infarction in men. *N Eng J Med* 1994, 330:1549-1554.

[23] Sesso HD, Paffenbarger RS Jr, Lee IM: Physical activity and coronary heart disease in men: The Harvard Alumni Health Study. *Circulation* 2000, 102:975-980.

[24] Lee IM, Sesso HD, Paffenbarger RS Jr: Physical activity and coronary heart disease risk in men: does the duration of exercise episodes predict risk? *Circulation* 2000, 102:981-986.

[25] Lakka TA, Laukkanen JA, Rauramaa R, et al.: Cardiorespiratory fitness and the progression of carotid atherosclerosis in middle-aged men. *Ann Intern Med* 2001, 134:12-20.

[26] National Institute of Health Consensus Development Panel on Physical Activity and Cardiovascular Health. Physical activity and cardiovascular health. *JAMA* 1996, 276:241-246.

[27] Pate RR, Pratt M, Blair SN, et al.: Physical activity and public health: a recommendation from the Centers for Disease Control and Prevention and the American College of Sport Medicine. *JAMA* 1995, 273:402-407.

[28] Bouchard C: Physical activity and health: introduction to the dose-response symposium. *Med Sci Sports Exerc.* 2001, 33:S347-350.

[29] Pearson TA: New tools for coronary risk assessment: what are their advantages and limitations? *Circulation* 2002,105:886-892.

[30] Smith SC, Jr., Amsterdam E, Balady GJ, et al.: Prevention Conference V: Beyond secondary prevention: identifying the high-risk patient for primary prevention: tests for silent and inducible ischemia: Writing Group II. *Circulation* 2000, 101:E12-6.

[31] Tavel ME: Stress testing in cardiac evaluation: current concepts with emphasis on the ECG. *Chest* 2001, 119:907-925.

[32] Bruce RA, Hossack KF, DeRouen TA, Hofer V: Enhanced risk assessment for primary coronary heart disease events by maximal exercise testing: 10years' experience of Seattle Heart Watch. *J Am Coll Cardiol* 1983, 2:565-573

[33] Mark DB, Shaw L, Harrell FE Jr, et al.: Prognostic value of a treadmill exercise score in outpatients with suspected coronary artery disease. *N Engl J Med* 1991, 325:849-853.

[34] Morrow K, Morris CK, Froelicher VF, et al.: Prediction of cardiovascular death in men undergoing noninvasive evaluation for coronary artery disease. *Ann Intern Med* 1993,118:689-695.

[35] Revill SM, Beck KE, Morgan MD: Comparison of the peak exercise response measured by the ramp and 1-min step cycle exercise protocols in patients with exertional dyspnea. *Chest* 2002, 121:1099-1105.

[36] Gibbons RJ, Balady GJ, Beasley JW, et al.: ACC/AHA Guidelines for Exercise Testing. A report of the American College of Cardiology/American Heart Association Task Force on Practice Guidelines (Committee on Exercise Testing). *J Am Coll Cardiol* 1997, 30:260-311.

[37] Paffenbarger RS Jr, Hyde RT, Wing AL, et al.: The association of changes in physical-activity level and other lifestyle characteristics with mortality among men. *N Engl J Med* 1993, 328:538-545.

[38] Myers J, Gullestad L, Vagelos R, et al.: Clinical, hemodynamic, and cardiopulmonary exercise test determinants of survival in patients referred for evaluation of heart failure. *Ann Intern Med* 1998, 129:286-293.

[39] Robbins M, Francis G, Pashkow FJ, et al.: Ventilatory and heart rate responses to exercise : better predictors of heart failure mortality than peak oxygen consumption. *Circulation* 1999, 100:2411-2417.

[40] Prakash M, Myers J, Froelicher V, et al.: Clinical and exercise test predictors of all-cause mortality. Results from > 6000 consecutive referred male patients. *Chest* 2001, 120:1003-1013.

[41] Weiner DA, Ryan TJ, McCabe CH, et al.: The role of exercise testing in identifying patients with improved survival after coronary artery bypass surgery. *J Am Coll Cardiol* 1986, 8:741–748.

[42] Gulati M, Pandey DK, Arnsdorf MF, et al.: Exercise capacity and the risk of death in women: the St James Women Take Heart Project. *Circulation* 2003, 108:1554-1559.

[43] Spin JM, Prakash M, Froelicher VF, et al.: The prognostic value of exercise testing in elderly men. *Am J Med* 2002, 112: 453–459.

[44] Mora S, Redberg RF, Cui Y, et al.: Ability of Exercise Testing to Predict Cardiovascular and All-Cause Death in Asymptomatic Women: A 20-Year Follow-up of the Lipid Research Clinics Prevalence Study, *JAMA* 2003, 24:1600-1607.

[45] Wannamethee SG, Shaper AG, Walker M: Changes in physical activity, mortality, and incidence of coronary heart disease in older men. *Lancet* 1998, 351:1603-1608.

[46] Wannamethee G, Shaper AG: Physical activity and stroke in British middle aged men. *BMJ* 1992, 304:597–601.

[47] Herman B, Schmitz PI, Leyten AC, et al.: Multivariate logistic analysis of risk factors for stroke in Tilburg, the Netherlands. *Am J Epidemiol* 1983, 118:514–525.

[48] Lee IM, Paffenbarger RS Jr: Physical activity and stroke incidence: the Harvard Alumni Health Study. *Stroke.* 1998;29:2049–2054.

[49] Kurl S, Laukkanen JA, Rauramaa R, et al.: Cardiorespiratory fitness and the risk for stroke in men. *Arch Intern Med* 2003, 163:1682-1688.

[50] Salonen JT, Puska P, Tuomilehto J.: Physical activity and the risk of myocardial Infarction, cerebral stroke and death: A longitudinal study in Eastern Finland. *Am J Epidemiol* 1982, 115:526-537.

[51] Sacco RL, Gan R, Boden-Albala B, et al.: Leisure-time physical activity and ischemic stroke risk: the Northern Manhattan Stroke Study. *Stroke* 1998, 29:380-387.

[52] Paffenbarger RS Jr: Factors predisposing to fatal stroke in longshoremen. *Prev Med* 1972, 1:522-528.

[53] Evenson KR, Rosamond WD, Cai J, et al.: Physical activity and ischemic stroke risk. The atherosclerosis risk in communities study. *Stroke* 1999, 30:1333-1339.

[54] Hambrecht R, Niebauer J, Marburger C, et al.: Various intensities of leisure time physical activity in patients with coronary artery disease: effects on cardiorespiratory fitness and progression of coronary atherosclerotic lesions. *J Am Coll Cardiol* 1993, 22:468-477.

[55] Lee CD, Folsom AR, Blair SN: Physical activity and stroke risk: a meta-analysis. *Stroke* 2003, 34:2475-2481.

[56] Lee IM, Rexrode KM, Cook NR, et al.: Physical activity and coronary heart disease in women: is "no pain, no gain" passe? *JAMA* 2001, 285: 1447-54.

[57] Lee CD, Blair SN, Jackson AS: Cardiorespiratory fitness, body composition, and all-cause and cardiovascular disease mortality in men. *Am J Clin Nutr* 1999, 69:373-380.

[58] Al Suwaidi J, Higano ST, Holmes DR Jr, et al.: Obesity is independently associated with coronary endothelial dysfunction in patients with normal mildly diseased coronary arteries. *J Am Coll Cardiol* 2001, 37:1523-1528.

[59] Cai H, Harrison DG: Endothelial dysfunction in cardiovascular diseases: the role of oxidant stress. *Circ Res* 2000, 87:840-844.

[60] Brown AA, Hu FB: Dietary modulation of endothelial function: implications for cardiovascular disease. *Am J Clin Nutr* 2001, 73: 673-686.

[61] Newby DE, McLeod AL, Uren NG, et al.: Impaired coronary tissue plasminogen activator release is associated with coronary atherosclerosis and cigarette smoking: direct link between endothelial dysfunction and atherothrombosis. *Circulation* 2001, 103: 1936-1941.

[62] Hambrecht R, Wolf A, Gielen S, et. al.: Effect of exercise on coronary endothelial function in patients with coronary artery disease. *N Engl J Med* 2000, 342:454-460.

[63] Gielen S, Schuler G, Hambrecht R: Exercise training in coronary artery disease and coronary vasomotion. *Circulation* 2001, 103:E1-6.

[64] McAllister RM, Hirai T, Musch TI: Contribution of endothelium-derived nitric oxide (EDNO) to the skeletal muscle blood flow response to exercise. *Med Sci Sports Exerc* 1995, 27:1145–1151.

[65] Sakamoto S, Minami K, Niwa Y, et al.: Effect of exercise training and food restriction on endothelium-dependent relaxation in the Otsuka Long-Evans Tokushima Fatty rat, a model of spontaneous NIDDM. *Diabetes* 1998, 47:82–86.

[66] Maiorana A, O'Driscoll G, Cheetham C, et al.: The effect of combined aerobic and resistance exercise training on vascular function in type 2 diabetes. *J Am Coll Cardiol* 2001, 38:860–866.

[67] Lavrencic A, Salobir BG, Keber I: Physical training improves flow-mediated dilation in patients with the polymetabolic syndrome. *Arterioscler Thromb Vasc Biol* 2000, 20:551–555.

[68] Paterick TE, Fletcher GF: Endothelial function and cardiovascular prevention: role of blood lipids, exercise, and other risk factors. *Cardiol Rev* 2001, 9:282–286.

[69] Salomaa V, Riley W, Kark JD, et al.: Non-insulin-dependent diabetes mellitus and fasting glucose and insulin concentrations are associated with arterial stiffness indexes: the ARIC Study. *Circulation* 1995, 91:1432–1443.

[70] Bank AJ, Shammas RA, Mullen K, Chuang PP: Effects of short-term forearm exercise training on resistance vessel endothelial function in normal subjects and patients with heart failure. *J Card Fail* 1998, 4:193-201.

[71] Pilegaard H, Ordway GA, Saltin B, Neufer PD: Transcriptional regulation of gene expression in human skeletal muscle during recovery from exercise. *Am J Physiol Endocrinol Metab* 2000, 279:E806-14.

[72] Febbraio MA, Koukoulas I: HSP72 gene expression progressively increases in human skeletal muscle during prolonged, exhaustive exercise. *J Appl Physiol* 2000, 89:1055-1060.

[73] Vaitkevicius PV, Fleg JL, Engel JH, et al.: Effects of age and aerobic capacity on arterial stiffness in healthy adults. *Circulation* 1993, 88:1456–1462.

[74] DeFronzo RA: Insulin resistance, hyperinsulinemia, and coronary artery disease: a complex metabolic web. *J Cardiovasc Pharmacol* 1992, 20:S1–S16.

[75] Watkins LL, Sherwood A, Feinglos M et al.: Effects of exercise and weight loss on cardiac risk factors associated with syndrome X. Arch Intern Med. 2003;163: 1889-1895.

[76] Kelley DE, Goodpaster BH. Effects of exercise on glucose homeostasis in type 2 diabetes mellitus. *Med Sci Sports Exerc* 2001, 33:S495–S501.

[77] Whelton SP, Chin A, Xin X, He J: Effect of aerobic exercise on blood pressure: a meta-analysis of randomized, controlled trials. *Ann Intern Med* 2002, 136:493–503.

[78] Kelley GA, Kelley KA, Tran ZV: Aerobic exercise and resting blood pressure: a meta-analytic review of randomized, controlled trials. *Prev Cardiol* 2001, 4:73–80.

[79] Hagberg JM, Ferrell RE, Dengel DR, Wilund KR: Exercise training-induced blood pressure and plasma lipid improvements in hypertensives may be genotype dependent. *Hypertension* 1999, 34:18-23.

[80] Tuomilehto J, Lindstrom J, Eriksson JG, et al.: Finnish Diabetes Prevention Study Group. Prevention of type 2 diabetes mellitus by changes in lifestyle among subjects with impaired glucose tolerance. *N Engl J Med* 2001; 344:1343-50.

[81] Rauramaa R, Salonen JT, Seppänen K, et al.: Inhibition of platelet aggregability by moderate intensity physical exercise: a randomized clinical trial in overweight men. *Circulation* 1986; 74:939-944.

[82] Stratton JR, Chandler WL, Scwartz RS, et al.: Effects of physical conditioning on fibrinolytic variables and fibrinogen in young and old healthy adults. *Circulation* 1991; 83: 1692-1697.

[83] Kramsch DM, Aspen AJ, Abramowitz BM, et al.: Reduction of coronary atherosclerosis by moderate conditioning exercise in monkeys on an atherogenic diet. *N Engl J Med* 1981, 305:1483-1489.

[84] Beere PA, Glagov S, Zarins CK: Retarding effect of lowered heart rate on coronary atherosclerosis. *Science* 1984, 226:180-182.

[85] Smith JK, Dykes R, Douglas JE, et al.: Long-term exercise and atherogenic activity of blood ischemic heart disease. *JAMA* 1999, 281:1722-1727.

[86] Hambrecht R, Niebauer J, Fihn E, et al.: Physical training in patients with stable chronic heart failure: effects on cardiorespiratory fitness and ultrastructural abnormalities of leg muscles. *J Am Coll Cardiol* 1995, 25:1239-1249.

[87] Langer A, Freeman MR, Josse RG, et al.: Detection of silent myocardial ischemia in diabetes mellitus. *Am J Cardiol* 1991, 67:1073–1078.

[88] Uusitupa MI, Mustonen JN, Airaksinen KE: Diabetic heart muscle disease. *Ann Med* 1990, 22:377–386.

[89] Kelemen MH, Effron MB, Valenti SA, Stewart KJ: Exercise training combined with antihypertensive drug therapy: effects on lipids, blood pressure, and left ventricular mass. *JAMA* 1990, 263:2766–2771.

[90] Levy WC, Cerqueira MD, Abrass IB, et al.: Endurance exercise training augments diastolic filling at rest and during exercise in healthy young and older men. *Circulation* 1993, 88:116–126.

[91] Bouchard C, Rankinen T, Chagnon YC, et al.: Genomic scan for maximal oxygen uptake and its response to training in the HERITAGE Family Study. *J Appl Physiol* 2000, 88:551-559.

[92] Perusse L, Rice T, Province MA, et al.: Familial aggregation of amount and distribution of subcutaneous fat and their responses to exercise training in the HERITAGE family study. *Obes Res* 2000, 8:140-150.

[93] Bouchard C, Rankinen T: Individual differences in response to regular physical activity. *Med Sci Sports Exerc* 2001, 33:S446-S451.

[94] Kesaniemi YA, Danforth E, Jensen MD, et al.: Dose-response issues concerning physical activity and health: an evidence-based symposium. Consensus statement. *Med Sci Sports Exerc* 2001, 33:S351-358.

[95] Dunn LA, Marcus BH, Kampert JB, et al.: Comparison of lifestyle and structured interventions to increase physical activity and cardiorespiratory fitness: a randomized trial. *JAMA* 1999, 281:327-334.

[96] Thompson PD, Buchner D, Pina IL, et al.: Exercise and physical activity in the prevention and treatment of atherosclerotic cardiovascular disease: a statement from the Council on Clinical Cardiology (Subcommittee on Exercise, Rehabilitation, and Prevention) and the Council on Nutrition, Physical Activity, and Metabolism (Subcommittee on Physical Activity) *Circulation* 2003, 107:3109-3116.

[97] Lemaitre RN, Siscovick DS, Raghunathan TE, et al.: Leisure-time physical activity and the risk of primary cardiac arrest. *Arch Intern Med* 1999, 159:686-690.

[98] Albert CM, Mittleman MA, Chae CU, et al.: Triggering of sudden death from cardiac causes by vigorous exertion. *N Engl J Med* 2000, 343:1355-1361.

[99] Froelicher VF, Fearon WF, Ferguson CM, et al. Lessons learned from studies of the standard exercise ECG test. *Chest* 1999, 116:1442-1451.

[100] Cohen-Solal A, Barnier P, Pessione F et al.: Comparison of the long-term prognostic value of peak exercise oxygen pulse and peak oxygen uptake in patients with chronic heart failure. *Heart* 1997, 78:572-576.

[101] Cohen-Solal A, Tabet JY, Logeart D, et al.: A non-invasively determined surrogate of cardiac power ('circulatory power') at peak exercise is a powerful prognostic factor in chronic heart failure. *Eur Heart* J 2002, 23:806-814.

In: Focus on Atherosclerosis Research
Editor: Leon V.Clark, pp. 63-72

ISBN 1-59454-044-6

Chapter III

Childhood Risk Factors Predict Adult Risk Associated with Subclinical Cardiovascular Disease and Strategies for Prevention: The Bogalusa Heart Study

Gerald S. Berenson * ***and Sathanur R. Srinivasan***
Tulane Center for Cardiovascular Health, Tulane School of Public Health and Tropical Medicine, New Orleans, LA

Abstract

Cardiovascular (C-V) risk factors can be identified beginning in childhood and these are predictive of C-V disease later in life. Observations in the Bogalusa Heart Study have shown an important correlation of clinical risk factors in early life with anatomic changes at autopsy in the aorta and coronary vessels with atherosclerosis and cardiac and renal changes related to hypertension. These observations have been extended by echo Doppler studies of carotid arteries and intima media thickness (IMT). A close association of risk factors in young adults, 20-38 years of age, occurs with IMT, and a marked increase in IMT is noted as the numbers of risk factors increase. Obesity and low density lipoprotein cholesterol (LDL-C) are strong predictors of subclinical disease. Observations of risk factors in young individuals and noninvasive studies of structure and function changes of the C-V system have strong implications for prevention by cardiologists.

* Address for correspondence: Gerald S. Berenson, M.D., Director, Tulane Center for Cardiovascular Health, 1440 Canal Street, Suite 1829, New Orleans, LA 70112, USA

Introduction

Cardiovascular (C-V) risk factors are identifiable in childhood and are predictive of future C-V risk in adulthood [1-4]. Although clinical C-V heart disease occurs in later life, evidence of atherosclerosis and traits of hypertension and diabetes mellitus are evident in childhood. The precise initiating, environmental and genetic factors of these diseases are still being investigated, but by identifying the earliest determinants of C-V risk we can achieve a great deal in determining susceptibility (risk) to complex conditions, such as coronary heart disease or diabetes. Risk factors as measured clinically are governed by both genes and environment; what is generally examined is the genetic-environmental interaction. The long-term burden of C-V risk factors and lifestyle behaviors developing in early childhood create a long-term burden ultimately being recognized as C-V events. Risk factor levels at an early age relate to beginning anatomic changes in the C-V-renal system. Unhealthy lifestyles, such as high fat-high calorie diet, tobacco use and physical inactivity leading to obesity, represent strong environmental factors related to the latent or "silent" C-V lesions in the C-V system. Cardiologists now have the opportunity to study C-V risk and determine subtle underlying cardiac and vascular changes in young individuals.

Population

The Bogalusa Heart Study is a comprehensive study of early natural history of C-V disease in a population of children and young adults from a biracial (65% white, 35% -black), semi-rural community [2,3]. The Study began in 1973 and has continued with multiple cross-sectional and longitudinal surveys. Extensive C-V risk factor data have been collected on approximately 16,000 individuals now providing a database from birth to 38 years of age. Many subjects have had multiple observations.

Tracking – Predictability of Childhood Risk Factors for Adult Levels

Levels of risk factor variables for an individual tend to remain over time in a given rank relative to peers. "Tracking", based on findings beginning in early life, is important for predicting adult C-V risk factors and clinical disease. Risk factors track to varying degrees, with measures of obesity tracking the most [5]. Obesity, which begins in infancy and childhood, persist into adulthood and levels in childhood are strongly predictive of adult obesity. Serum total cholesterol and low density lipoprotein cholesterol (LDL-C) track almost as well. The magnitude of tracking of triglycerides, high density lipoprotein cholesterol (HDL-C), and blood pressure is relatively low, probably due to biologic and measurement variability. Of note, clusters of multiple risk factors persist strongly from childhood to adulthood, especially among those with increased body fatness [6]. Multiple risk factors can be shown to track and subjects with one or more risk factors are likely to have three or four additional risk factors. It is of interest that baseline adiposity in childhood precedes hyperinsulinemia at older age periods, independently of race, sex, and baseline insulin levels. Adiposity is an independent predictor of the risk of developing a cluster of risk variables consistent with the metabolic syndrome or insulin resistance and it begins in childhood. The exposure to a burden of multiple risk factors over time enhances the development of coronary atherosclerosis and hypertensive C-V disease and this persistence of risk factors from childhood creates a burden as a precursor of C-V changes.

Secular Trends – Changes over Time

Repeated surveys in this biracial community have allowed observations of secular trends through examination of subjects at comparable ages following a common protocol. Of particular importance is the dramatic increase of obesity over time, especially in black females. Since 1973 children on average gained 5 kilograms of body weight without increases in height [7]. This upward secular trend in childhood obesity reflects the national trend seen in the general population of children and associated with a similar increase of obesity in adults [8]. With this increasing obesity there is a notable increase in the incidence of type 2 diabetes mellitus, with an increase in the early onset of diabetes, even during adolescence. The persistence of risk factors from childhood indicates that the precursors of atherosclerosis and essential hypertension begin early and progress into adulthood with evidence of underlying subclinical C-V lesions.

In addition, a secular trend in diet has also occurred with an increase of P/S ratio from 0.30 to approximately 0.45 in the mid 70's with the introduction of liquid oils for cooking. The total fat consumption has decreased from 38-40% to a level of approximately 33% of the diet with a reduction in saturated fat. In addition there has been a more marked decrease of cholesterol intake, likely related to a decreased consumption of eggs. Although increased body weight has occurred without an increase of body height no detectable changes of dietary caloric intake has been noted. However, the variability in the method of quantitating diet is probably inadequate to accurately document slight caloric changes. Dietary imbalance with physical inactivity has probably accounted for the dramatic increase of body weight and body fatness. Dietary patterns have changed, however, with a large increase of high carbohydrate soft drinks, and high intake of corn starch.

Clustering of Multiple Risk Factors

The adverse impact of obesity on individual C-V risk variables has been found in both children and adults. Of interest is the coexistence of multiple risk factors beginning in childhood and persisting into adulthood. Individual risk factors correlate with other risk factors in both childhood and adulthood alike, for example, obesity and high blood pressure levels. Consequently, risk factors tend to occur in constellations. A condition that includes dyslipidemia, hypertension, hyperinsulinemia/insulin resistance, and central obesity has been termed Syndrome X, metabolic syndrome, or insulin resistance syndrome. The degree of clustering of risk variables has been found to increase with age and in preadolescents extending to young adulthood. The age-related changes in obesity and the attendant insulin resistance seem to govern the degree of clustering. Clustering is quite evident in adults as illustrated by the findings in adult epidemiologic studies, such as the Framingham Study and the Multiple Risk Factor Intervention Trial (MRFIT). The coexistence of multiple risk factors in the young has strong implications for adult heart disease. Central obesity and the attendant insulin resistance/hyperinsulinemia occurring in childhood are the driving force of the metabolic syndrome and this combination of risk factors strongly predicts the risk of developing a constellation of metabolic, hemodynamic, thrombotic and inflammatory disorders [9]. Of considerable importance, the burden of multiple risk factors from childhood on the C-V system can be evaluated by non-invasive imaging procedures. These procedures provide the opportunity of addressing "silent" C-V disease and identifying individuals at high risk requiring aggressive prevention methods.

Autopsy Studies: Evidence of Underlying Cardiovascular Disease

Autopsy studies in young individuals in the Bogalusa Heart Study have provided the most compelling evidence that an unusually high prevalence (90%) of coronary atherosclerosis occurs by the third decade of life. The early occurrence of atherosclerosis was found in autopsy studies of the International Atherosclerosis Study emphasized by McGill [10] and in soldiers dying in the Korean and Vietnamese wars [11,12]. Fatty streaks begin to appear in the aorta even earlier than 3 years of age. It has been recently reported that anatomical changes in the fetal arterial wall are conditioned by maternal hypercholesterolemia during pregnancy and adversely influence the rate of progression of early lesions throughout childhood [13]. Lesions appear in other vascular beds early and beginning in adolescence progressive lesions involve the coronary arteries [14]. Coronary vessels demonstrate both fatty streaks and fibrous plaques of raised, collagen-capped lesions that are precursors of more advanced disease. There are also race-sex differences with a definite lag in the development of lesions in females vs males at this young age. Blacks show more aorta fatty streaks and white males more coronary artery fibrous plaques [15-16]. Of importance is the striking relationship of antemortem C-V risk factors to the extent of lesions. With the severity of atherosclerosis increasing there is a curvilinear increase in lesions as the number of multiple risk factors increase, Figure 1 [17]. The Pathobiologic Determinants of Atherosclerosis in Youth (PDAY) Study has confirmed the extent of these autopsy findings showing the importance of multiple risk factors in young adulthood [18]. These observations are in concert with the multiple risk factor concept of the Framingham Risk Score for predicting morbidity and mortality. It is also important to note that autopsy studies have shown atherosclerosis developing at various vascular sites, including carotid, cerebral, and renal vessels at a young age, indicating the systemic effects of atherosclerosis [19].

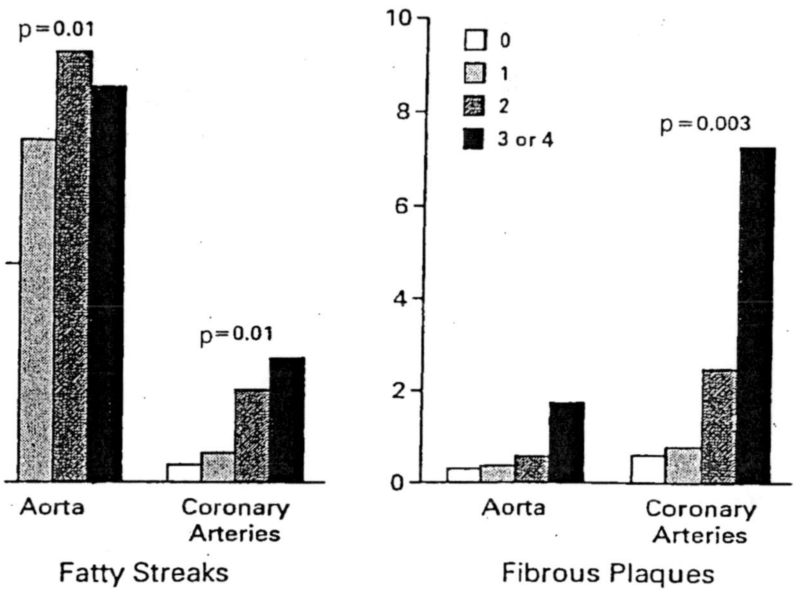

Figure 1. The effect of multiple risk factors on the extent of atherosclerosis in the aorta and coronary arteries in children and young adults. The severity of intimal surface involved with lesions in subjects with 0, 1, 2, and 3 risk factors is shown. Elevated risk factors included body-mass index, systolic blood pressure, and serum triglycerides and LDL cholesterol concentrations were defined as values above the 75th percentile for the study group (specific for study period, race, sex, and age). There were 52 subjects with no risk factors, 20 with one, 14 with two, and 7 with three or four. The P value is based on the analysis of trend. A marked increase in the percentage of the intimal surface covered by fibrous plaques, raised, collagen-capped lesions, is evident in the coronary vessels associated with multiple risk factors. (Berenson, Ref. 17)

In addition, there is an increase of left ventricular mass associated with higher levels of blood pressure. Increased left ventricular mass has been shown to be an independent risk factor for future cardiac events in adults. We found that obesity and insulin resistance, as important risk factors, are predictive of left ventricular hypertrophy [20]. Our echocardiographic studies in children indicate that although linear growth is a major determinant of cardiac growth, overweight is an important determinant of the acquisition of increased left ventricular mass preceding the development of elevated blood pressure in the future.

Carotid Artery Studies

Although a profile of C-V risk factors is useful in predicting C-V events, inclusion of measurements of subclinical atherosclerosis by non-invasive imaging techniques can be a powerful approach to determine the extensiveness and severity of asymptomatic disease and potential future risk – evidence of "silent" disease.

Recently there has been considerable interest in imaging for calcification of coronary arteries, but these observations are often limited in determining the extensiveness of coronary atherosclerosis of non-calcified lesions, even in individuals with known coronary artery disease. The Muscatine Study showed a relation of obesity from childhood with coronary calcification [20], although coronary calcification showed a limited relationship to C-V risk factors [21]. Of the methods, such as coronary arteriography and intravascular coronary ultrasound studies (IVUS) even though these methods confirm disease at a young age, these are not applicable for large population studies nor can they be recommended for asymptomatic individuals. On the other hand, echoDoppler studies of the carotid arteries do provide a non-invasive and readily available tool for evaluating large numbers of asymptomatic individuals.

We measured carotid intima-media thickness ((IMT) of carotid arteries by B-mode duplex ultrasonography in 517 black and white subjects, aged 20 to 38 years, enrolled in the Bogalusa Heart Study. The risk factor variables included parental history, blood pressure, obesity measures, lipoproteins, insulin, glucose and cigarette smoking [22]. The means of three values each for right and left farwall was used for the common, the bulb or bifurcation, and internal carotid segments. IMT in the common carotid and the bulb segments were significantly greater in blacks, males, and older individuals.

Lumen narrowing by encroaching atherosclerotic plaques is not expected to occur at a young age. However, marked individual differences occur in IMT. The top 5% of carotid artery at the common segment shows a range of 0.84 to 1.1 mm IMT, with a mean of 0.68 mm, and for the carotid bulb 0.90 to 2.1 mm, with a mean of 0.87 mm. Earlier, in a related study regarding vascular carotid artery distensibility measured in 10- to 17-year olds, we were able to show an association of increased stiffness in terms of pressure-strain elastic modulus (Petersen elastic modulus, Ep) with elevated levels of both serum total cholesterol and systolic blood pressure, and a positive parental history of myocardial infarction [23]. It is possible that studies of compliance and distensibility may show functional changes even preceding a detectable structural disease of the arterial wall. We have been able to show that Ep and Young's Elastic Modulus (YEM) also relate to increasing numbers of risk factors similar to the changes of coronary atherosclerosis.

The association among different segments of carotid IMT and risk factor variables occurs in the expected direction. Systolic blood pressure, race, age, LDL cholesterol and HDL cholesterol explain 16% variance in the common carotid IMT; age, systolic blood pressure, HDL cholesterol, LDL cholesterol, race and, importantly, insulin explain 19.4% variance in the carotid bulb IMT; and sex and body mass index explain 4.7% variance in the internal carotid IMT. Increases in IMT with increasing numbers of risk factors (0,1,2,3 or more) similar to the Framingham multi-variate risk score were noted for the common carotid and the carotid bulb segments, Figure 2. These observations are consistent with the notion that the carotid bulb or bifurcation is more susceptible for atherosclerotic damage than the common carotid or or internal carotid segments. Zarins, et al showed the importance of pulsatile hemodynamic forces on arterial wall structure, which led Glagov and his group to the concept of remodeling [24]. This may be a basis for thicker IMT at the bulb segment and stronger association of risk factors similar to that noted for coronary arteries; both areas of the vascular system are also musculoelastic in structural composition.

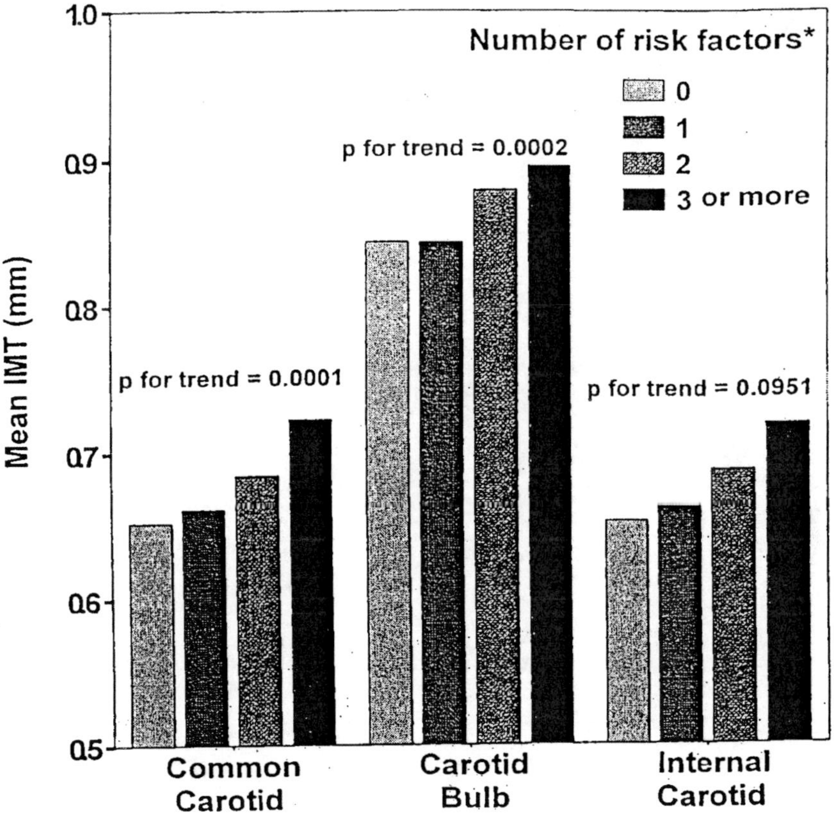

Figure 2. Mean intima media thickness (IMT) of three segments of the carotid artery related to the number of risk factors. *Total cholesterol to HDL-C ratio, waist circumference, syst. BP, and insulin (>75th percentile specific for age, race, and gender), and smoking. (Urbina, Ref. 22)

The subclinical underlying lesion in the carotid arteries, especially at the bulb site, by relating to multiple C-V risk factors in young individuals should help identify individuals of particular risk for future C-V events. The carotid ultrasound measurements are relatively easy to obtain and are a marker for the intensity of the long-term burden of C-V risk factors beginning at a young age. Studies of such changes may provide a history of vascular age in contrast to chronological age. Understanding risk factors at a young age and observing their effects on the C-V system can guide early prevention modalities.

Strategies to Prevent Cardiovascular Disease

A High Risk Model

A Family Health Promotion program was developed for individuals detected with abnormal risk of developing or having coronary artery disease [25,26]. One example is to involve the family, spouse and children, of the patient who has by-pass surgery, or angioplasty or stent implacement into a training program. Training is conducted over 10-12 weeks by a multidisciplinary team of cardiologists and a nurse specialist and including a nutritionist, exercise specialist and psychologist (behavior oriented specialist). A C-V risk profile is conducted on each family member and periodic training is continued. Although such prevention is not insurance-funded at present, this kind of program can be very successful. More specific is a model of preventing early hypertension by use of low dose of hypertensive medication, diet and exercise for children, tracking at a level of 90th percentile, a level shown to be associated with target organ C-V changes [27].

A Public Health or Population Model

The importance of beginning prevention in early life has to be emphasized by cardiologists being confronted with the daily problem of morbidity and mortality from heart diseases. Although the effectiveness of drug therapy has been demonstrated in many clinical trials, primary intervention by altering lifestyles plays a major role in managing patients and families at C-V risk [25,26]. To establish effectiveness of primary prevention alone at a population level is almost impossible, since it would require a large scale population study over many years. As a population approach for the young, we have developed an effective comprehensive health education program (Health Ahead/Heart Smart) for elementary school children [28,29]. This includes the classroom, cafeteria, physical education, lifestyles of teachers and it involves parents. Lifestyles and behaviors that influence C-V risk factors are learned in childhood. Healthy lifestyles need to be adopted early because they are critical to modulating risk factors in adult life. Cardiologists and primary care physicians have a tremendous opportunity to guide and propagate prevention programs for the young, destined to be our next generation of cardiac patients [30].

Acknowledgments

The Bogalusa Heart Study is a joint effort of many individuals whose cooperation is gratefully acknowledged. We are especially grateful for the children who grew up contributing to this research. Supported by grants HL38844 from the National Heart, Lung and Blood Institute, AG16592 from the National Institute on Aging, and HD 43820 from the National Institute of Child Health and Human Development.

References

[1] Lauer RM, Shekelle RB, eds. *Childhood Prevention of Atherosclerosis and Hypertension*, Raven Press, New York; pp. 484, 1980.

[2] Berenson GS, McMahan CA, Voors AW, Webber LS, Srinivasan SR, Frank GC, Foster TA, Blonde CV. *C-V Risk Factors in Children—The Early Natural History of Atherosclerosis and Essential Hypertension* (editorial assistance: Andrews C, Hester HE). New York: Oxford University Press, pp 450, 1980.

[3] Berenson GS (ed). *Causation of C-V Risk Factors in Children: Perspectives on Cardiovascular Risk in Early Life*. New York: Raven Press, pp 408, 1986.

[4] Akerblom HK, Ukari M, Pesonen E, Dahl M, Kaprio EA, Nuutinen EM, et al. Cardiovascular Risk in Young Finns. *Annals of Med* 23:35-39, 1991.

[5] Srinivasan SR, Bao W, Wattigney WA, Berenson GS: Adolescent overweight is associated with adult overweight and related multiple cardiovasuclar risk factors. The Bogalusa Heart Study. *Metabolism* 45(2):235-240, 1996..

[6] Boa W, Srinivasan Sr, Wattigney WA, Berenson GS. Persistence of multiple cardiovascular risk clustering related to syndrome X from childhood to young adulthood. *Arch Intern Med*, 154:1842-47, 1994.

[7] Gidding SS, Bao W, Srinivasan SR, Berenson GS. Effects of secular trends in obesity on coronary risk factors in children. The Bogalusa Heart Study. *J Pediatr* 127:868-74, 1995.

[8] Mohdad AH, Serdula MK, Dietz WH, Bowman BA, Marks JS, Koplan JP. The spread of the obesity epidemic in the United States, 1991 1998. *JAMA* 282:1519-22, 1999.

[9] Srinivasan SR, Myers L, Berenson GS. Temporal Association between obesity and hyperinsulinemia in children, adolescents, and young adults: The Bogalusa Heart Study. *Metabolism*, 48:928-34, 1999.

[10] McGill HC Jr. Persistent problems in the pathogenesis of atherosclerosis. *Arteriosclerosis*, 4:443-51, 1984.

[11] Enos WF, Beyer JC, Holmes RH. Pathogenesis of coronary disease in American soldiers killed in Korea. *JAMA* 158:912-14, 1955.

[12] McNamara JD, Molot MA, Stremple JF, Cutting RT. Coronary artery disease in combat casualties in Vietnam. *JAMA* 216:1185-87, 1971.

[13] Napoli C, Glass CK, Witztum JL, Deutsch R, D'Armiento FP. Influences of maternal hypercholesterolemia during pregnancy on progression of early atherosclerotic lesions in childhood: Fate of early lesions in children (FELIC) Study. *Lancet* 354:1234-41, 1999.

[14] Stary HC, Blankenhorn DH, Chandler AB, Glagov S, Insull W Jr, Richardson M, et al. A definition of the intima of human arteries and of its atherosclerosis-prone regions. *Circ* 85:391-405, 1992.

[15] Freedman DS, Newman WP III, Tracy RE, Voors AW, Srinivasan SR, Webber LS, et al. Black/white differences in aortic fatty streaks in adolescence and early childhood: The Bogalusa Heart Study. *Circ* 77:856-64, 1988.

[16] Berenson GS, Wattigney WA, Tracy RE. Atherosclerosis of aorta and coronary arteries and cardiovascular risk factors in persons aged 6-30 years studied at necropsy: The Bogalusa Heart Study. *Am J Card*, 70:851-58, 1992.

[17] Berenson GS, Srinivasan SR, Bao W, Newman WP III, Tracy, RE, Wattigney W, for the Bogalusa Heart Study. Association between multiple cardiovascular risk factors and atherosclerosis in children and young adults. *N Eng J Med,* 338:1650-56, 1998.

[18] McGill HC Jr, McMahan CA, Malcom RT, Oalmann MC, Strong JP. Effect of serum lipoproteins and smoking on atherosclerosis in young men and women. *Arterioscl Thromb Vasc Biol* 17:95-106, 1997.

[19] McGarry P, Solberg LA, Guzman MA, Strong JP. Cerebral atherosclerosis in New Orleans. Comparison of lesions by age, sex, and race. *Lab Invest* 52:533-39, 1985.

[20] Urbina EM, Gidding SS, Bao W, Pickoff AS, Berdusis K, Berenson GS: Effect of body size ponderosity and blood pressure on left ventricular growth in children and young adults, The Bogalusa Heart Study. *Circulation* 91(9):2400-2406, 1995.

[21] Mahoney LT, Burns TL. Stanford W, Thompson BH, Witt JD, Rost CA, Lauer RM. Coronary risk factors measured in childhood and young adult life is associated with coronary artery calcification in young adults: the Muscatine Study. *J Am Coll Cardiol* 27:277-84, 1996.

[22] Urbina EM, Srinivasan SR, Tang R, Bond MG, Kieltyka L, Berenson GS. Impact of Multiple Coronary Risk Factors on the Intima-Media Thickness of Different Segments of Carotid Artery in Healthy Young Adults (The Bogalusa Heart Study). *Am J Cardiol* 90:953-958, 2002.

[23] Riley WA, Freedman DS, Higgs NA, Barnes RW, Zinkgraf SA, Berenson GS. Decreased arterial elasticity associated with C-V disease risk factors in the young: The Bogalusa Heart Study. *Arteriosclerosis* 6:377-86, 1986.

[24] Zarins CK, Giddens DP, Bharadvaj BK, Sottiurai VS, Mabon RF, Glagov S. Carotid bifurcation atherosclerosis. Quantitative correlation of plaque localization with flow velocity profiles and with shear stress. *Circ Res* 53:502-11, 1983.

[25] Johnson CC, Nicklas TA, Arbeit ML, Harsha DW, Serpas DC, Hunter SM, Wattigney W, Berenson GS: C-V intervention for high-risk families: The Heart Smart Program. *South Med J* 84(11):1305-1312, 1991.

[26] Berenson GS, Harsha DW, Johnson CC, Nicklas TA: Teach families to be Heart Smart. *Patient Care* June 15, pp 135-145, 1993.

[27] Berenson GS, Shear CL, Chiang Y-K, Webber LS, Voors AW: Combined low-dose medication and primary intervention over a 30 month period for sustained high blood pressure in childhood. *Am J Med Sci* 299:79-86, 1990.

[28] Downey AM, Greenberg JS, Virgilio SJ, Berenson GS: Health promotion model for "Heart Smart": The medical school, university, and community. *Am J Hlth Promotion* 13:31-46, 1989.

[29] Arbeit ML, Johnson CC, Mott DS, Harsha DW, Nicklas TA, Webber LS, Berenson GS. The Heart Smart cardiovascular school health program. Behavior correlates of risk factor change. *Prev Med* 21:18-32, 1992.

[30] Downey AM, Cresanta JL, Berenson GS. Cardiovascular health promotion: "Heart Smart" and the changing role of the physician. *Am J Prev Med* 5:279-95, 1989.

In: Focus on Atherosclerosis Research
Editor: Leon V.Clark, pp. 73-90

ISBN 1-59454-044-6

Chapter IV

A Role for *Chlamydia pneumoniae* in Atherosclerosis: Fact or Fiction?

Vicky Y. Hoymans[1,*], Johan M. Bosmans[1], Margareta M. Ieven[2], Christiaan J. Vrints[1]

[1] The Division of Cardiology, University of Antwerp, UA, Belgium
[2] The Division of Microbiology, University of Antwerp, UA, Belgium

Abstract

Chlamydia pneumoniae, a human respiratory pathogen, has been repeatedly linked to atherosclerotic disease based on sero-epidemiologic studies, direct detection of the organism in atherosclerotic lesions, animal experiments, in vitro data, and to a lesser extent tissue culture. Later on, meta-analysis showed no differences in antibody titers between cases and healthy controls. In addition, the detection of *C. pneumoniae* DNA within atherosclerotic lesions by PCR varied from 0% to 100% between different laboratories, with evidence of contamination being gathered. Furthermore, the correlation between *C. pneumoniae* DNA or antigen detection in the vessel wall and antibody titres in the serum remains controversial. Recent data have also revealed potential crossreaction with nonchlamydial plaque constituents, indicating that immunohistostaining results obtained with antibodies against *C. pneumoniae* have to be interpreted with caution. Besides methodological difficulties, not all investigators have used the same tests, antigens, strains, primers and protocol conditions. Whether patients with atherosclerosis or acute coronary syndromes benefit from antibiotics against *C. pneumoniae* is still an evolving issue. Animal models and in vitro studies, however, continue to accumulate convincing data for an association. In this chapter, a literature overview is given of the main studies published on *C. pneumoniae* and its potential relation to atherosclerotic disease.

* Correspondence: Vicky Hoymans, Division of Cardiology, University of Antwerp (UA), Wilrijkstraat 10, B-2600 Edegem Belgium, Tel. +32-3-821 46 93, Fax +32-3-830 23 05, e-mail: vicky.hoymans@uza.be

Introduction

Classical risk factors of atherosclerosis, coronary heart disease (CHD) and myocardial infarction (MI), e.g. hyperlipidemia, hypertension, obesity, diabetes, smoking, lack of physical exercise and genetic predisposition, do not completely explain for all new cases of cardiovascular disease. In 1988, Saikku et al [1] proposed a provoking relationship between positive serology for *Chlamydia pneumoniae* (*C. pneumoniae*) and atherosclerosis. Since this initial report, many studies have been published investigating a possible (causative) linkage between *C. pneumoniae* and atherosclerotic disease.

Biology and Epidemiology of *C. pneumoniae*

The genus Chlamydia is currently composed of four species: *C. psittaci*, *C. trachomatis*, *C. pneumoniae* (in a recent paper renamed *Chlamydophila pneumoniae* [2]) and *C. pecorum*. All but the latter cause disease in humans. *C. trachomatis* primarily causes ocular and urogenital infections, *C. psittaci* causes ornithosis-psittacosis, while *C. pneumoniae* is an important cause of respiratory illness. *C. pecorum* only infects sheep and cattle [3].

Chlamydiae are obligate intracellular gram-negative bacteria (no peptidoglycan layer and no own ATP production) with a unique biphasic life-cycle, involving an extracellular infectious elementary body (EB, 0.3 - 0.4 μm) and an intracellular metabollically active reticulate body (RB, ± 1μm). After endocytosis of an EB by a susceptible host cell, the EB differentiates into a RB. The RBs grow by binary fission and are susceptible to antibiotics. During the entire intracellular developmental cycle, Chlamydiae remain within the host cell endosome, and inhibit the fusion of the endosome with lysosomes, thereby resisting intracellular killing. Late in the growth cycle, the RBs redifferentiate into EBs, and are released upon exocytosis or lysis of the host cell, starting a new infectious cycle (figure 1) One inclusion may contain about 100 - 1000 Chlamydia particles. [4].

To date, only one serovar of *C. pneumoniae* named TWAR is known, meaning that every isolate of the organism has a nearly identical antigenic make-up. This should make the development of a vaccine against *C. pneumoniae* much more easier than for the other *Chlamydia* species, that have numerous serologically distinct variants. In 1989, the TWAR organism was established as the third species of the genus Chlamydia and named *C. pneumoniae* on the basis of DNA homology, ultrastructural and immunologic studies [5].

Besides respiratory disease, *C. pneumoniae* has been associated with atherosclerosis, myocarditis, endocarditis, sarcoidosis, multiple sclerosis, asthma and Alzheimer's disease. Infection with the organism is common. Antibodies are infrequent before the age of 5 years, reach 10 % at the age of 5 – 10 years, and above age 20, approximately 50 % of adults possess antibodies to *C. pneumoniae*. This antibody rate continues to rise through lifehood, reaching 80 % in men and 70 % in women in old age. Seroprevalence is equal in both sexes until the age of 15 years old, but seropositivity rates in adults are higher in men than in women. Antibodies from first infection are usually lost within 3 to 5 years, suggesting that reinfection occurs frequently. Humans are the only known reservoir of *C. pneumoniae* and transmission is believed to occur from person to person via infected respiratory secretions. Infection with the organism is endemic, but epidemics also tend to occur (table 1) [5][6].

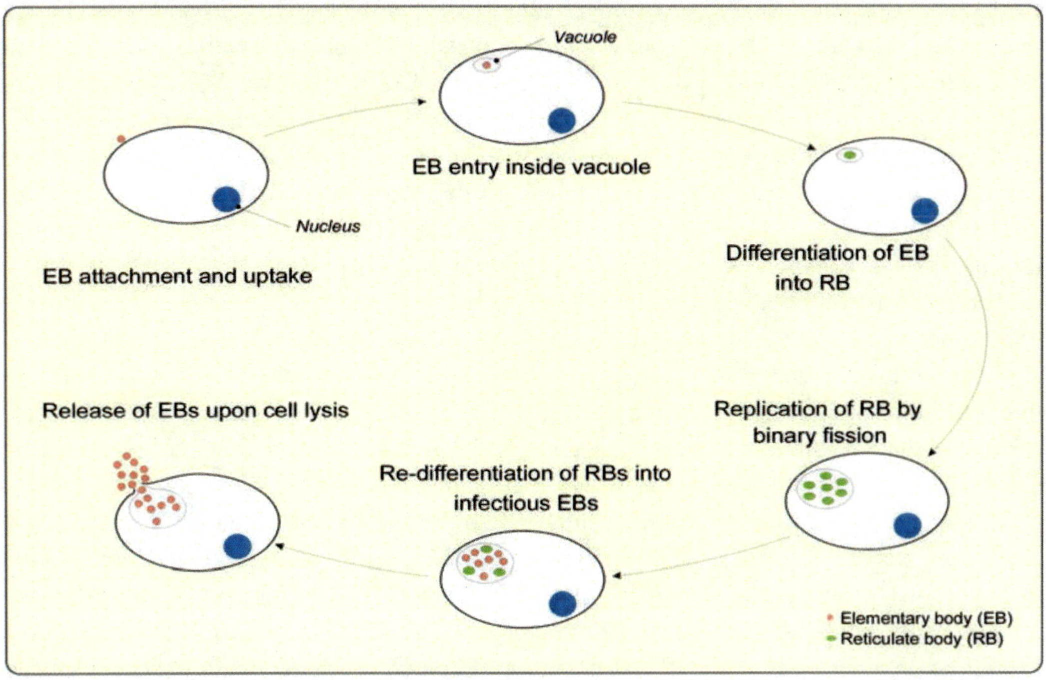

Figure 1. Life-cycle of *C. pneumoniae*

Table 1. *Chlamydia pneumoniae* Characteristics

Identified	**1986**
Type of organism	Gram-negative intracellular bacterium
Host	Humans
Transmission	Person to person by respiratory secretions
Seroprevalence	Increases with age, reaching 70-80 % in elderly
Cause of	Pneumonia, pharyngitis, sinusitis, bronchitis
Associated diseases	Atherosclerosis, myocarditis, endocarditis, sarcoidosis, multiple sclerosis, asthma, Alzheimer's disease?
Antibiotics	Macrolides, tetracyclines, quinolones

Seroepidemiologic Studies

In 1988, Saikku et al. [1] demonstrated that patients with coronary artery disease (CAD) were significantly more likely to carry anti - *C. pneumoniae* IgG and IgA antibodies than healthy controls. Since this initial report, many cross-sectional and case-control seroepidemiologic studies on the association of *C. pneumoniae* and atherosclerosis have followed and several but not all reported similar findings. Results from prospective studies – the more preferred methodology for examining a possible association – however appear to minimise the association between baseline *C. pneumoniae* IgG titers and subsequent risk for major coronary events. The Physicians' Health Study [7] for example, found no association between IgG seropositivity to *C. pneumoniae* and risk of future MI over a follow-up period of 12 years. The ARIC Study [8] found higher levels of serum IgG to *C. pneumoniae* among individuals free of clinical evidence of CHD at baseline that subsequently developed the disease over a follow-up period of approximately 3.3 years. However, when other risk factors for cardiovascular disease were controlled for in this study, this association was weakened and became statistically nonsignificant. The relationship only persisted in non-smokers. Another 8-year prospective study [9] measured IgG, IgA and IgM antibodies to *C. pneumoniae* and circulating immune complexes containing *C. pneumoniae* antigen in baseline serum samples from cases experiencing a subsequent coronary event during follow-up and from their matched controls, but no case-control differences could be demonstrated. The Caerphilly prospective heart disase study [10] did find an association between baseline *C. pneumoniae* IgA titres and subsequent risk of death from ischaemic heart disease during a 13-year follow-up period. An overview of these studies is presented in table 2.

Mainly due to the variability in results, no consensus has been reached concerning the validity of the hypothesis associating serologic markers of *C. pneumoniae* infection with cardiovascular disease. The microimmunofluorescence (MIF) test is considered as the gold standard and many seroepidemiologic studies have relied on this test for antibody determination. Currently, the MIF test is used as a guideline for treatment in several ongoing antibiotic trials. The test allows simultaneous detection of antibodies to all three Chlamydia species that can be found in humans, and is able to differentiate between IgG, IgA and IgM antibody classes, which is useful for identifying the type of infection (primary infection, acute infection, chronic infection or reinfection). However, the MIF test has some major drawbacks: it is a labor intensive method, requiring an experienced reader for proper interpretation of slides, and the majority of MIF tests used are in-house developed tests lacking standardisation. Therefore, it is very difficult to compare the published findings of different laboratories and to evaluate the validity of these results. In a recent study [18], reproducibility for the MIF test among different research laboratories was assessed and found to be 60 - 80 % depending on the criteria used. Furthermore, the specificity of MIF has been questioned as cross-reactions between the major outer membrane proteins (MOMP) of different *Chlamydia* species were reported [19][20]. Commercially available serological tests in an enzyme linked immunoassay (EIA) format detecting *C. pneumoniae* species-specific antibodies were introduced, which might offer a solution for all the problems related to the MIF test. EIAs are more objective, relatively simple to perform, easier to standardise and therefore suitable for large-scale screening and use in diagnostic clinical laboratories. However, we have recently demonstrated that commonly used serological enzyme immunoassays for *C. pneumoniae* IgG antibody detection vary greatly in sensitivity and specificity. The choice of serological assay significantly affected antibody seroprevalence, and consequently the assumed relationship between seropositivity for *C. pneumoniae* and coronary artery disease. [21].

Table 2. Results of Prospective Studies of *Chlamydia pneumoniae* Seropositivity and Cardiovascular Disease

Study	N° cases	N° controls	Serologic test	Mean follow-up	Association (Y/N)
Ridker et al[7]	343	343	MIF (IgG)	12 years	N
Nieto et al[8]	246	550	MIF (IgG)	3.3 years	N
Tavendale et al[9]	217	217	ELISA (IgG, IgA & IgM)	7.5 years	N
Strachan et al[10]	278	1057	MIF (IgG & IgA)	13 years	IgA – Y IgG – N
Ridker et al[11]	122	244	MIF (IgG)	3 years	N
Danesh et al[12]	496	989	TRF (IgG)	16 years	N
Wald et al[13]	647	1294	TRF (IgG & IgA)	15.6 years	N
Siscovick et al[14]	213	405	MIF (IgG)	4.5 years	Low/moderate titer-N High titer–Y
Von Hertzen et al[15]	67	NS	MIF (IgG & IgA)	6.5 years	N
Glader et al[16]	78	156	MIF (IgG) rELISA (IgA)	max. 9 years	N
Saikku et al[17]	102	102	MIF (IgG & IgA)	6 months	IgA – Y IgG – N

MIF: microimmunofluorescence test.
ELISA: enzym linked immunosorbent assay.
TRF: time resolved fluoroscopic immunoassay.
NS: not specified.
Source: Only Siscovick and colleages found a high *C. pneumoniae* IgG titer (IgG $\geq$ 1:1024, and not IgG $\geq$ 1:512) to be associated with an increased risk of myocardial infarction and coronary heart disease death.

C. pneumoniae Detection in Vascular Tissue and Peripheral Blood Mononuclear Cells (PBMC)

In 1992, Shor and colleagues [22] were the first to demonstrate *C. pneumoniae* in coronary atherosclerotic plaques by electronmicroscopy (EM) and immunocytochemistry (ICC). They extended their study [23] in 1993, by performing polymerase chain reaction (PCR) on the tissues, and detected that 20 out of 36 patients contained *C. pneumoniae* DNA, while none of the non-atherosclerotic tissues were found positive. Until now, few studies failed to detect the organism in human atheromata, and those that failed, focused on DNA detection by PCR. In general, DNA detection of *C. pneumoniae* has always resulted in a lower prevalence of positives compared to studies aiming at the detection of *C. pneumoniae* antigens by immunohistochemistry [24-28]. This is rather astonishing since the detection of DNA by PCR is supposed to be more sensitive than immunohistochemical staining. The reason for this discrepancy in results has often been attributed to the presence of PCR inhibitors in the tissues examined. Our study group failed to demonstrate *C. pneumoniae* DNA in a wide spectrum of 205 atherosclerotic lesions in different human arteries by performing 4 different PCR assays targeting 3 different genes of *C. pneumoniae*. This after controlling for the presence of PCR inhibitors through construction and inclusion of internal controls, thereby excluding false negative results.

Interestingly, *C. pneumoniae* has been detected in atherosclerotic tissue of seronegative persons and was not always found in seropositive individuals [29]. Furthermore, there seems to be no association between the detection of *C. pneumoniae* components and the disease stage [30]. Some research groups reported higher detection rates in severe lesions than in mild lesions, but few studies found statistically significant differences [31][32]. Finally, several studies investigated the in-vivo presence of *C. pneumoniae* DNA in human leukocytes. If *C. pneumoniae* causes or contributes to the atherogenic process, it must have a cellular vehicle transporting it from the respiratory system or other infected sites of the body to the vessel wall, since EBs – the extracellular infectious metabolically inactive form – have never been found circulating freely in the blood stream. The prevalence of *C. pneumoniae* DNA in patients with CAD ranged from 0% to 59 % [33], and DNA of *C. pneumoniae* has also been detected in PBMC of healthy blood donors.

Hence, detection rates from different laboratories are highly variable. Possible reasons for these conflicting results are the patchy distribution of the organism in the lesions, examination of only a small amount of tissue and application of non-standardised laboratory procedures. In one respect, these variable laboratory results gave rise to great scepticism about the role of *C. pneumoniae* in atherosclerosis. On the other hand, some researchers became even more convinced of this "overwhelming evidence" for the presence of *C. pneumoniae* in atherosclerotic tissue, thereby suggesting that failure to find the organism were the result of technical failures. Therefore, the Centers for Disease Control and Prevention (USA) and the Laboratory Centre for Disease Control (Canada) have recently published recommendations [34] for standardising *C. pneumoniae* diagnostic assays, including serological testing, culture, DNA amplification and tissue diagnostics.

The results obtained concerning the relation of *C. pneumoniae* with cardiovascular disease should be interpreted in line with the biology of Chlamydiae, which is characterised by alternate phases of activity and latency followed by fibrosis and scarring, when the organism is no longer viable (e.g. ocular trachoma and post *C. trachomatis* adnexitis). As it is difficult to study early atherosclerotic lesions, the majority of cases examined were advanced lesions, in which DNA of *C. pneumoniae* could have been partially or totally degraded. As illustrated by quality control studies [35], many published reports on *C. pneumoniae* DNA detected by PCR may reflect laboratory contaminations. These findings rather suggest persistence of *C. pneumoniae* antigens and absence of viable organisms after the acute infection has subsided. Or are nonchlamydial plaque constituents responsible for the positive histological staining with anti-chlamydial antibodies in diseased arteries? [36]. These interpretations also offer an explanation for the very low recovery of *C. pneumoniae* from atherosclerotic arteries by cell culture.

Animal Experiments

Rabbits and mice have been the main animal species used to investigate whether *C. pneumoniae* infection plays a causative role in the atherosclerotic process. These studies have demonstrated that *C. pneumoniae* respiratory infection induces mild aortic lesions in New Zealand White Rabbits fed a normal diet [37][38]. New Zealand White rabbits normally do not develop atherosclerosis while being fed a regular diet. When these animals were given a cholesterol-enriched diet, repeated intranasal *C. pneumoniae* infection accelerated the formation of intimal thickening in the aorta [39], an effect that could be diminished by a 7 week treatment of the acute infection with azithromycin starting 5 days after inoculation. Interestingly, antibiotic treatment did not eliminate chlamydial antigen as was demonstrated by immunofluorescence testing with a genus-specific monoclonal antibody. PCR for detection of specific DNA was not performed.

In mice with low-density lipoprotein receptor deficiency and a cholesterol-enriched diet, monthly *C. pneumoniae* inoculation was found to exacerbate the hypercholesterolemia induced atherosclerosis after a period of 9 months [40]. However, no histopathologic changes were seen in the aortas of LDL-receptor deficient mice fed a regular diet. The latter is in agreement with the study of Blessing et al.[41], who demonstrated that repeated intranasal inoculations of *C. pneumoniae* induced inflammatory changes in the aorta of a small number of C57BL/6J mice fed a regular diet, but did not initiate atherosclerotic lesions. In addition, *C. pneumoniae* specific immune complexes were demonstrated in the sera of 42 % of 19 infected animals, but no correlation was found between histopathologic examination of arterial tissue and these circulating immune complexes. Circulating immune complexes can induce endothelial injury and have been associated with coronary heart disease [42][43].

The risk for arteriosclerosis is modulated by the apoE allele. ApoE-deficient mice are knockout mice that have high levels of plasma cholesterol and spontaneously develop atherosclerosis that is morphologically very similar to that in humans. Aalto-Setala et al.[44] reported the repeated *C. pneumoniae* infections not being capable of accelerating atherogenic changes in the aortic root of apoE-deficient mice that were maintained on either a low- or a high- fat diet. In contrast, the study of Moazed et al.[45] demonstrated that chronically infected ApoE-deficient mice, fed a regular diet, had a 2.4 fold increase in lesion area in the aortic arch at 16 weeks of age compared to sham-inoculated mice. At 20 weeks of age, this difference was reduced to a 1.5 fold increase. Rothstein et al.[46] extended these observations by reporting an increase in aortic lesion size in 26 week-old ApoE-deficient mice infected at 12 and 14 weeks of age. Remarkebly, they did not find a significant difference in aortic lesion index of ApoE null mice infected at 10 and 12 weeks of age and killed at 20 weeks of age. Gieffers et al.[47] assessed the genotype of 30 patients with coronary heart disease that had PCR-proven *C. pneumoniae* infection of coronary arteries, and found that the apoE-deficient genotype was not associated with an increased risk for vascular chlamydia infection. Finally, Wright et al.[48] bred apoE-deficient mice with animals incapable of responding to bacterial LPS due to a deficiency in toll-like receptor 4, and demonstrated that atherogenesis was unaffected in these double deficient mice. The development of atherosclerotic lesions and plasma cholesterol levels in double deficient mice fed a high-fat diet were not altered by the absence of toll-like receptor 4. Further, atherosclerosis in germ-free apoE-deficient mice was not measurably different from that in control apoE-deficient mice. As a consequence, they speculated that infectious agents are not critical for murine atherogenesis, but may still exacerbate the disease process through indirect effects on metabolism and known risk factor.

In summary, it seems that in some rabbit and mouse models *C. pneumoniae* infection may play a contributory role in the atherogenic process. The observed discrepancies in results between experiments may be attributed to differences in strains, inoculum or methodology. However, it should be kept in mind that no ideal animal model for human atherosclerosis exists.

Antibiotic Treatment Trials

Treatment of *C. pneumoniae* in Animal Models of Atherosclerosis

Fong et al.[49] infected one-month old New Zealand White specific-pathogen-free rabbits, fed a cholesterol-free chow, on 3 separate occasions, 2 weeks apart via the nasopharynx with *C. pneumoniae*. Early treatment with azithromycin - 30 mg/kg orally, daily for 3 days, and then once every 6 days until sacrifice at 10 or 11 weeks after initial inoculation – starting 5 days after initial inoculation was very effective in preventing early atherosclerotic lesions. Delayed treatment starting 6 weeks after initial inoculation, was ineffective in reducing or reversing atherosclerosis. Interestingly, treatment with azithromycin did not significantly decrease the prevalence of chlamydial antigen in the aorta. Muhlestein et al.[39] showed that treatment with azithromycin – 30 mg/kg intramuscularly daily the first week and twice-weekly for another 6 weeks – starting 3 days after final triple *C. pneumoniae* inoculation prevented accelerated aortic intimal thickening. Again, immunofluorescence staining showed that chlamydial antigen was not eliminated by antibiotic therapy. In another study [50], 10 weeks old ApoE null mice were treated for 2 weeks after intranasal infection with azithromycin or placebo. Azithromycin was given orally at 24 mg/kg starting at 2 weeks and at 3 weeks after the second and final inoculation. The therapy did not have any effect on aortic lesion size, not in the infected mice, nor in the noninfected animals.

As illustrated by these animal experiments, antibiotic therapy starting late after a *C. pneumoniae* infection seems to have no effect on atherogenesis. Only early treatment of an acute infection may be beneficial.

Antibiotic Treatment Trials in Humans

Retrospective Case Control Studies

Meier et al.[51] demonstrated that in a patient population with a first-time MI, cases were significantly less likely to have used tetracyclines or quinolones during the previous 3 years. No such protective effect was found for prior use of macrolides. Another case-control study [52] reported a lack of association between first risk MI and 5– year prior use of erythromycin, tetracycline or doxycycline. In agreement, Pilote et al.[53] studied exposure to antichlamydial antibiotics during the 6 months before acute MI and did not find any association.

Prospective Placebo Controlled Treatment Studies: Antibiotics and Cardiovascular Events

Three small trials studying the effect of macrolide treatment of *C. pneumoniae* in patients with CAD, found a reduction in coronary events. Gupta et al.[54] randomized 60 male survivors of acute MI with persisting seropositivity by MIF (IgG titer $\geq$ 1:64) to receive either placebo or oral azithromycin 500 mg/day for 3 days or for 6 days. Antibiotic treated patients showed a decrease in their MIF titers and had a lower risk of adverse cardiovascular events over a follow-up period of 18 months. The results of the CLARIFY (Clarithromycin in Acute Coronary Syndrome Patients in Finland) study [55] showed that a 3-month treatment with clarithromycin versus placebo, reduced cardiovascular events in patients (n = 148) with acute unstable angina or non-Q-wave coronary syndromes, followed-up for 1.5 years. In the study of Gurfinkel et al. [56], 202 patients with unstable angina or non-Q-wave MI were randomized to roxithromycin 150 mg orally twice daily or placebo for 30 days. At 1 month, a significant reduction in events was seen in the patients treated with roxithromycin compared with the control group, but after six months of follow-up, the difference was no longer statistically significant. Interestingly, anti-*C. pneumoniae* IgG titers were unchanged in both groups. In analogy, a moderately large sized (n = 302) study [57] in CAD patients with serologic evidence of prior *C. pneumoniae* exposure followed-up for 2 years, did not demonstrate a significant reduction in cardiovascular events after a three month course of azithromycin compared to the placebo group. In the STAMINA (South Thames Trial of Antibiotics in Myocardial Infarction and Unstable Angina) trial [58], 324 patients with AMI or unstable angina were randomized to receive a 1-week antibiotic treatment regime of azithromycin (500 mg once daily), metronidazole (400 mg twice daily), and omeprazole (20 mg twice daily), or placebo. An improvement in event-free survival was observed in patients treated with the azithromycin-containing regimen ($p = 0.061$) after a follow-up period of 1 year. This beneficial effect of antibiotic treatment on events was independent of *C. pneumoniae* serological infection status at study entry. In the ANTIBIO (Antibiotic therapy After an Acute Myocardial Infarction) trial [59] (n = 872), antibiotic therapy with roxithromycin (300 mg daily) for 6 weeks in patients with AMI resulted in no better clinical outcome after 1 year of follow-up compared to the placebo group. Finally, in the larger (n = 1439) AZACS (Azithromycin in Acute Coronary Syndromes) study [60], which was very recently published, it was demonstrated that a 5-day course of azithromycin did not reduce the number of recurrent events in patients with unstable angina or AMI.

At present, several large and adequately powered antibiotic secondary prevention trials are underway. The WIZARD (Weekly Intervention with Zithromax for Atherosclerosis and its related Disorders) trial [61] is the only study that has been completed. The patient population consisted of 7000 patients stable after a myocardial infarction, who were seropositive for *C. pneumoniae*. The study showed no clinical benefit of a 3 month treatment with azithromycin after 1-4 years of follow-up: the 7 % reduction in events was not statistically significant.

Considering the results of animal studies, one might expect that such limited courses of antibiotic therapy effective against *C. pneumoniae* in patients with cardiovascular disease will not be beneficial since *C. pneumoniae* infection is most probably not acute in this population. Furthermore, Anderson et al. [62] and Gurfinkel et al.[56] provided evidence that macrolides exert an anti-inflammatory effect, while other studies [63] have shown that tetracyclines and macrolides possess both antioxidative actions and inhibitory effects on metalloproteinases, which could all be responsible for a better clinical outcome of patients with CHD irrespective of the antibacterial efficacy of these antibiotics.

Mechanisms

In vitro, *C. pneumoniae* is able to infect and replicate within a number of cell types, including human monocytes/macrophages, peripheral blood monocyte (PBMC)-derived macrophages, vascular endothelial cells and arterial smooth muscle cells [64][65], which are all important components of the atherosclerotic plaque. *C. pneumoniae* infection of monocytes/macrophages induced the production of pro-inflammatory cytokines, such as tumor necrosis factor α (TNFα), interleukin-6 (Il-6), interleukin-1β (Il-1β), and interferon α (IFNα), as well as upregulation of CD14 [66]. Furthermore, a marked increase in the number of foam cells and accumulation of cholesteryl esters occured, when macrophages were exposed to *C. pneumoniae* in the presence of low-density lipoprotein (LDL) [67]. Foam cell formation was not inhibited by the antioxidants butylated hydroxytoluene or fucoidan, which block scavenger receptors, suggesting that lipid accumulation did not result from the uptake of modified LDL and did not involve scavenger receptors. In contrast, *C. pneumoniae*-induced foam cell formation was inhibited by heparin, which blocks the binding of LDL to the LDL-receptor. These results suggest that high levels of native LDL can be ingested by human macrophages following exposure to *C. pneumoniae*. The same authors further demonstrated that the heat-stable chlamydial lipopolysaccharide (LPS) is the principal *C. pneumoniae* component that induces macrophage native lipoprotein uptake and foam cell formation [68], and that heat-labile *C. pneumoniae* components, such as the chlamydial heat shock protein 60 (hsp60), are capable of inducing monocyte LDL oxidation [69].

Remarkebly, Blessing et al. [70] showed that lipid loading of macrophages in vitro with oxidized and acetylated LDL rendered them less susceptible to infection with *C. pneumoniae* compared to cells pretreated with native LDL or without lipid. Oxidized LDL inhibited infection to a greater extent than did acetylated LDL. However, once *C. pneumoniae* infection was established, the burst size was equivalent between infected macrophages treated with oxidized and native LDL. No significant differences were observed in the rate of attachment and internalisation of the organism between lipid-loaded and non-lipid loaded cells, nor could a direct toxic effect of oxidized lipids to *C. pneumoniae* explain the suppressed growth of the organism in lipid-loaded macrophages. On the basis of the inverse correlation that was found between the degree of infection and the intracellular cholesterol concentration of the lipid-loaded cells, the authors speculated that the intracellular milieu of foam cells was less conducive for growth of *C. pneumoniae*. Noteworthy, despite the resistance to infection, preincubation of macrophages with native or oxidized LDL did not prevent *C. pneumoniae* from inducing a pro-inflammatory cytokine (Il-1β, Il-6, TNFα) and chemokine (monocyte chemotactic protein, MCP-1) release within 24 hours after infection. Furthermore, both heat-killed and UV-inactivated *C. pneumoniae* stimulated the secretion of

TNFα. These findings are consistent with the observations of Netea et al. [71], who found evidence that live *C. pneumoniae* are not necessarely required for stimulation of cytokine (TNFα, Il-1β and Il-6) and chemokine (Il-8, MCP-1 and MIP-1α) production by PBMC, but that molecular components of dead *C. pneumoniae* organisms are also potent cell activators.

In contrast to the results observed by Blessing et al., Knoebel and colleagues [72] had previously reported an increase in infectivity of oxLDL-treated smooth muscle cells with *C. pneumoniae*. Possible explanations for this contrasting observation were suggested like the expression of a smaller number of scavenger receptors and the reduced capacity of smooth muscle cells to accumulate lipids. In addition, the authors pointed to the use of cycloheximide in the study with smooth muscle cells, an antibiotic known to enhance the susceptibility of many cell lines to chlamydial infections.

Besides the induction of cellular LDL oxidation, chlamydial hsp60 may promote atherogenesis by other mechanisms. Kol and colleagues [73] showed that chlamydial and human hsp60 colocalize within plaque macrophages in human atheroma, and induce the production of macrophage TNFα and matrix metalloproteinase-9 (MMP-9). Furthermore, both chlamydial and human hsp60 induce the expression of adhesion molecules on endothelial cells, but not on smooth muscle cells, stimulate Il-6 production and trigger activation of NF-κB in human endothelial cells, smooth muscle cells and macrophages in vitro [74]. In addition, Mayer et al. [75] reported the presence of endothelial cytotoxicity mediated by serum antibodies to chlamydial hsp60.

In vitro infection of endothelial cells induces the expression of adhesion molecules such as endothelial-leukocyte adhesion molecule-1 (E-selectin), intercellular adhesion molecule-1 (ICAM-1) and vascular cell adhesion molecule-1 (VCAM) [76], induces the production of the pro-inflammatory cytokine interleukin-6 (Il-6) [74], enhances tissue factor procoagulant activity [77], promotes platelet adhesion [71] and MCP-1 secretion [78], and stimulates transendothelial migration of neutrophils and monocytes [79]. A functional role for NF-κB was detected in the induction of MCP-1 gene expression by *C. pneumoniae* [78]. Furthermore, data was acquired implicating a role for a heat-labile component of *C. pneumoniae*, since UV-inactivated but not heat-inactivated organisms retained the ability to stimulate MCP-1 secretion from human umbilical vein endothelial cells (HUVEC) [80].

Surface-exposed outer membrane proteins such as Omp4 and Omp5 may be involved in attachment of *C. pneumoniae* to endothelial cells [81]. Gaydos et al. [82] demonstrated that transfer of *C. pneumoniae* infection to endothelial cells may proceed by cell-to-cell spread from adherent mononuclear phagocytes. A recent report [83] showed that protein tyrosine phosphorylation, mitogen-activated protein kinase (MAPK) stimulation, and activation/translocation of NF-κB occured within minutes of *C. pneumoniae* addition to endothelial cells. Within hours, an increased mRNA and surface expression of E-selectin, ICAM-1 and VCAM-1 were noted, resulting in an enhanced leukocyte-endothelial cell interaction. These data suggest that bacterial attachment might be sufficient to trigger a cascade of events, which could lead to endothelial damage, inflammation and thrombosis.

Experiments have shown that *C. pneumoniae* infection induces a procoagulant protein and pro-inflammatory cytokine expression in smooth muscle cells and endothelial cells by upregulation of tissue factor, plasminogen activator inhibitor (PAI-1), and interleukin-6 (Il-6) via NF-κB activation [84].

Gieffers et al. [85] studied the growth and activity of *C. pneumoniae* within isolated circulating monocytes of two healthy volunteers in the presence of rifampin and azithromycin, and found that inoculation of these cells resulted in the development of inclusion bodies after 3 days, whether these antibiotics were added (simultaneously) with the chlamydiae or not. In addition, monocytes isolated after a standard course of azithromycin (500 mg/day for 3 days) or rifampin (600 mg/day for 6 days) could still acquire *C. pneumoniae* infection when cultured in the presence of antibiotics. The viability and metabolic activity of *C. pneumoniae* were proven after 2 hours and 1, 3 and 10 days by Trypan blue staining and the detection of chlamydial mRNA transcripts. Moreover, *C. pneumoniae* was culturally retrieved from circulating CD14+ cells of patients with unstable angina that had previously undergone oral azithromycin treatment, indicating that chlamydiae can survive an antichlamydial therapy within monocytes in vitro and in vivo.

Furthermore, Geng et al. [86] demonstrated that human PBMCs exposed to *C. pneumoniae* respond to infection by secreting Il-12 and Il-10, and that infected PBMCs were resistant to apoptosis induced by 8-methoxypsoralen, hypericin or FAS ligand-mediated apoptosis. The *C. pneumoniae* infection– induced secretion of Il-10, but not Il-12, would at least be partially responsible for the anti-apoptotic activity of the pathogen. *C. pneumoniae* infection in vitro also elicits the externalisation of the lipid phosphatidylserine on the surface of human epithelial, endothelial, granulocytic and monocytic cells, a phenomenon associated with cellular development, differentiation and death. Infection-induced phosphatidylserine exposure was immediate (< 5 minutes), transient, calcium dependent, infectious dose dependent, and unaffected by a broad-spectrum caspase inhibitor [87]. Cytokines such as IFNγ, TNFα [88] and lymphotoxin (TNFβ) [89] were found to inhibit the intracellular replication of *C. pneumoniae* in HEp-2 cells, and thereby can alter the intracellular fate of the organism.

Conclusion

The role of *C. pneumoniae* in atherosclerosis is controversial and unresolved. There is a lack of consistent serologic data, the in vivo results are extremely variable, isolation of *C. pneumoniae* by culture has only succeeded in a very limited number of studies, and antibiotic therapy effective against *C. pneumoniae* in patients with cardiovascular disease seems not beneficial. Animal experiments and in vitro studies tend to support a contributory role for *C. pneumoniae* infection in atherosclerosis.

References

[1] Saikku P, Mattila K, Nieminen MS, Makela PH, Huttunen JK, Valtonen V. Serological evidence of an association of a novel Chlamydia, TWAR, with chronic coronary heart disease and acute myocardial infarction. *Lancet* 1988;2:983-6.

[2] Everett KDE, Bush RM, Andersen AA. Emended description of the order *Chlamydiales*, proposal of *Parachlamydiaceae* fam. nov., and *Simkaniaceae* fam. Nov., each containing one monotypic genus, revised taxonomy of the family *Chlamydiaceae*,

including a new genus and five new species, and standards for the identification of organisms. *Int J Syst Bacteriol.* 1999;49:415-40.

[3] Blanchard TJ, Mabey DCW. Chlamydial infections. *In:* Emmerson AM, Hawkey PM, Gillespie SH eds. Principles and practice of clinical bacteriology. John Wiley & sons, 1997:723-43.

[4] Hoymans VY, Bosmans JM, Ieven MM, Vrints CJ. *Chlamydia pneumoniae* and atherosclerosis. *Acta Chirurgica Belgica* 2002;102:317-22.

[5] Grayston JT. Background and current knowledge of *Chlamydia pneumoniae* and atherosclerosis. *J Infect Dis* 2000;181:S402-10.

[6] Grayston JT. Does *Chlamydia pneumoniae* cause atherosclerosis? *Arch Surg* 1999;134:930-4.

[7] Ridker PM, Kundsin RB, Stampfer MJ, Poulin S, Hennekens CH. Prospective study of *Chlamydia pneumoniae* IgG seropositivity and risks of future myocardial infarction. *Circulation* 1999;99:1161-4.

[8] Nieto FJ, Folsom AR, Sorlie PD, Grayston JT, Wang SP, Lloyd E. *Chlamydia pneumoniae* infection and incident coronary heart disease. *Am J Epidemiol* 1999;150:149-56.

[9] Tavendale R, Parratt D, Pringle SD, Brook RA, Tunstall-Pedoe H. Serological markers of *Chlamydia pneumoniae* infection in men and women and subsequent coronary events. *Eur Heart J* 2002;23:301-7.

[10] Strachan DP, Carrington D, Mendall MA, et al. Relation of *Chlamydia pneumoniae* serology to mortality and incidence of ischaemic heart disease over 13 years in the Caerphilly prospective heart disease study. *BMJ* 1999;318:1035-40.

[11] Ridker PM, Hennekens CH, Buring JE, Kundsin R, Shih J. Baseline IgG antibody titers to *Chlamydia pneumoniae*, *Helicobacter pylori*, Herpes Simplex Virus, and Cytomegalovirus and the risk for cardiovascular disease in women. *Ann Intern Med* 1999;131:573-7.

[12] Danesh J, Whincup P, Walker M, et al. *Chlamydia pneumoniae* IgG titres and coronary heart disease: prospective study and meta-analysis. *BMJ* 2000;321:208-13.

[13] Wald NJ, Law MR, Morris JK, Zhou X, Wong Y, Ward ME. *Chlamydia pneumoniae* infection and mortality from ischaemic heart disease: large prospective study. *BMJ* 2000;321:204-7.

[14] Siscovick DS, Schwartz SM, Corey L, et al. *Chlamydia pneumoniae*, Herpes Simplex Virus type 1, and Cytomegalovirus and incident myocardial infarction and coronary heart disease in older adults. *Circulation* 2000;102:2335-40.

[15] Von Hertzen L, Isoaho R, Kivelä SL, Saikku P. Relation of *C. pneumoniae* antibodies to ischaemic heart disease. *BMJ* 1999;319:1575-6.

[16] Glader CA, Boman J, Saikku P, et al. The proatherogenic properties of lipoprotein(a) may be enhanced through the formation of circulating immune complexes containing *Chlamydia pneumoniae*-specific IgG antibodies. *Eur Heart J* 2000;21:639-46.

[17] Saikku P, Leinonen M, Tenkanen L, et al. Chronic *Chlamydia pneumoniae* infection as a risk factor for coronary heart disease in the Helsinki Heart Study. *Ann Intern Med* 1992;116:273-8.

[18] Peeling RW, Wang SP, Grayston JT, et al. *Chlamydia pneumoniae* serology: interlaboratory variation in microimmunofluorescence assay results. *J Infect Dis* 2000;181(Suppl 3):S426-9.

[19] Kern DG, Neill MA, Schachter J. A seroepidemiologic study of *Chlamydia pneumoniae* in Rhode Island. Evidence of serologic cross-reactivity. *Chest* 1993;104:208-13.

[20] Wong YK, Sueur JM, Fall CH, Orfila J, Ward ME. The species specificity of the microimmunofluorescence antibody test and comparisons with a time resolved fluoroscopic immunoassay for measuring IgG antibodies against *Chlamydia pneumoniae*. *J. Clin. Pathol.* 1999;52:99-102.

[21] Hoymans VY, Bosmans JM, Van Renterghem L, et al. Importance of methodology in determination of *Chlamydia pneumoniae* seropositivity in healthy subjects and in patients with coronary atherosclerosis. *J Clin Microbiol* 2003; 41(9):4049-53.

[22] Shor A, Kuo CC, Patton DL. Detection of *Chlamydia pneumoniae* in coronary arterial fatty streaks and atheromatous plaques. *S Afr Med J* 1992;82:158-61.

[23] Kuo CC, Shor A, Campbell LA, Fukushi, Patton DL, Grayston JT. Demonstration of *Chlamydia pneumoniae* in atherosclerotic lesions of coronary arteries. *J Infect Dis* 1993;167:841-9.

[24] Bishara J, Pitlik S, kazakov A, Sahar G, Haddad M, Vojdani A, Rosenberg S, Samra Z. Failure to detect *Chlamydia pneumoniae* by cell culture and polymerase chain reaction in major arteries of 93 patients with atherosclerosis. *Eur J Clin Microbiol Infect Dis* 2003;22(5):300-2.

[25] Chierichetti F, Arbustini E, Arici V, Moghadam SP, Conti B, Bagliani A. Identification of *Chlamydia pneumoniae* DNA in carotid plaques. *Angiology* 2000;51(10):827-30.

[26] LaBiche R, Koziol D, Quinn TC, Gaydos C, Azhar S, Ketron G, Sood S, Degraba TJ. Presence of *Chlamydia pneumoniae* in human symptomatic and asymptomatic carotid atherosclerotic plaque. *Stroke* 2001;32:855-60.

[27] Meijer A, Roholl PJM, Gielis-Proper SK, Ossewaarde JM. *Chlamydia pneumoniae,* rather than viable bacteria, persist in atheroscleotic lesions. *J Clin Pathol* 2000;53:911-6.

[28] Ericson K, Saldeen TGP, Lindquist O, Påhlson C, Mehta JL. Relationship of *Chlamydia pneumoniae* infection to severity of human coronary atherosclerosis. *Circulation* 2000;101:2568-71.

[29] Puolakkainen M, kuo CC, Shor A, Wang SP, Grayston JT, Campbell LA. Serological response to *Chlamydia pneumoniae* in adults with coronary arterial fatty streaks and fibrolipid plaques. *J Clin Microbiol* 1993;31(8):2212-4.

[30] Davidson M, Kuo CC, Middaugh JP, et al. Confirmed previous infection with *Chlamydia pneumoniae* (TWAR) and its presence in early coronary atherosclerosis. *Circulation* 1998;98:628-33.

[31] Thomas M, Wong Y, Thomas D, et al. Relationship between direct detection of *Chlamydia pneumoniae* DNA in human coronary arteries at postmortem examination and histological severity (Stary grading) of associated atherosclerotic plaque. *Circulation* 1999;99:2733-6.

[32] Ericson K, Saldeen TGP, Lindquist O, Påhlson C, Mehta JL. Relationship of *Chlamydia pneumoniae* infection to severity of human coronary atherosclerosis. *Circulation* 2000;101:2568-71.

[33] Hoymans V, Bosmans J, Ursi D, et al. Systemic Inflammation, *Chlamydia pneumoniae* DNA in circulating leukocytes and coronary atherosclerosis'. *Acta Cardiologica* 2002,57(3):213-9.

[34] Dowell SF, Peeling RW, Boman J, et al. Standardizing *Chlamydia pneumoniae* assays: Recommendations from the Centers for Disease Control and Prevention (USA) and the Laboratory Centre for Disease Control (Canada). *Clin Iinfect Dis* 2001;33:492-502.

[35] Apfalter P, Blasi F, Boman J, et al. Multicenter comparison trial of DNA extraction methods and PCR assays for detection of *Chlamydia pneumoniae* in endarterectomy specimens. *J Clin Microbiol* 2001;39:519-24.

[36] Hoymans VY, Bosmans JM, ursi D, et al. *Chlamydia* pneumoniae constituents in human atheroma colocalize with ceroid deposits. 4th Annual Conference on Arteriosclerosis, Thrombosis and Vacsular Biology, May 2003. *Abstract.*

[37] Fong IW, Chiu B, Viira E, Fong MW, Jang D, Mahony J. Rabbit model for *Chlamydia pneumoniae* infection. *J Clin Microbiol* 1997;35:48-52.

[38] Laitinen K, Laurila A, Pyhala L, Leinonen M, Saikku P. *Chlamydia pneumoniae* infection induces inflammatory changes in the aortas of rabbits. *Infect Immun* 1997;65:4832-5.

[39] Muhlestein JB, Anderson JL, Hammond EH, et al. Infection with *Chlamydia pneumoniae* accelerates the development of atherosclerosis and treatment with azithromycin prevents it in a rabbit model. *Circulation* 1998;97:633-6.

[40] Hu H, Pierce GN, Zhong G. The atherogenic effects of Chlamydia are dependent on serum cholesterol and specific to *Chlamydia pneumoniae. J Clin Invest* 1999;103:747-53.

[41] Blessing E, Lin TM, Campbell LA, Rosenfeld ME, Lloyd D, Kuo C. *Chlamydia pneumoniae* induces inflammatory changes in the heart and aorta of normocholesterolemic C57BL/6J mice. *Infect Immun* 2000;68:4765-8.

[42] Lopez-Virella MF, Virella G. Immunological and microbiological factors in the pathogenesis of atherosclerosis. *Clin Immunol Immunopathol* 1985; 37:377-86.

[43] Linnanmaki E, Leinonen M, mattila K, Nieminen MS, Valtonen V, Saikku P. *Chlamydia pneumoniae*-specific circulating immune complexes in patients with chronic coronary heart disease. *Circulation* 1993;4:1130-4.

[44] Aalto-Setala K, Laitinen K, Erkkila M, et al. *Chlamydia pneumoniae* does not increase atherosclerosis in the aortic root of apolipoprotein E-deficient mice. *Arterioscler Thromb Vasc Biol* 2001;21:578-84.

[45] Moazed TC, Campbell LA, Rosenfield ME, Grayston JT, Kuo CC. *Chlamydia pneumoniae* infection accelerates the progession of atherosclerosis in apolipoprotein E-deficient mice. *J Infect Dis* 1999;180:238-41.

[46] Rothstein NM, Quinn TC, Madico G, Gaydos CA, Lowenstein CJ. Effect of azithromycin on murine arteriosclerosis exacerbated by *Chlamydia pneumoniae. J Infect Dis* 2001;183:232-8.

[47] Gieffers J, Solbach W, Maass M. Cardiovascular infection by *Chlamydia pneumoniae* is not related to apolipoprotein E genotype. *Clin Infect Dis* 2000;31:1515-6.
[48] Wright SD, Burton C, Hernandez M, et al. Infectious agents are not necessary for murine atherogenesis. *J Exp Med* 2000;191:1437-41.
[49] Fong IW, Chiu B, Viira E, et al. Can an antibiotic (macrolide) prevent *Chlamydia pneumoniae*-induced atherosclerosis in a rabbit model? *Clin Diagn Lab Immunol* 1999;6:891-4.
[50] Rothstein NM, Quinn TC, Madico G, Gaydos CA, Lowenstein CJ. Effect of azithromycin on murine arteriosclerosis exacerbated by *Chlamydia pneumoniae. J Infect Dis* 2001;183:232-8.
[51] Meier CR, Derby LE, Jick SS, Vasilakis C, Jick H. Antibiotics and risk of subsequent first-time acute myocardial infarction. *JAMA* 1999;281:427-31.
[52] Jackson LA, Smith NL, Heckbert SR, Grayston JT, Siscovick DS, Psaty BM. Lack of association between first myocardial infarction and past use of erythromycin, tetracycline, or doxycycline. *Emerg Infect Dis* 1999;5:281-4.
[53] Pilote L, Green L, Joseph L, Richard H, Eisenberg MJ. Antibiotics against *Chlamydia pneumoniae* and prognosis after acute myocardial infarction. *Am Heart J* 2002;143:294-300.
[54] Gupta S, Leatham EW, Carrington D, Mendall MA, Kaski JC, Camm AJ. Elevated *Chlamydia pneumoniae* antibodies, cardiovascular events, and azithromycin in male survivors of myocardial infarction. *Circulation* 1997;96:404-7.
[55] Sinisalo J, mattila K, Valtonen V, et al. Effect of 3 months of antimicrobial treatment with clarithromycin in acute non-Q-wave coronary syndrome. *Circulation* 2002;105:1555-60.
[56] Gurfinkel E, Bozovich G, Beck E, Testa E, Livellara B, Mautner B. Treatment with the antibiotic roxithromycin inpatients with acute non-Q-wave coronary syndromes. The final report of the ROXIS study. *Eur Heart J* 1999;20:121-7.
[57] Muhlestein JB, Anderson JL, Carlquist JF, et al. Randomized secondary prevention trial of azithromycin in patients with coronary artery disease. *Circulation* 2000;102:1755-60.
[58] Stone AFM, Mendall MA, Kaski J-C, et al. Effect of treatment for *Chlamydia pneumoniae* and *Helicobacter pylori* on markers of inflammation and cardiac events in patients with acute coronary syndromes. *Circulation* 2002;106:1219-23.
[59] Zahn R, Schneider S, Frilling B, et al. Antibiotic therapy after acute myocardial infarction. *Circulation* 2003;107:1253-9.
[60] Cercek B, Shah PK, Noc M, et al. Effect of short-term treatment with azithromycin on recurrent ischaemic events in patients with acute coronary syndrome in the Azithromycin in Acute Coronary Syndrome (AZACS) trial: a randomised controlled trial. *Lancet* 2003;361:809-13.
[61] Grayston JT. Antibiotic treatment of atherosclerotic cardiovascular disease. *Circulation* 2003;107:1228-30.
[62] Anderson JL, Muhlestein JB, Carlquist J, et al. Randomized secondary trial of azithromycin in patients with coronary artery disease and serological evidence for *Chlamydia pneumoniae* infection. *Circulation* 1999;99:1540-7.

[63] Meier CR. Antibiotics in the prevention and treatment of coronary heart disease. *J Infect Dis* 2000;181:S558-62.

[64] Gaydos CA, Summersgill JT, Sahney NN, Ramirez JA, Quinn TC. Replication of *Chlamydia pneumoniae* in vitro in human macrophages, endothelial cells, and aortic artery smooth muscle cells. *Infect Immun* 1996;64:1614-20.

[65] Suvi-Sirkku E, Teppo AM, Laitinen K, Saikku P, Linnavuori K, Leinonen M. Growth of *Chlamydia pneumoniae* in cultured human PBMC and induction of a cytokine response. *Microbial Pathogenesis* 1996;21:215-21.

[66] Heinemann M, Susa M, Simnacher U, Marre R, Essig A. Growth of *Chlamydia pneumoniae* induces cytokine production and expression of CD14 in a human monocytic cell line. *Infect Immun* 1996;64(11):4872-5.

[67] Kalayoglu MV, Byrne GI. Induction of macrophage foam cell formation by *Chlamydia pneumoniae*. *J Infect Dis* 1998;177:725-9.

[68] Kalayoglu MV, Byrne GI. A *Chlamydia pneumoniae* component that induces macrophage foam cell formation is chlamydial lipopolysaccharide. *Infect Immun* 1998;66:5067-72.

[69] Kalayoglu MV, Hoerneman B, La Verda D, Morrison SG, Morrison RP, Byrne GI. Cellular oxidation of low-density lipoprotein by *Chlamydia pneumoniae*. *J Infect Dis* 1999;180:780-90.

[70] Blessing E, Kuo CC, Lin TM, et al. Foam cell formation inhibits growth of *Chlamydia pneumoniae* but does not attenuate *Chlamydia pneumoniae*-induced secretion of proinflammatory cytokines. *Circulation* 2002;105:1976-82.

[71] Netea MG, Selzman CH, Kullberg BJ, et al. Acellular components of *Chlamydia pneumoniae* stimulate cytokine production in human blood mononuclear cells. *Eur J Immunol* 2000;30:541-9.

[72] Knoebel E, Vijayagopal P, Figueroa JE 2nd, Martin DH. In vitro infection of smooth muscle cells by *Chlamydia pneumoniae*. *Infect Immun* 1997;65:503-5.

[73] Kol A, Sukhova GK, Lichtman AH, Libby P. Chlamydial heat shock protein 60 localizes in human atheroma and regulates macrophage tumor necrosis factor-α and matrix metalloproteinase expression. *Circulation* 1998;98:300-7.

[74] Kol A, Bourcier T, Lichtman AH, Libby P. Chlamydial and human heat shock protein 60s activate human vascular endothelium, smooth muscle cells, and macrophages. *J Clin Invest* 1999;103:571-7.

[75] Mayr M, Metzler B, Kiechl S, et al. Endothelial cytotoxicity mediated by serum antibodies to heat shock proteins of Escherichia coli and *Chlamydia pneumoniae*. *Circulation* 1999;99:1560-6.

[76] Kaukoranta –Tolvanen SS, Ronni T, Leinonen M, Saikku P, Laitinen K. Expression of adhesion molecules on endothelial cells stimulated by *Chlamydia pneumoniae*. *Microb Pathog* 1999;21:407-11.

[77] Fryer RH, Schwobe EP, Woods ML, Rodgers GM. Chlamydia species infect human vascular endothelial cells and induce procuagulant activity. *J Invest Med* 1997;45:168-74.

[78] Molestina RE, Miller RD, Lentsch AB, Ramirez JA, Summersgill JT. Requirement for NF-κB in transcriptional activation of monocyte chemotactic protein 1 by *Chlamydia pneumoniae* in human endothelial cells. *Infect Immun* 2000;68(7):4282-8.

[79] Molestina RE, Miller RD, Ramirez JA, Summersgill JT. Infection of human endothelial cells with *Chlamydia pneumoniae* stimulates transendothelial migration of neutrophils and monocytes. *Infect Immun* 1999;67:1323-30.

[80] Molestina RE, Dean D, Miller RD, Ramirez JA, Summersgill JT. Characterization of a strain of *Chlamydia pneumoniae* isolated from a coronary atheroma by analysis of the *Omp1* gene and biological activity in human endothelial cells. *Infect Immun*;66(4):1370-6.

[81] Knudsen K, Madsen AS, Mygind P, Christiansen G, Birkelund S. Identification of two novel genes encoding 97- to 99-kilodalton outer membrane proteins of *Chlamydia pneumoniae. Infect Immun* 1999;67:375-83.

[82] Gaydos CA, Pham DG, Quinn TC. *Chlamydia pneumoniae* infections in macrophages and coronary artery endothelial cells. *In:* STARY A ed. Proceedings of the Third Meeting of the European Society for *Chlamydia* Research. Società Editrice Esculapio, Bologna, Italy, 1996:223.

[83] Krüll M, Klucken AC, Wuppermann FN, et al. Signal transduction pathways activated in endothelial cells following infection with *Chlamydia pneumoniae. J Immunol* 1999;162:4834-41.

[84] Dechend R, Maass M, Gieffers J, et al. *Chlamydia pneumoniae* infection of vascular smooth muscle and endothelial cells activates NF-κB and induces tissue factor and PAI-1 expresion. *Circulation* 1999;100:1369-73.

[85] Gieffers J, Fullgraf H, Jahn J, et al. *Chlamydia pneumoniae* infection in circulating human monocytes is refractory to antibiotic treatment. *Circulation* 2001;103:351-6.

[86] Geng Y, Shane RB, Berencsi K, et al. *Chlamydia pneumoniae* inhibits apoptosis in human peripheral blood mononuclear cells through induction of Il-10. *J Immunol* 2000;164:5522-9.

[87] Goth SR, Stephens RS. Rapid, transient phosphatidylserine externalization induced in host cells by infection with *Chlamydia* spp. *Infect Immun* 2001;69(2):1109-19.

[88] Summersgill JT, Sahney NN, Gaydos CA, Quinn TC, Ramirez JA. Inhibition of *Chlamydia pneumoniae* growth in HEp-2 cells pretreated with gamma interferon and tumor necrosis factor alpha. *Infect Immun* 1995;63(7):2801-3.

[89] Matsushima H, Shirai M, Ouchi K, et al. Lymphotoxin inhibits *Chlamydia pneumoniae* growth in HEp-2 cells. *Infect Immun* 1999;67(6):3175-9.

In: Focus on Atherosclerosis Research
Editor: Leon V.Clark, pp. 91-113
ISBN 1-59454-044-6

Chapter V

Oxidized Low-Density Lipoprotein Complexed with β2-Glycoprotein I as a Common Metabolic Form in Atherogenesis and Autoimmune-Mediated Atherosclerosis

Eiji Matsuura*
Department of Cell Chemistry, Okayama University Graduate School of Medicine and Dentistry, Okayama 700-8558, Japan
Luis R. Lopez
Corgenix, Inc., Westminster, CO 80234, USA

Abstract

Oxidative stress is thought to be etiologically related to atherosclerosis. Experimental evidence clearly demonstrates the presence of oxidized LDL (oxLDL) in the intima of arteries and that it contributes to the initiation and progression of atherosclerotic lesions. We have recently demonstrated that oxLDL interacts with an endogenous plasma protein, β2-glycoprotein I (β2GPI), to form complexes and that these complexes circulate in the blood stream of patients with systemic inflammatory diseases. β2GPI/oxLDL complexes have been implicated as autoantigens relevant in atherogenesis. A role of antiphospholipid antibodies (i.e. anti-β2GPI) in arterial thrombosis has been proposed, and circulating IgG immune complexes containing β2GPI and oxLDL have been detected in patients with autoimmune diseases such as systemic lupus erythematosus (SLE) and antiphospholipid syndrome (APS). Our *in vitro* experiments showed that oxLDL quickly interacts with β2GPI via specific ligands generated by Cu^{2+}-oxidation in

* Correspondence: Eiji Matsuura, Ph.D., Associate Professor, Department of Cell Chemistry, Okayama University Graduate School of Medicine and Dentistry, 2-5-1 Shikata-cho, Okayama 700-8558, Japan, Tel: +81-86-235-7402, Fax: +81-86-235-7404, E-mail: eijimatu@md.okayama-u.ac.jp

which the negative charge of oxLDL is neutralized by the complex formed. Novel ELISA systems allowed the measurement of circulating β2GPI/oxLDL complexes and autoantibodies to these complexes to study their association with various thrombotic manifestations. Even though many questions still remain, β2GPI/oxLDL complex can be described as a common metabolic form in atherogenesis and a significant participant in autoimmune-mediated atherosclerosis.

Key words: antiphospholipid syndrome, antiphospholipid antibody, β2-glycoprotein I, oxidized LDL, atherosclerosis

Background

Atherosclerosis is a major health concern in developed countries worldwide. During the last 20 years, evidence has accumulated that atherosclerosis is closely related to lipoprotein metabolism, and that it is an inflammatory/immune-mediated disease. In particular, oxidized low-density lipoprotein (oxLDL) is involved in early development of atherosclerotic lesions through the formation of macrophage-derived foam cells (1-4).

Antiphospholipid syndrome (APS) is characterized by the presence of heterogeneous autoantibodies and by the occurrence of thromboembolic complications in the arterial and/or venous vasculatures (5, 6). This prothrombotic state is associated with the appearance of anticardiolipin antibodies (aCL), lupus anticoagulants, and others. It has been commonly accepted that β2-glycoprotein I (β2GPI) is a major antigen for aCL present in patients with APS (7-11).

OxLDL and β2GPI are present in the human atheroma, and in particular, oxLDL is the principal source of lipid that accumulates within cells of the atherosclerotic lesions (12). In 1993, it was first reported that aCL cross-react with malondialdehyde (MDA)-modified LDL in patients with systemic lupus erythematosus (SLE) (13). It was also shown that anti-β2GPI antibodies could be good markers for arterial thrombosis in SLE patients (14, 15). More recently, we have demonstrated that oxLDL binds to endogenous β2GPI, and that these complexes can be found in the blood stream of patients with various chronic diseases, such as SLE, APS, chronic renal disease, diabetes mellitus, and also in patients with "acute" myocardial infarction (AMI). These complexes are occasionally present as IgG immune complexes with anti-β2GPI antibodies in SLE and APS patients (16). Our recent *in vitro* experiments also showed that oxLDL-β2GPI complexes are taken up by macrophages via anti-β2GPI autoantibody-mediated phagocytosis (16-19). Thus, IgG immune complexes with oxLDL and β2GPI may be atherogenic. In contrast, a set of reports indicated that natural antibodies (mainly of the IgM class) derived from hyperlipidemic mice reduced the incidence of atherosclerosis in experimental models (20-23).

In this chapter, we review recent new insights into atherosclerosis, namely, "autoimmune-mediated mechanisms participating in the development of atherosclerosis and our recent observations on the clinical significance of oxLDL complexed with β2GPI.

Atherogenesis

Atherosclerosis is a pathological condition in which arteries undergo thickening of the intima region causing a decrease in their elasticity. The coronary and cerebral arteries, as well as the aorta are the sites most commonly affected by the atherosclerotic changes. The appearance of lipid laden foam cells is a characteristic early event in the development of atherosclerotic lesions. The diagram in Figure 1A represents a current consensus on the events leading to the earliest stage of atherosclerosis. Hypercholesterolemia is most commonly associated with an elevation of LDL, and this lipoprotein is the ultimate source of the cholesterol that accumulates in foam cells. The increment of circulating LDL levels in the lumen and/or the shear stress lead to an increase in the adherence of circulating monocytes to arterial endothelial cells, followed by the entry of these elements (LDL, oxLDL and monocytes) into the intima. The oxidative modification of LDL is thought to be catalyzed by these cells at the site of the arterial lesions, resulting in foam cell formation (oxLDL-loaded macrophages). Some foam cells are also generated from smooth muscle cells. Atherosclerosis develops under complicated interrelated conditions in which numerous pro-inflammatory molecules and/or adhesion molecules participate, i.e., monocyte chemo-attractant protein-1 (MCP-1), macrophage colony-stimulating factor (M-CSF), interferon-γ (IFN-γ, tumor necrosis factor-α (TNF-α), interleukine-4 (IL-4), platelet-derived growth factor (PDGF), heparin-binding EGF-like growth factor (HB-EGF), intercellular adhesion molecules (ICAMs), vascular cell adhesion molecules (VCAM-1), and endothelial selectin (E-selectin) and so on (24-27). Also, scavenger receptors, and some cell-cell interactions, possibly via CD40 and CD40 ligands, are involved in the development of atheroma (28).

When the luminal surface of the atherosclerotic lesion is damaged and ruptures, the activation of blood coagulation accompanied by the aggregation of platelets generates thrombi. Thrombus formation can lead to complete occlusion of the blood vessel, the primary cause of AMI and cerebral infarction.

Scavenger Receptors

Goldstein and Brown first discovered LDL specific receptors which contribute to the uptake of LDL by macrophages (29, 30). They also indicated the presence of other type(s) of receptors for chemically modified LDLs, namely, scavenger receptors (29, 31). These scavenger receptors are not down-regulated and may lead to the accumulation of massive amounts of intracellular neutral lipids in macrophages resulting in the formation of foam cells. Initially, acetylated LDL has been used as a ligand for scavenger receptors but such acetylation is not seen in physiological conditions. On the contrary, oxLDL is a physiological ligand for scavenger receptors and it is generated by peroxidation when LDL is co-cultured with endothelial cells or when incubated with a metal ion such as Cu^{2+} or Fe^{2+}.

Kodama and his colleagues first cloned the scavenger receptor SR-A (32, 33), specific for acetylated-LDL and Cu^{2+}-oxLDL. This followed the description of several different types of scavenger receptors, i.e. MARCO (a novel macrophage receptor with collagenous structure), SR-B1, CD36, Macrosialin, CD68, LOX-1, SREC, SRPSOX, and so on (34-41).

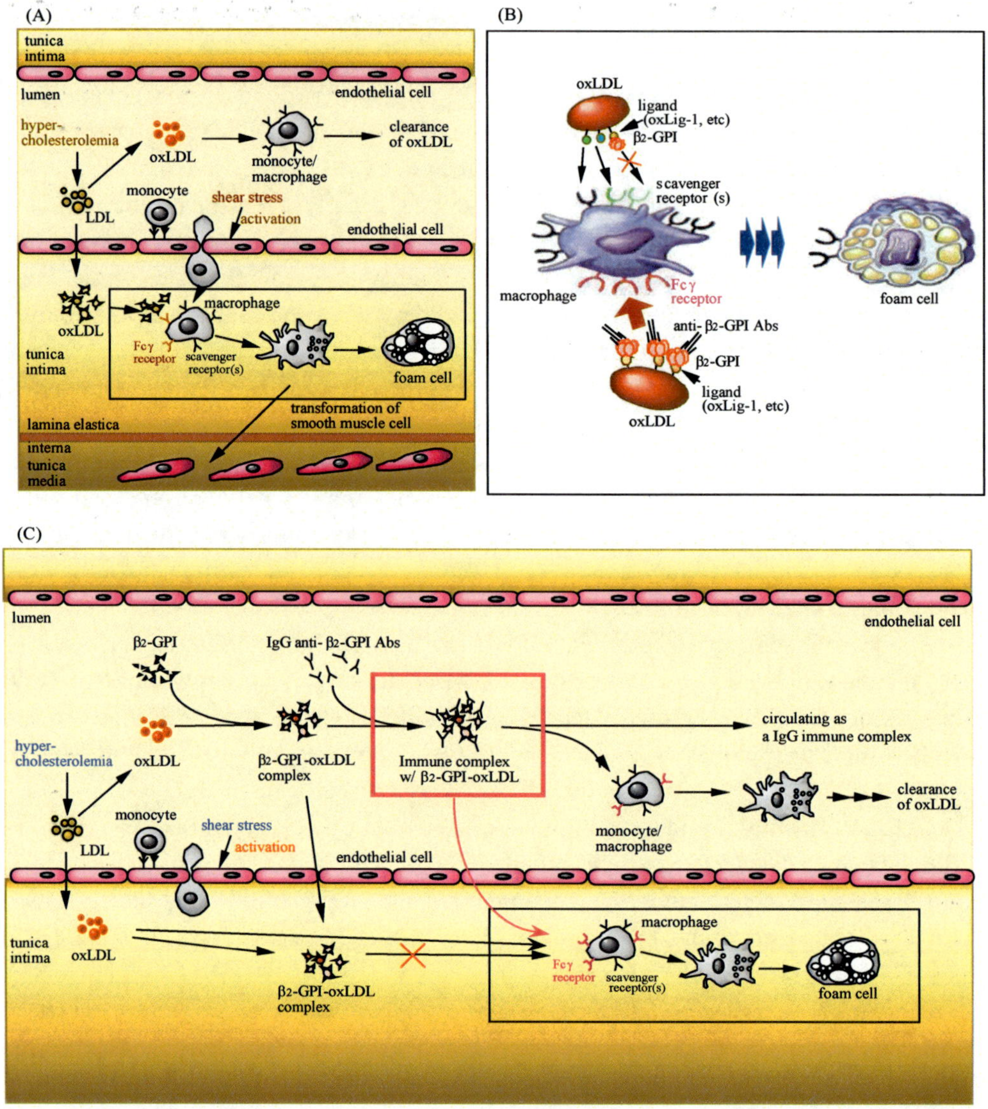

Figure 1. Mechanisms on development of atherosclerosis. A: The general consensus on atherogenesis. B: A possible mechanism on anti-β2GPI autoantibody-mediated oxLDL uptake by macrophages in APS. C: A proposed mechanism on development of thrombosis in APS.

Oxidation of LDL

The LDL particle contains 700 molecules of phospholipids, 600 of free cholesterols, and 1600 of cholesteryl esters, 185 of triglycerides, and 1 of apolipoprotein B (apoB) composed of 4536 amino acid residues (3).

Both the lipids and apoB are subjected to oxidation, and apoB breaks down into fragments of different sizes by oxidative attack, from 14 kDa to over 550 kDa (42). It is well known that a key feature of LDL's oxidation is the breakdown of these polyunsaturated fatty acids to yield a broad array of smaller fragments including aldehydes and ketones that can become conjugated to amino lipids or to apoB (43). The polyunsaturated fatty acids in cholesterol esters, phospholipids, and triglycerides are subject to free radical-initiated oxidation and can participate in chain reactions that amplify the extent of damage. Recently, it was reported that two oxidized fatty acid components, i.e., 9- or 13-hydroxyoctadecadienoic acid (9-HODE and 13-HODE) activate peroxisome proliferator-activator receptor γ (PPARγ), a transcriptional regulator of genes linked to lipid metabolism, and result in up-regulation of the scavenger receptor, CD36 (44). Thus, particular lipid components of oxLDL via the PPARγ activation can promote foam cell formation. Linoleic acid is a predominant polyunsaturated fatty acid in LDL and is present mainly as a cholesterol ester (45). In mildly oxidized LDL, cholesteryl hydroperoxyoctadecadienoic acid (Chol-HPODE) and cholesteryl hydroxyoctadecadienoic acid (Chol-HODE) were detected as main constituents of oxidation products (46). Chol-HPODE was reported to inactivate platelet-derived growth factor (47).

Thus, the oxidative breakdown of either free or esterified polyunsaturated fatty acids result in fatty acid-hydroperoxides to form highly reactive products containing aldehyde and ketone functions. Such active functions can form Schiff base adducts with lysine residues of the apoB moiety of LDL or other proteins, and with primary amine containing phospholipids such as phosphatidylserine and phosphatidylethanolamine.

Cholesterol is also converted to oxysterols, and it is especially oxidized at the 7-position. 7-Hydroxycholesterol (both free and esterified) is the major oxysterol formed during early events in LDL oxidation, with 7-ketocholesterol dominating at later stages (48). Recent studies indicated that elevated plasma levels of 7β-hydroxycholesterol may be associated with an increased risk of atherosclerosis (49). At later stage in LDL oxidation, cholesterol or 7-ketocholesteryl esters of 9-oxononanoate derived from cholesteryl linoleate, were detected as the most abundant fraction of oxidized cholesteryl linoleate (50-52). As a result of oxidation, huge numbers of oxidative structures are literally generated.

Chemically modified LDL, such as MDA-modified LDL, acetylated LDL and Cu^{2+}-mediated oxLDL were widely examined as experimental models of denatured LDL for studying the mechanisms of the development of atherosclerosis. Among these models, trace amounts of Cu^{2+} can induce LDL oxidation, resulting in highly reproducible LDL damage (53). This process leads to oxidized LDL that shares many structural and functional properties with LDL oxidized by cells or oxLDL extracted from arterial atherosclerotic plaques. Incubation of LDL with any of several different types of cells, or with Cu^{2+} even in the absence of cells, results in oxidatively modified LDL with similar properties (54). In 1989, there was already general agreement about the use of Cu^{2+}-oxidized LDL as an autoantigen because the oxLDL found in atheromatous lesions and oxLDL extracted from atherosclerotic lesions exhibit nearly all of the physicochemical and immunological properties of Cu^{2+}-oxLDL (55). Thus, Cu^{2+}-mediated oxLDL might be a more suitable model for physiological LDL rather than other chemically modified LDL, such as MDA-LDL. *In vivo*, LDL might be alternatively oxidized by released Cu^{2+} from ceruloplasmin, the major copper-containing component of mammalian plasma (56, 57).

Common Metabolic Form of oxLDL (Complex Formation of oxLDL and β2GPI)

β2GPI is a 50 kDa single-chain polypeptide composed of 326 amino acid residues, arranged in 5 homologous repeats known as complement control protein domains. β2GPI's fifth domain contains a patch of positively charged amino acids that likely represents the binding region for phospholipids (58, 59). β2GPI binds strongly to negatively charged molecules, such as phospholipids, heparin, and certain lipoproteins as well as to activated platelets and apoptotic cell membranes. This binding may aid the clearance of apototic cells from circulation (60). Further, β2GPI may have anticoagulant properties, as it has been shown to inhibit the intrinsic coagulation pathway, prothrombinase activity, and ADP-dependent platelet aggregation (61). It has also been reported to interact with several components of the protein C, protein S anticoagulant system (62).

The appearance of antiphospholipid antibodies such as aCL and lupus anticoagulants were found to be associated with thromboembolic manifestations such as cerebral or myocardial infarctions, venous and/or arterial thrombosis, recurrent fetal loss, and occasional thrombocytopenia in patients with systemic lupus erythematosus (5, 6). A subgroup of patients with these features was classified as having antiphospholipid syndrome (APS). Antiphospholipid antibodies were initially considered to be directed to phospholipids, but now it is widely considered that some phospholipid-binding proteins are true antigens for antiphospholipid antibodies. In 1990, three groups of investigators (7, 9) independently reported that aCL were directed to β2-GPI complexed with cardiolipin. We also reported that aCL recognize a neo-epitope on β2-GPI adsorbed onto hydrophobic surfaces with a negative charge such as an oxygenated polystyrene plate (11).

In our recent study (16-19), an *in vitro* specific interaction between Cu^{2+}-oxLDL and β2GPI was successfully demonstrated by ELISA, biosensor analysis (Figure 2), and ligand blot analysis on a silica gel plate for thin-layer chromatography (TLC) (Figure 3). Thus, oxLDL but not native LDL is subsequently targeted by β2GPI and anti-β2GPI autoantibodies. The ligands providing the specific interaction between oxLDL and β2GPI are chloroform-extractable lipids. Two major β2GPI ligands, i.e., oxLig-1 and oxLig-2, were further purified by reverse-phase HPLC and their structures was identified to be 7-ketocholesteryl-9-carboxynonanoate [9-oxo-9-(7-ketocholest-5-en-3β-yloxy) nonanoic acid (IUPAC)] and 7-ketocholesteryl-12-carboxy (keto) dodecanoate, respectively (Figure 4). Cholesteryl linoleate which is present in LDL as a major core lipid, represents the most probable candidate for a precursor of these ligands.

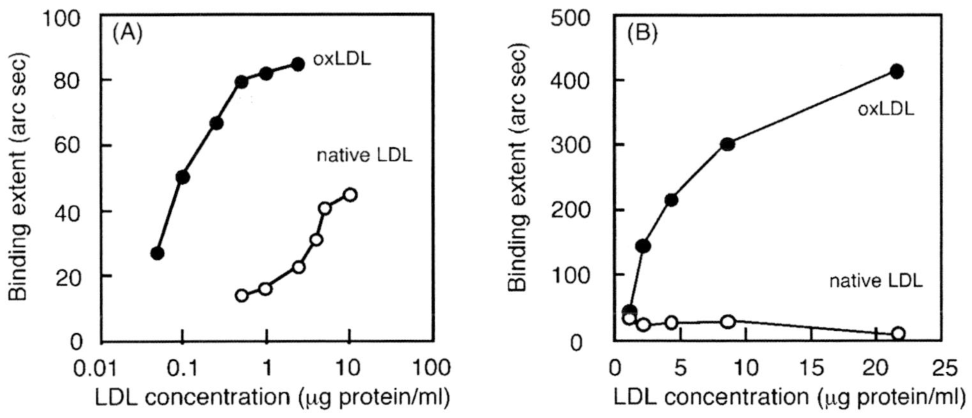

Figure 2. Molecular interactions among LDLs, β2GPI, and anti-β2GPI autoantibody, detected by an optical biosensor (IAsys). (A) Native LDL or oxLDL binding to solid phase β2GPI. (B) Native LDL or oxLDL binding to solid phase WB-CAL-1 antibody (anti-β2GPI autoantibody) in the presence of β2GPI.

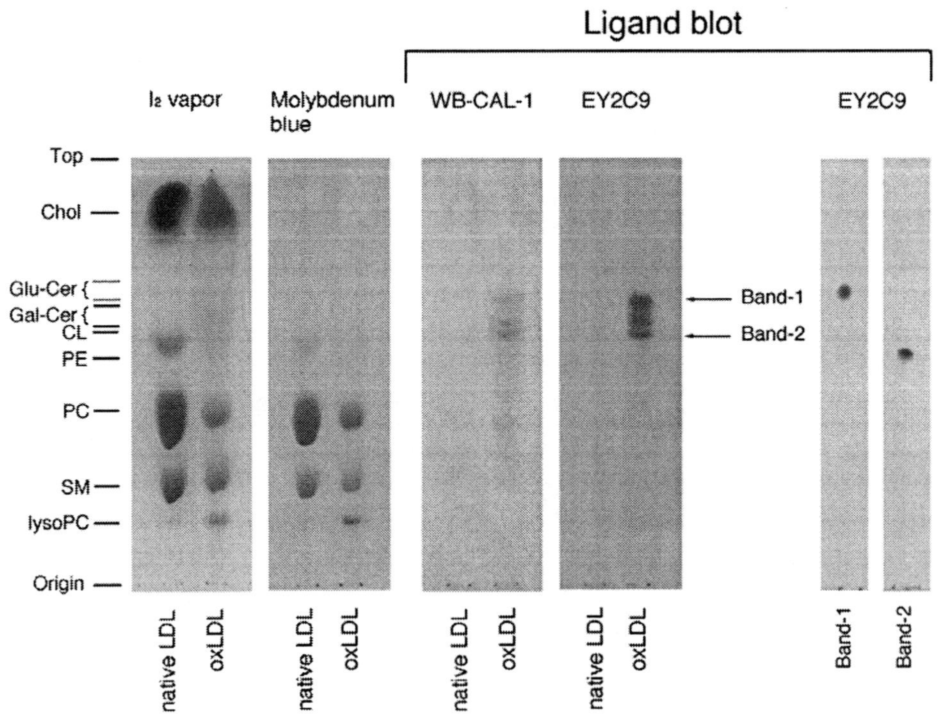

Figure 3. Thin-layer chromatography (TLC) and ligand blot of lipid extracts from LDLs. Lipids were spotted on a TLC plate, developed in chloroform/methanol/30% ammonia/water (120: 80: 10: 5, v/v/v/v). Plates were stained with I2 vapor, molybdenum blue. Ligand blot was performed with β2GPI and anti-β2GPI antibody (WB-CAL-1 and EY2C9). WB-CAL-1 and EY2C9 are monoclonal anti-β2GPI autoantibodies derived from NZW x BXSB F1 mouse (an animal model of APS) and APS patient, respectively.

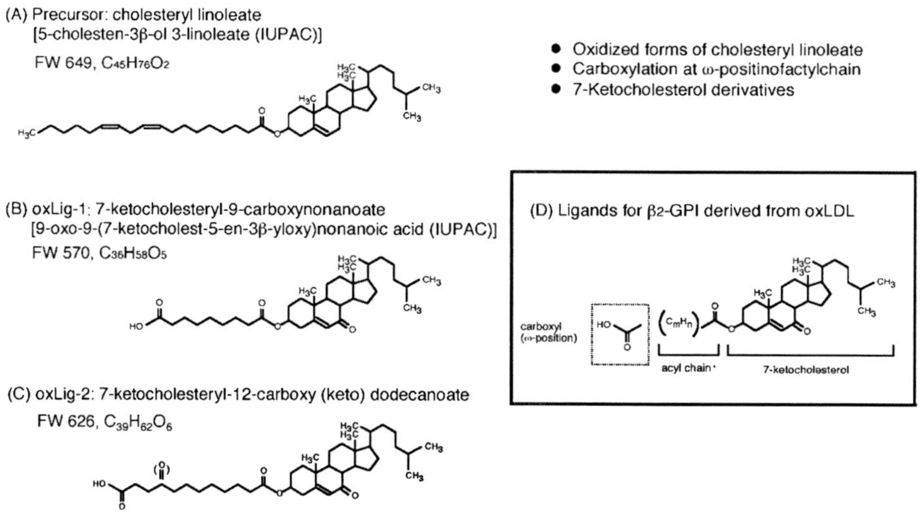

Figure 4. Structures of oxLDL derived ligands, specific for β2GPI. A: Cholesteryl linoleate (as a precursor), B and C: Major ligands for β2GPI (oxLig-1 and oxLig-2, respectively), C: A common structure of the β2GPI ligands. oxLig-1: ^{1}H-NMR (300.1 MHz, CDCl3):δ=5.71 (s, 1 H, H-6), 4.78-4.69 (m, 1 H, H-3); ^{13}C-NMR (75.5 MHz, CDCl3): δ=202.5, 179.7, 173.4, 164.5, 127.1, 72.4, 55.2, 50.4, 50.2, 45.8, 43.5, 39.9, 38.7, 36.6, 36.1, 29.2, 28.9, 28.4, 25.3, 25.0, 24.2, 23.2, 23.0, 19.3, 17.7, 12.4; m/z (FD-MS): 571 [$(M+H)^+$, C36H59O5 requires 571].

The *in vitro* interaction between β2GPI and Cu^{2+}-oxLDL is initially reversible by Mg^{2+} treatment due to the electrostatic interaction between an ω-carboxyl function and a lysine residue of the β2GPI which later progresses to a much more stable interaction such as Schiff base formation with an ω-aldehyde (Figure 5). Interestingly, the increased negative charge in Cu^{2+}-oxLDL was neutralized by the interaction with β2GPI (Figure 6). As described later, we detected the stable oxLDL/β2GPI complexes in sera of patients with various chronic diseases, such as APS, SLE, chronic renal diseases, and diabetes mellitus, or also "acute" myocardial infarction (AMI) and that the complexes are occasionally present as IgG immune complexes with anti-β2GPI antibodies in APS and SLE patients (16). The strength of the binding and the neutralized charge of the complexes may contribute to their stability in the blood stream.

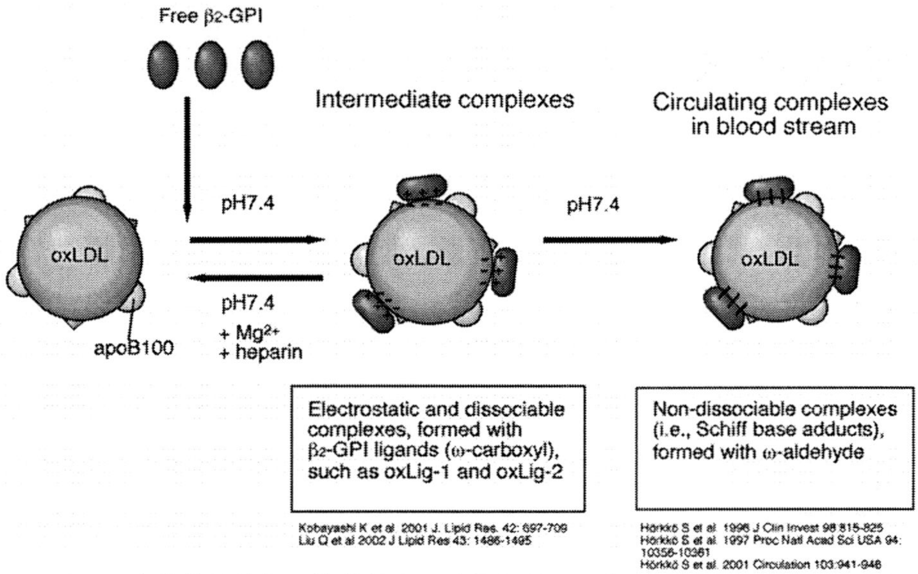

Figure 5. Mechanisms on complex formation between oxLDL and β2GPI.

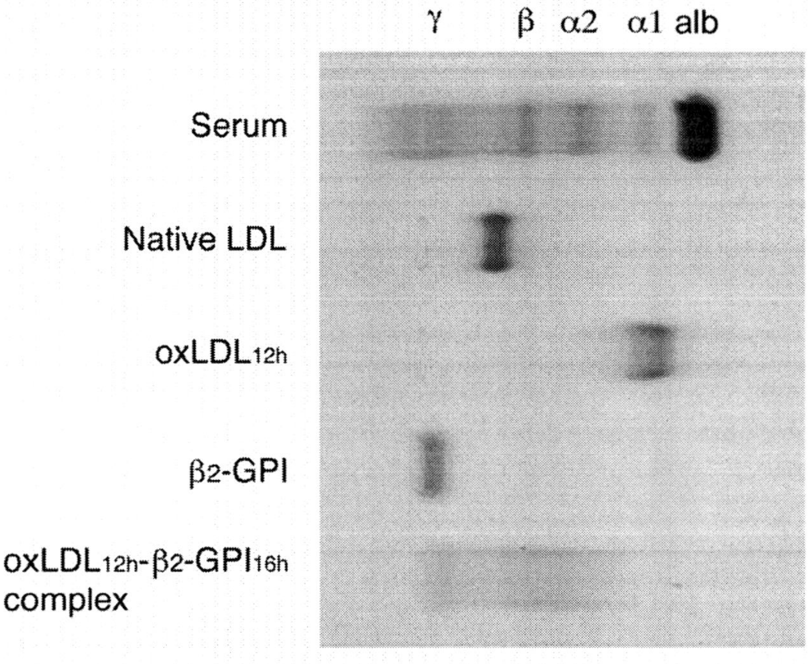

Figure 6. Agarose-electrophoresis of modified LDLs.

β2GPI and Autoantibody-Dependent Uptake of oxLDL by Macrophage

In 1997, we first demonstrated that $^{125}I\text{-}Cu^{2+}$-oxLDL was significantly taken up by macrophages in the presence of β2GPI and anti-β2GPI autoantibody (17), and the uptake by macrophages of liposomes containing β2GPI ligands, such as oxLig-1 and oxLig-2 as models of oxLDL ligands, and confirmed the previous results (16, 18, 19). Thus, as indicated by *in vitro* experiments, IgG anti-β2GPI autoantibodies are atherogenic. The oxLDL uptake is probably mediated by Fcγ receptors rather than scavenger receptors (i.e., by phagocytosis) (Figure 1B). In contrast, Fcμ receptors have poor phagocytic properties, possibly making IgM class of autoantibodies and/or natural antibodies anti-atherogenic (protective).

Both a ketone function at 7 position (one at 22 position is not) on the sterol backbone and ω-carboxyl function on the acyl chain of these β2GPI ligands are responsible either for the interaction between oxLDL and β2GPI and for the β2GPI and anti-β2GPI mediated uptake of oxLDL by macrophages (16).

Novel ELISA for oxLDL/β2GPI Complexes and for Antibodies to the Complexes

Utilizing an anti-β2GPI autoantibody, WB-CAL-1, derived from NZW x BXSB F1 mouse (63), we established a novel detection system for β2GPI/oxLDL complexes circulating in sera (Figure 7). Microtiter plates were pre-coated with WB-CAL-1, diluted serum samples applied and bound β2GPI/oxLDL complexes captured with enzyme-labeled anti-apoB antibody. The WB-CAL-1 antibody is specific for complex forms of β2GPI but not for free forms of β2GPI in solution. Significant amounts of stable and non-dissociable complexes between oxLDL and β2GPI were detected in sera of patients with chronic diseases. Anti-β2GPI/oxLDL antibodies can also be detected in the presence of β2GPI in ELISA using oxLig-1 pre-coated microtiter plate (16, 18, 19, 64).

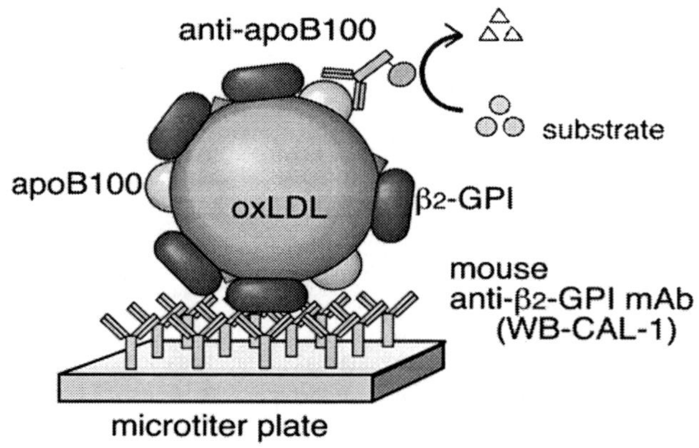

Figure 7. A principle of detection system (ELISA) for oxLDL complexes with β2GPI.

Antiphospholipid Antibodies and Hypercoagulability

Thromboembolic disease is a significant cause of morbidity and mortality of public health importance. The medical literature reports that 25% to 40% of the general population will die from a thrombotic event, and that one-fourth of the first thrombotic events will be fatal (65, 66). Hypercoagulability (thrombophilia) refers to a broad group of inherited or acquired molecular defects that may result in hyperactivity of the coagulation system and the development of thrombosis (67). Hypercoagulability is not an uniform disease, and appears to be characterized by complex interactions of various predisposing factors. Current laboratory investigations will likely identify the underlying coagulation defect(s) in 40 to 60% of patients, a number substantially higher than a 10% reported two decades ago (68). Defining the exact cause(s) of hypercoagulability may determine the type and duration of antithrombotic treatments. The antiphospholipid syndrome (APS) is probably the most common cause of acquired hypercoagulability in the general population (69, 70).

Elevated serum levels of antiphospholipid antibodies along with thrombotic events of both the venous and arterial vasculature, or with pregnancy morbidity (miscarriages and fetal loss) represent the major features of the APS. APS may be present in the context of a systemic autoimmune disorder i.e. systemic lupus erythemtosus (SLE) and referred to as secondary APS, or in the absence of an obvious underlying disease (primary APS) (5, 71). Antiphospholipid antibodies appear to increase the risk of thrombosis by at least two-fold in the context of an autoimmune disease (72). In both primary and secondary APS, recurrence rates of up to 30% for thrombosis with a mortality of up to 10% over a 10-year follow-up period have been reported (73). Understanding the pathogenic mechanism(s) involved in immune-mediated thrombosis, the complexities of the serologic diagnosis, and the heterogeneity of the clinical features of APS remain the focus of many research groups. Accurate diagnosis of APS is critically important as long-term anticoagulation is the only treatment proven to reduce the risk of thrombosis (74).

Antiphospholipid antibodies, aCL or lupus anticoagulants, are a heterogeneous group of autoantibodies with a possible pathogenic role in the development of the clinical manifestations of APS. These antibodies are characterized by their reactivity to negatively charged phospholipids, phospholipid/protein complexes, and certain proteins presented on suitable surfaces (i.e. activated cell membranes, oxygenated polystyrene) (11, 75). Several plasma proteins that participate in coagulation and interact with phospholipids have been described as antiphospholipid cofactors, i.e. β2GPI, prothrombin, and annexin V. β2GPI has been shown to be a relevant antigenic target for antiphospholipid antibodies (7, 9-11).

Binding of antiphospholipid antibodies to β2GPI may interfere with the anticoagulant functions promoting thrombosis. Anti-β2GPI antibodies have been reported as more specific for thrombosis and APS than aCL antibodies (76), and recent prospective studies have shown that aCL antibodies, particularly β2GPI-dependent, or anti-β2GPI antibodies are important predictors for arterial thrombosis (myocardial infarction and stroke) in men (15, 77, 78). These studies suggest that β2GPI and anti-β2GPI antibodies play a pathogenic role in the development of thrombosis, particularly in arterial thrombosis in SLE and APS patients.

Autoimmunity, Antiphospholipid Syndrome and Atherosclerosis

The premature (or accelerated) development of atherosclerosis has been recently recognized in patients with systemic autoimmune diseases (79, 80, 81). The traditional risk factors for atherosclerosis as well as autoimmune disease treatments (i.e. steroids) failed to account for these changes (82). Today's SLE survival rate (>80% over a 10-year follow-up) may have uncovered hidden causes of mortality and morbidity. SLE mortality rates due to cardiovascular disease have surpassed that from SLE disease itself or from complications such as infections in most studies (83). In addition, abnormal myocardial perfusion results by Tc99m emission tomography have been reported in about 43% of asymptomatic SLE patients, and increased carotid intima medial thickness (IMT) with atherosclerotic plaques have also been demonstrated by B-mode ultrasound in over 33% of these patients. Autoantibodies against oxLDL, phospholipids, and Lp(a), have been proposed as alternative mechanisms as well as certain biochemical and genetic abnormalities (84). Although venous thromboembolic complications are the most common clinical findings in APS patients (85-87), about 25% of the APS patients enrolled into a European cohort of 1,000 patients presented an arterial thrombotic event (myocardial infarction, cerebrovascular accident, angina, etc.) as the initial clinical manifestation. When all the arterial thrombotic events were considered, up to 31% of the patients presented these complications (88).

The cholesterol that accumulates in macrophage-derived foam cells of atherosclerotic lesions in the arterial intima is derived from circulating lipoproteins, mainly from LDL. However, LDL must be modified before is ingested by macrophages via scavenger receptors and the oxidative modification of LDL represent such a mechanism (89). Several studies have demonstrated that atherosclerosis is an inflammatory disease, involving the dysregulation of cholesterol homeostasis by aberrant interactions between lipid-modulating elements and mediators of inflammation (90). Although the initiating inflammatory factor(s) remain to be elucidated, likely candidates include oxLDL, immunological injury, homocysteine and possibly infectious agents. An active role of antibodies in this process has been proposed (91).

It seems that oxLDL plays an important pathogenic role in early events leading to atherosclerosis (3, 92), acting as a pro-inflammatory chemotactic agent for macrophages and T lymphocytes (93). In addition, oxLDL has been found in human and rabbit atherosclerotic lesions (55), and shown to be an immunogen producing autoantibodies in patients with autoimmune disorders, such as SLE and APS (13, 94, 95). β2GPI has also been localized in human atherosclerotic lesions by immunohistochemical staining (12), which suggests a role of β2GPI (and antiphospholipid antibodies, i.e. anti-β2GPI antibodies) in atherosclerosis. The participation of the immune system in the development of atherosclerosis is becoming apparent and some antiphospholipid antibodies may also be possible participants.

We have demonstrated that β2GPI binds to oxLDL, not to native LDL, initially forming dissociable electrostatic complexes followed by more stable complexes bound by covalent interactions (59, 16). In addition, we also reported that oxidation of LDL is a common occurrence in both APS and SLE patients without APS, and have demonstrated stable β2GPI/oxLDL complexes in more than 60% of these patients (Figure 8A). However, serum levels of these complexes were not associated with the activity of SLE or with any major clinical manifestation of APS. Although it can be hypothesized that this interaction might be related to chronic inflammation of the vasculature that occurs in autoimmune patients, the mechanism(s) for the increased oxidation of LDL are not fully understood. It is possible that the interaction between β2GPI and oxLDL may promote oxLDL clearance from circulation to prevent thrombus formation. Serum levels of β2GPI/oxLDL complexes fluctuated widely when measured in samples obtained at different time intervals over a 12-month follow-up from 6 SLE patients. This suggests that oxidation and formation of complexes are very active processes under unknown regulatory mechanism(s). Stable β2GPI/oxLDL complexes may be clinically relevant as they have been implicated as atherogenic autoantigens, and their presence may represent a risk factor or an indirect but significant contributor for thrombosis and atherosclerosis in patients with an autoimmune background (16).

The high-density lipoprotein (HDL)-associated enzyme paraoxonase (PON) has antioxidant activity that protects LDL from oxidation (96). Patients with high levels of aCL antibodies have decreased PON activity (97). Furthermore, increased levels of IgG anti-β2GPI antibodies have been associated with reduced PON activity in patients with SLE and primary APS (98). PON activity is also known to increase with lipid-lowering drugs (99, 100), and in one study, cholesterol-lowering statins prevented the *in vitro* endothelial cell activation normally induced by anti-β2GPI antibodies (101). Antioxidant treatment for 4 to 6 weeks has been observed to decrease the titer of circulating aCL antibodies in patients with SLE and APS (102). Vascular (endothelial) injury as seen in autoimmune patients may affect PON activity or any other antioxidant mechanism, triggering LDL oxidative changes. Taken together, these findings provide additional support to the hypothesis that oxidative stress plays an important role in antiphospholipid antibody production and development of thrombosis in APS.

We have also measured serum levels of IgG anti-β2GPI/oxLig-1 autoantibodies in patients with APS, and evaluated their association with the major clinical manifestations of APS (16, 64, in revision). IgG autoantibodies against β2GPI/oxLig-1 were significantly elevated in about 40% of patients with APS as compared to SLE patients without APS. Further, these antibodies showed a stronger correlation with arterial thrombosis than venous thrombosis and pregnancy morbidity (Figure 8B). Anti-β2GPI/oxLig-1 antibodies were present in 3 of 4 SLE patients with active disease followed over a 12-month period, while 2 patients with inactive disease and β2GPI/oxLDL complexes did not have these antibodies. The coexistence of β2GPI/oxLig-1 autoantibodies with β2GPI/oxLDL complexes, suggest that these two elements interact perhaps forming circulating immune complexes (β2GPI/oxLDL/antibody). Actually, we detected such immune complexes in serum samples of SLE and/or APS in the most recent study (16) (Figure 1C). This observation along with the increased macrophage uptake of β2GPI/oxLDL complexes in the presence of anti-β2GPI/oxLig-1 antibodies (14, 17-19), provide a possible explanation for the accelerated development of atherosclerosis in autoimmune patients. The ELISA system used in our study seems to detect only free (unbound) antibodies to β2GPI/oxLig-1 complexes. Although preliminary, our results also suggest that IgG anti-β2GPI/oxLig-1 antibodies may represent a distinct subset of antiphospholipid antibodies (to anti-β2GPI), and that they may coexist with other antibodies. Thus, IgG anti-β2GPI/oxLDL (oxLig-1) antibodies appear to be a useful serologic marker with high specificity for APS and might possibly have a pathogenic role in atherosclerosis risk in autoimmune patients.

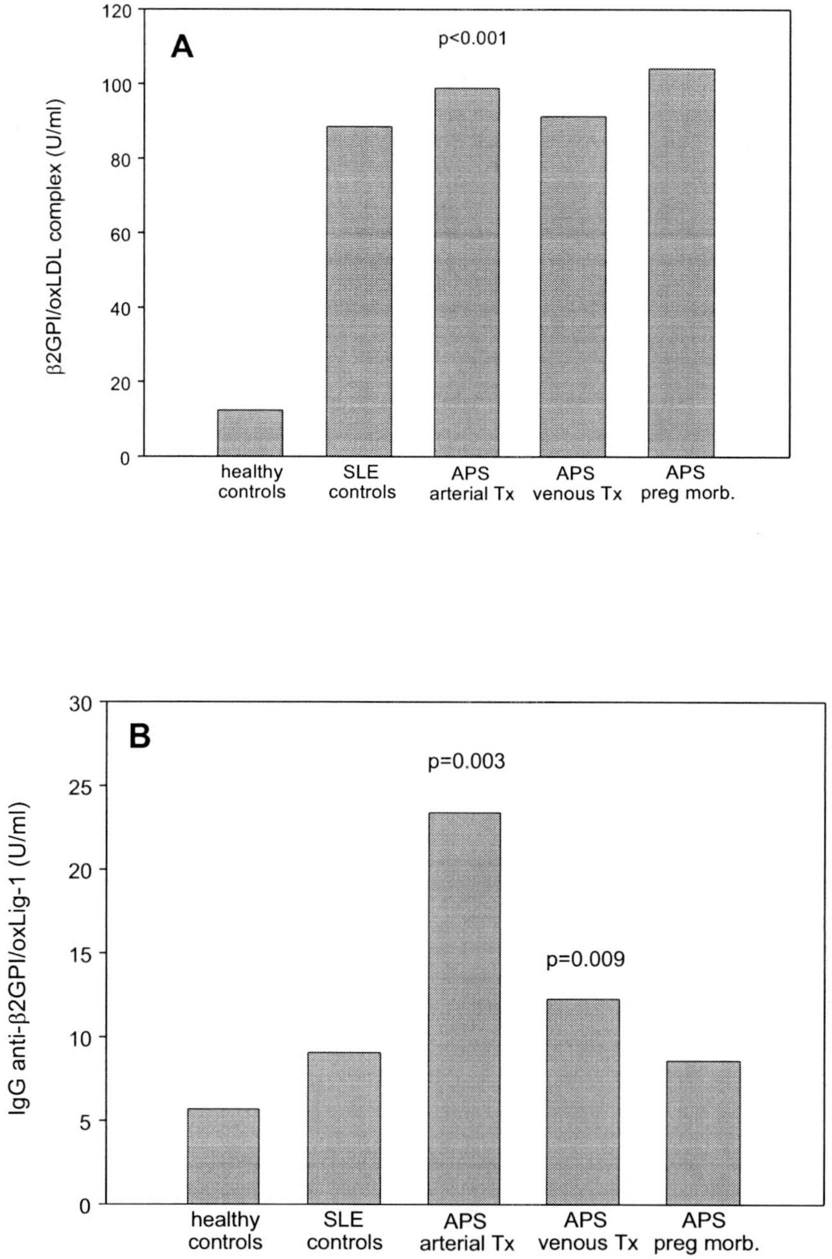

Figure 8. Mean serum levels of (A, top) β2GPI/oxLDL complexes and (B, bottom) IgG anti-β2GPI/oxLig-1 antibodies measured by ELISA in healthy controls (n=40), systemic lupus erythematosus (SLE) patients without APS (n=50) and antiphospholipid syndrome (APS) patients classified into arterial thrombosis (n=45), venous thrombosis (n=40) or pregnancy morbidity (n=15) subgroups. P values shown versus healthy controls.

Summary and Future Direction

Current atherosclerosis research at different levels i.e. histopathological, biochemical and immunological, continues to unravel the complex nature of this process. New factors are being investigated and unique interactions among diverse factors described, all resulting in a better understanding of atherogenic mechanism(s). In addition to the oxidative modification of LDL, the action of local soluble mediators of inflammation and chemotactic factors, the accumulation of immune cells (macrophages and T lymphocytes) in response to various external and/or internal stimuli, an active humoral immune response has been recently incorporated into the process. A significant role of antiphospholipid antibodies in the development of atherosclerosis was proposed when some groups reported in SLE patients the presence of antibodies to oxLDL, some with cross-reactivity with aCL antibodies, as well as the localization of β2GPI in human atherosclerotic plaques. The interaction between β2GPI and oxLDL to form circulating complexes strongly suggests that this complex is an atheroantigen. This interaction has been further characterized and the oxLDL-derived ligand (oxLig-1) specific for β2GPI has been identified and synthesized. APS patients may produce antibodies to this complex and the resulting circulating immune complexes trigger atherosclerotic changes. The physiologic relevance of this finding has been demonstrated *in vitro* by the enhanced macrophage uptake of β2GPI/oxLDL/antibody complexes. The antibodies used in these experiments were aCL antibodies β2GPI-dependent or anti-β2GPI antibodies. The possible participation of Fcγ receptors in the increased macrophage uptake of oxLDL-containing complexes seem to be particularly important in the development of foam cells and atherosclerotic plaques in the intima of arteries.

The development of ELISA systems to measure β2GPI/oxLDL complexes and anti-β2GPI/oxLDL (oxLig-1) antibodies provided additional tools to further study the role of the humoral immune response in the atherosclerotic process. Stable and possibly pathogenic β2GPI/oxLDL complexes were demonstrated in the serum of both SLE and APS patients, but anti-β2GPI/oxLDL (oxLig-1) antibodies seem to be restricted to SLE patients with APS or primary APS. Further, the association of these antibodies with arterial thrombosis was stronger than with venous thrombosis in these patients. At this point, these results should be interpreted in the context of an autoimmune disease. Studies are underway to determine whether β2GPI/oxLDL complexes and/or anti-β2GPI/oxLDL (oxLig-1) antibodies exist in non-autoimmune conditions i.e., diabetes mellitus, and what would be the effect of aging in normal individuals. We have detected β2GPI/oxLDL complexes in patients with syphilis but not in rheumatoid arthritis, indicating that oxidation of LDL and the formation of complexes with β2GPI is not restricted to SLE. In contrast, both patient groups syphilis and rheumatoid arthritis did not develop anti- β2GPI/oxLDL (oxLig-1) antibodies. These antibodies seem restricted to SLE and APS and we hypothesize that the presence of these antibodies accelerates the development of atherosclerosis.

Are there other molecules, antigens, antibodies, or other interactions playing a role in the development of atherosclerosis? It has been reported that β2GPI may interact with various negatively charged molecules. An inflammatory reaction may modify certain molecules making them reactive with β2GPI. The resulting complex may exert unwanted effect and even trigger an immune response. For example, the acute-phase C-reactive protein (CRP), fibrinogen, tumor necrosis factor-α (TNF-α), heat-shock protein (HSP), homocysteine, etc. are being use to assess the risk for cardiovascular disease, however, the exact mode of action(s) are not complete understood. It is possible that β2GPI may participate in additional, yet unknown, interactions some acting directly on arterial intima or inducing the production of antibodies which may form immune complexes that promote atherosclerosis. Current research is yielding better, more sensitive and specific markers for atherosclerosis. In addition, having a more complete picture of the atherogenic mechanisms will not only favor the development of more effective targeted therapeutic strategies but the implementation of prevention programs for acquired hypercoagulable states and atherosclerosis.

References

[1] Steinbrecher UP, Parthasarathy S, Leake DS, Witztum JL, Steinberg D. Modification of low density lipoprotein by endothelial cells involves lipid peroxidation and degradation of low density lipoprotein phospholipids. *Proc Natl Acad Sci USA*. 1984; 81: 3883-7.

[2] Steinberg D, Parthasarathy S, Carew TE, Khoo JC, Witztum JL. Beyond cholesterol. Modifications of low-density lipoprotein that increase its atherogenicity. *N Engl J Med* 1989; 320: 915-24.

[3] Steinberg D. Low density lipoprotein oxidation and its pathobiological significance. *J Biol Chem* 1997; 272: 20963-6.

[4] Heinecke JW. Mechanisms of oxidative damage of low density lipoprotein in human atherosclerosis. *Curr Opin Lipid.* 1997; 8: 268-74.

[5] Hughes GRV, Harris EN, Gharavi AE. The anticardiolipin syndrome. *J Rheumatol* 1986; 13: 486-9.

[6] Harris EN, Gharavi AE, Boey ML, Patel BM, Mackworth-Young CG, Loizou S, Hughes GRV. Anticardiolipin antibodies: detection by radioimmunoassay and association with thrombosis in systemic lupus erythematosus. *Lancet* 1983; 2: 1211-4.

[7] McNeil HP, Simpson RJ, Chesterman CN, Krilis SA. Anti-phospholipid antibodies are directed against a complex antigen that includes a lipid-binding inhibitor of coagulation: β2-glycoprotein I (apolipoprotein H). *Proc Natl Acad Sci USA* 1990; 87: 4120-4.

[8] Galli M, Comfurius P, Maassen C, Hemker HC, de Baets MH, Van Breda-Vriesman PJC, Barbui T, Zwaal RFA, Bevers EM. Anticardiolipin antibodies (ACA) directed not to cardiolipin but to a plasma protein cofactor. *Lancet* 1990; 335:1544-7.

[9] Matsuura E, Igarashi Y, Fujimoto M, Ichikawa K, Koike T. Anticardiolipin cofactor(s) and differential diagnosis of autoimmune disease. *Lancet* 1990; 336: 177-8.

[10] Matsuura E, Igarashi Y. Fujimoto M, Ichikawa K, Suzuki T, Sumida T, Yasuda T, Koike T. Heterogeneity of anticardiolipin antibodies defined by the anticardiolipin cofactor. *J Immunol* 1992; 148: 3885-91.

[11] Matsuura E, Igarashi Y, Yasuda T, Triplett DA, Koike T. Anticardiolipin antibodies recognize β2-glycoprotein I structure altered by interacting with an oxygen modified solid phase surface. *J Exp Med* 1994; 179: 457-62.

[12] George J, Harats D, Gilburd B, Afek A, Levy Y, Schneiderman J, Barshack I, Kopolovic J, Shoenfeld Y. Immunolocalization of β2-glycoprotein I (apolipoprotein H) to human atherosclerotic plaques: potential implications for lesion progression. *Circulation* 1999; 99: 2227-30.

[13] Vaarala O, Alfthan G, Jauhiainen M, Leirisalo-Repo M, Aho K, Palosuo T. Crossreaction between antibodies to oxidised low-density lipoprotein and to cardiolipin in systemic lupus erythematosus. *Lancet* 1993 341: 923-5.

[14] Tinahones FJ, Cuadrado MJ, Khamashta MA, Mujic F, Gomez-Zumaquero JM, Collantes E, Hughes GRV. 1998. Lack of cross reaction between antibodies to β2-glycoprotein-I and oxidized low-density lipoprotein in patients with antiphospholipid syndrome. *Br J Rheumatol* 1998; 37: 746-9.

[15] Romero FI, Amengual O, Atsumi T, Khamashta MA, Tinahones FJ, Hughes GRV. Arterial disease in lupus and secondary antiphospholipid syndrome: Association with anti-β2-glycoprotein I antibodies but not with antibodies against oxidized low-density lipoprotein. *Br J Rheumatol* 1998; 37: 883-8.

[16] Kobayashi K, Kishi M, Atsumi T, Bertolaccini ML, Makino H, Sakairi N, Yamamoto I, Yasuda T, Khamashta MA, Hughes GRV, Koike T, Voelker DR, Matsuura E. Circulating oxidized LDL forms complexes with β2-glycoprotein I: implication as an atherogenic autoantigen. *J Lipid Res* 2003; 44: 716-26.

[17] Hasunuma Y, Matsuura E, Makita Z, Katahira T, Nishi S, Koike T. Involvement of β2-glycoprotein I and anticardiolipin antibodies in oxidatively modified low-density lipoprotein uptake by macrophages. *Clin Exp Immunol* 1997; 107: 569-73.

[18] Kobayashi K, Matsuura E, Liu Q, Furukawa J, Kaihara K, Inagaki J, Atsumi T, Sakairi N, Yasuda T, Voelker DR, Koike T. A specific ligand for β2-glycoprotein I mediates autoantibody-dependent uptake of oxidized low density lipoprotein by macrophages. *J Lipid Res* 2001; 42: 697-709.

[19] Liu Q, Kobayashi K, Inagaki J, Sakairi N, Iwado A, Yasuda T, Koike T, Voelker DR, Matsuura E. ω-carboxyl variants of 7-ketocholesteryl esters are ligands for β2-glycoprotein I and mediate antibody-dependent uptake of oxidized LDL by macrophages. *J Lipid Res* 2002; 43: 1486-95.

[20] Chang MK, Bergmark C, Laurila A, Horkko S, Han KH, Friedman P, Dennis EA, Witztum JL. Monoclonal antibodies against oxidized low-density lipoprotein bind to apoptotic cells and inhibit their phagocytosis by elicited macrophages: evidence that oxidation-specific epitopes mediate macrophage recognition. *Proc Natl Acad Sci USA* 1999; 96: 6353-8.

[21] Horkko S, Bird DA, Miller E, Itabe H, Leitinger N, Subbanagounder G, Berliner JA, Friedman P, Dennis EA, Curtiss LK, Palinski W, Witztum JL. Monoclonal autoantibodies specific for oxidized phospholipids or oxidized phospholipid-protein adducts inhibit macrophage uptake of oxidized low-density lipoproteins. *J Clin Invest* 1999; 103: 117-28.

[22] Binder CJ, Horkko S, Dewan A, Chang MK, Kieu EP, Goodyear CS, Shaw PX, Palinski W, Witztum JL, Silverman GJ. Pneumococcal vaccination decreases atherosclerotic lesion formation: molecular mimicry between Streptococcus pneumoniae and oxidized LDL. *Nat Med* 2003; 9: 736-43.

[23] Rose N, Afanasyeva M. Autoimmunity: busting the atheorsclerotic plaque. *Nat Med* 2003; 9: 641-2.

[24] Cushing SD, Berliner JA, Valente AJ, Territo MC, Navab M, Parhami F, Gerrity R, Schwartz CJ, Fogelman AM. Minimally modified low density lipoprotein induces monocyte chemotactic protein 1 in human endothelial cells and smooth muscle cells. *Proc Natl Acad Sci USA* 1990; 87: 5134-8.

[25] Rajavashisth TB, Andalibi A, Territo MC, Berliner JA, Navab M, Fogelman AM, Lusis AJ. Induction of endothelial cell expression of granulocyte and macrophage colony-stimulating factors by modified low-density lipoproteins. *Nature* 1990; 344: 254-7.

[26] Nakata A, Miyagawa J, Yamashita S, Nishida M, Tamura R, Yamamori K, Nakamura T, Nozaki S, Kameda-Takemura K, Kawata S, Taniguchi N, Higashiyama S, Matsuzawa Y. Localization of heparin-binding epidermal growth factor-like growth factor in human coronary arteries. Possible roles of HB-EGF in the formation of coronary atherosclerosis. *Circulation* 1996; 94: 2778-86.

[27] Frostegard J, Wu R, Haegerstrand A, Patarroyo M, Lefvert AK, Nilsson J. Mononuclear leukocytes exposed to oxidized low density lipoprotein secrete a factor that stimulates endothelial cells to express adhesion molecules. *Atherosclerosis* 1993; 103: 213-9.

[28] Mach F, Schonbeck U, Sukhova GK, Atkinson E, Libby P. Reduction of atherosclerosis in mice by inhibition of CD40 signalling. *Nature* 1998; 394: 200-3.

[29] Goldstein JL, Ho YK, Basu SK, Brown MS. Binding site on macrophages that mediates uptake and degradation of acetylated low density lipoprotein, producing massive cholesterol deposition. *Proc Natl Acad Sci USA* 1979; 76: 333-7.

[30] Yamamoto T, Davis CG, Brown MS, Schneider WJ, Casey ML, Goldstein JL, Russell DW. The human LDL receptor: a cysteine-rich protein with multiple Alu sequences in its mRNA. *Cell* 1984; 39: 27-38.

[31] Brown MS, Goldstein JL. Scavenger cell receptor shared. *Nature* 1985; 316: 680-1.

[32] Kodama T, Reddy P, Kishimoto C, Krieger M. Purification and characterization of a bovine acetyl low density lipoprotein receptor. *Proc Natl Acad Sci USA* 1988; 85: 9238-42.

[33] Kodama T, Freeman M, Rohrer L, Zabrecky J, Matsudaira P, Krieger M. Type I macrophage scavenger receptor contains alpha-helical and collagen-like coiled coils. *Nature* 1990; 343: 531-5.

[34] Elomaa O, Kangas M, Sahlberg C, Tuukkanen J, Sormunen R, Liakka A, Thesleff I, Kraal G, Tryggvason K. Cloning of a novel bacteria-binding receptor structurally related to scavenger receptors and expressed in a subset of macrophages. *Cell* 1995; 80: 603-9.

[35] Elomaa O, Sankala M, Pikkarainen T, Bergmann U, Tuuttila A, Raatikainen-Ahokas A, Sariola H, Tryggvason K. Structure of the human macrophage MARCO receptor and characterization of its bacteria-binding region. *J Biol Chem* 1998; 273:4530-8.

[36] Ramprasad MP, Fischer W, Witztum JL, Sambrano GR, Quehenberger O, Steinberg D. The 94- to 97-kDa mouse macrophage membrane protein that recognizes oxidized low density lipoprotein and phosphatidylserine-rich liposomes is identical to macrosialin, the mouse homologue of human CD68. *Proc Natl Acad Sci USA* 1995; 92: 9580-4.

[37] Sambrano GR, Steinberg D. Recognition of oxidatively damaged and apoptotic cells by an oxidized low density lipoprotein receptor on mouse peritoneal macrophages: role of membrane phosphatidylserine. *Proc Natl Acad Sci USA* 1995; 92: 1396-1400.

[38] Rigotti A, Acton SL, Krieger M. The class B scavenger receptors SR-BI and CD36 are receptors for anionic phospholipids *J Biol Chem* 1995; 270: 16221-4.

[39] Sawamura T, Kume N, Aoyama T, Moriwaki H, Hoshikawa H, Aiba Y, Tanaka T, Miwa S, Katsura Y, Kita T, Masaki T. An endothelial receptor for oxidized low-density lipoprotein. *Nature* 1997; 386: 73-7.

[40] Minami M, Kume N, Shimaoka T, Kataoka H, Hayashida K, Akiyama Y, Nagata I, Ando K, Nobuyoshi M, Hanyuu M, Komeda M, Yonehara S, Kita T. Expression of SR-PSOX, a novel cell-surface scavenger receptor for phosphatidylserine and oxidized LDL in human atherosclerotic lesions. *Arterioscler Thromb Vasc Biol* 2001; 21: 1796-800.

[41] Shimaoka T, Kume N, Minami M, Hayashida K, Kataoka H, Kita T, Yonehara S. Molecular cloning of a novel scavenger receptor for oxidized low density lipoprotein, SR-PSOX, on macrophages. *J Biol Chem*. 2000; 275: 40663-6.

[42] Fong LG, Parthasarathy S, Witztum JL, Steinberg D. Nonenzymatic oxidative cleavage of peptide bonds in apoprotein B-100. *J Lipid Res* 1987; 28:1466-77.

[43] Esterbauer H, Jurgens G, Quehenberger O, Koller E. Autoxidation of human low density lipoprotein: loss of polyunsaturated fatty acids and vitamin E and generation of aldehydes. *J Lipid Res* 1987; 28: 495-509.

[44] Nagy L, Tontonoz P, Alvarez JG, Chen H, Evans RM. Oxidized LDL regulates macrophage gene expression through ligand activation of PPAR□. *Cell* 1998; 93: 229-40.

[45] Weidtmann A, Scheithe R, Hrboticky N, Pietsch A, Lorenz R, Siess W. Mildly oxidized LDL induces platelet aggregation through activation of phsopholipase A2. *Arterioscler Thromb Vasc Biol* 1995; 15: 1131-8.

[46] Kritharides L, Jessup W, Gifford J, Dean RT. A method for defining the stages of low density lipoprotein oxidation by the separation of cholesterol- and cholesteryl ester-oxidation products using HPLC. *Anal Biochem* 1993; 213: 79-89.

[47] van Heek M, Schmitt D, Toren P, Cathcart MK, DiCorleto PE. Cholesteryl hydroperoxyoctadecadienoate from oxidized low density lipoprotein inactivated platelet-derived growth factor. *J Biol Chem* 1998; 273: 19405-10.

[48] Brown AJ, Leong SL, Dean RT, Jessup W. 7-Hydroperoxycholesterol and its products in oxidized low density lipoprotein and human atherosclerotic plaque. *J Lipid Res* 1997; 38: 1730-45.

[49] Brown AJ, Jessup W. Oxysterols and atherosclerosis. *Atherosc*lerosis 1999; 142: 1-28.

[50] Kamido H, Kuksis A, Marai L, Myher JJ. Identification of cholesterol-bound aldehydes in copper-oxidized low density lipoprotein. *FEBS Lett* 1992; 304: 269-72.

[51] Kamido H, Kuksis A, Marai L, Myher JJ. Lipid ester-bound aldehydes among copper-*catalyzed* peroxidation products of human plasma lipoproteins. *J Lipid Res* 1995; 36: 1876-86.

[52] Hoppe G, Ravandi A, Herrera D, Kuksis A, Hoff HF. Oxidation products of cholesteryl linoleate are resistant to hydrolysis in macrophages, form complexes with proteins, and are present in human atherosclerotic lesions. *J Lipid Res* 1997; 38: 1347-60.

[53] Kleinveld HA, Hak-Lemmers HLM, Stalenhoef AFH, Demacker PNM. Improved measurement of low-density-lipoprotein susceptibility to copper-induced oxidation: application of a short procedure for isolating low-density lipoprotein. *Clin Chem* 1992; 38, 2066-72.

[54] Parthasarathy S, Fong LG, Quinn MT, Steinberg D. Oxidative modification of LDL: comparison between cell-mediated and copper-mediated modification. *Eur Heart J* 1990; 11 (Suppl.) E: 83-7.

[55] Yla-Herttuala S, Palinski W, Rosenfeld ME, Parthasarathy S, Carew TE, Butler S, Witztum JL, Steinberg D. Evidence for presence of oxidatively modified low density lipoprotein in atherosclerotic lesions of rabbit and man. *J Clin Invest* 1989; 85: 1086-95.

[56] Ehrenwald E, Chisolm GM, Fox PL. Intact ceruloplasmin oxidatively modifies low density lipoprotein. *J Clin Invest* 1994; 93: 1493-1501.

[57] Lamb D, Leake DS. Acidic pH enables caeruloplasmin to catalyse the modification of low-density lipoprotein. *FEBS Lett* 1994; 338: 122-6.

[58] Bouma B, de Groot PG, van den Elsen JM, Ravelli RB, Schouten A, Simmelink MJ, Derksen RH, Kroon J, Gros P. Adhesion mechanism of human β2-glycoprotein I to phospholipids based on its crystal structure. *EMBO J* 1999; 18: 5166-74.

[59] Hoshino M, Hagihara Y, Nishii I, Yamazaki T, Kato H, Goto Y. Identification of the phospholipid-binding site of human β2-glycoprotein I domain V by heteronuclear magnetic resonance. *J Mol Biol* 2000; 304: 927-39.

[60] Inanc M, Radway-Bright EL, Isenberg DA, β2-glycoprotein I and anti-β2-glycoprotein I antibodies: where are we now? *Br J Rheumatol* 1997; 36: 1247-57.

[61] Sheng Y, Kandiah DA, Krilis SA. β2-glycoprotein I: Target antigen for 'antiphospholipid' antibodies. Immunological and molecular aspects. *Lupus* 1998; 7: S5-9.

[62] Merrill JT, Zhang HW, Shen C, Butman BT, Jeffries EP, Lahita RG, Myones BL. Enhancement of Protein S anticoagulant function by β2-glycoprotein I, a major target antigen of antiphospholipid antibodies: β2-glycoprotein I interferes with binding of Protein S to its plasma inhibitor, C4b-binding protein. *Thromb Haemost* 1999; 81: 748-757.

[63] Hashimoto Y, Kawamura M, Ichikawa K, Suzuki T, Sumida T, Yoshida S, Matsuura E, Ikehara S, Koike T. Anticardiolipin antibodies in NZW x BXSB F1 mice. A model of antiphospholipid syndrome. *J Immunol* 1992;149:1063-8.

[64] Lopez D, Kobayashi K, Merrill JT, Matsuura E, Lopez LR. IgG Autoantibodies against β2-glycoprotein I complexed with a lipid ligand derived from oxidized low-density lipoprotein are associated with arterial thrombosis in antiphospholipid syndrome. Clin Dev Immunol (in press).

[65] De Stefano V, Finazzi G, Mannucci PM. Inherited thrombophilia: pathogenesis, clinical syndromes, and management. *Blood* 1996; 87: 3531-44.

[66] Bertina RM. Molecular risk factors for thrombosis. *Thromb Haemost* 1999; 82:601-9.

[67] Lane DA, Mannucci PM, Bauer KA, Bertina RM, Bochkov NP, Boulyjenkov V. Chandy M, Dahlback B, Ginter EK, Miletich JP, Rosendaal FR, Seligsohn U. Inherited thrombophilia: part 1. *Thromb Haemost* 1996; 76: 651-62.

[68] Francis JL. Laboratory investigation of hypercoagulability. *Semin Thromb Hemostas* 1998; 24: 111-26.

[69] Petri M. In: The antiphospholipid syndrome. II: Autoimmune thrombosis. Asherson RA, Cervera R, Piette J, Schoenfeld Y editors. 2nd edition. Amsterdam: Elsevier Science; 2002. p11-20.

[70] Thomas RH. Hypercoagulability syndromes. *Arch Intern Med* 2001; 161: 2433-39.

[71] Gharavi AE, Harris EN, Asherson RA, Hughes GRV. Anticardiolipin antibodies–isotype distribution and phospholipid specificity. *Ann Rheum Dis* 1987; 46: 1-6.

[72] Wahl DG, Guillemin F, de Maistre E, Perret C, Lecompte T, Thibaut G. Risk for venous thrombosis related to antiphospholipid antibodies in systemic lupus erythematosus: a meta analysis. *Lupus* 1997; 6: 467-73.

[73] Shah NM, Khamashta MA, Atsumi T, Hughes GR. Outcome of patients with anticardiolipin antibodies: a 10 year follow-up of 52 patients. *Lupus* 1998; 7: 3-6.

[74] Khamashta MA, Cuadrado MJ, Mujic F, Taub NA, Hunt BJ, Hughes GRV. The management of thrombosis in the antiphospholipid-antibody syndrome. *N Engl J Med* 1995; 332: 993-7.

[75] Roubey RA. Autoantibodies to phospholipid-binding plasma proteins: a new view of lupus anticoagulants and other "antiphospholipid" autoantibodies. *Blood* 1994; 84: 2854-67.

[76] Tsutsumi A, Matsuura E, Ichikawa K, Fujisaku A, Mukai M, Kobayashi S, Koike T. Antibodies to β2-glycoprotein I and clinical manifestations in patients with systemic lupus erythematosus. *Arthritis Rheum* 1996; 39: 1466-74.

[77] Vaarala O. Antiphospholipid antibodies in myocardial infarction. *Lupus* 1998; 7: S132-4.

[78] Brey RL, Abbott RD, Curb JD, Sharp DS, Ross GW, Stallworth CL, Kittner SJ. β2-glycoprotein I dependent anticardiolipin antibodies and the risk of ischemic stroke and myocardial infarction. *Stroke* 2001; 32: 1701-6.

[79] Van Doornum S, McColl G, Wicks IP. Accelerated atherosclerosis. An extraarticular feature of Rheumatoid Arthritis? *Arthritis Rheum* 2002; 46: 862-73.

[80] Ward MM. Premature morbidity from cardiovascular and cerebrovascular diseases in women with systemic lupus erythematosus. *Arthritis Rheum* 1999; 42: 338-46.

[81] Aranow C, Ginzler EM. Epidemiology of cardiovascular disease in systemic lupus erythematosus. *Lupus* 2000; 9: 166-9.

[82] Esdaile JM, Abrahamowicz M, Grodzicky T, Li Y, Panaritis C, du Berger R, Cote R, Grover SA, Fortin PR, Clarke AE, Senecal JL. Traditional Framingham risk factors fail to fully account for accelerated atherosclerosis in systemic lupus erythematosus. *Arthritis Rheum* 2001; 44: 2331-7.

[83] Schattner A, Liang MH. The cardiovascular burden of lupus; a complex challenge. *Arch Intern Med* 2003; 163: 1507-10.

[84] Lockshin MD, Salmon JE, Roman MJ. Atherosclerosis and Lupus: A work in progress. *Arthritis Rheum* 2001; 44: 2215-7.

[85] Harris EN, Chan JKH, Asherson RA, Hughes GRV. Thrombosis, recurrent fetal loss and thrombocytopenia–predictive value of the anticardiolipin antibody test. *Arch Intern Med* 1986; 146: 2153-6.

[86] Ginsburg KS, Liang MH, Newcomer L, Goldhaber SZ, Schur PH, Hennekens CH, Stampfer MJ. Anticardiolipin antibodies and the risk for ischemic stroke and venous thrombosis. *Ann Intern Med* 1992; 117: 997-1002.

[87] Bick RL, Baker WF. Antiphospholipid syndrome and thrombosis. *Semin Thromb Hemost* 1999; 25: 333-50.

[88] Cervera R, Piette JC, Font J, Khamashta MA, Shoenfeld Y, Camps MT, Jacobsen S, Lakos G, Tincani A, Kontopoulou-Griva I, Galeazzi M, Meroni PL, Derksen RH, de Groot PG, Gromnica-Ihle E, Baleva M, Mosca M, Bombardieri S, Houssiau F, Gris JC, Quere I, Hachulla E, Vasconcelos C, Roch B, Fernandez-Nebro A, Boffa MC, Hughes GR, Ingelmo M; Euro-Phospholipid Project Group. Antiphospholipid syndrome. Clinical and immunologic manifestations and patterns of disease expression in a cohort of 1,000 patients. *Arthritis Rheum* 2002; 46: 1019-27.

[89] Ross R. Atherosclerosis: an inflammatory disease. *N Engl J Med* 1999; 340: 115-126.

[90] Steinberg D. Atherogenesis in perspective: Hypercholesterolemia and inflammation as partners in crime. *Nature Med* 2002; 8: 1211-7.

[91] Virella G, Atchley DH, Koskinen S, Zheng D, Lopes-Virella M. Pro-atherogenic and pro-inflammatory properties of immune complexes prepared with purified human oxLDL antibodies and human oxLDL. *Clin Immunol* 2002; 105: 81-92.

[92] Berliner JA, Heinecke JW. The role of oxidized lipoproteins in atherogenesis. *Free Radic Biol Med* 1996; 20: 707-27.

[93] McMurray HF, Parthasarathy S, Steinberg D. Oxidatively modified low density lipoprotein is a chemoattractant for human T lymphocytes. *J Clin Invest* 1993; 92: 1004-8.

[94] Salonen JT, Yla-Herttuala S, Yamamoto R, Butler S, Korpela H, Salonen R, Nyyssonen K, Palinski W, Witztum JL. Autoantibodies against oxidized LDL and progression of carotid atherosclerosis. *Lancet* 1992; 339: 883-7.

[95] Vaarala O. Antiphospholipid antibodies and atherosclerosis. *Lupus* 1996; 5: 442-7.

[96] Durrington PN, Mackness B, Mackness MI. Paraoxonase and atherosclerosis. *Arterioscler Thromb Vasc Biol* 2001; 21: 473-80.

[97] Lambert M, Boullier A, Hachulla E, Fruchart JC, Teissier E, Hatron PY, Duriez P. Paraoxonase activity is dramatically decreased in patients positive for anticardiolipin antibodies. Lupus 2000; 9: 299-300.

[98] Delgado Alves J, Ames PR, Donohue S, Stanyer L, Nourooz-Zadeh J, Ravirajan C, Isenberg DA, Noorouz-Zadeh J. Antibodies to high-density lipoprotein and β2-glycoprotein I are inversely correlated with Paraoxonase activity in Systemic lupus erythematosus and primary antiphospholipid syndrome. *Arthritis Rheum* 2002; 46: 2686-94.

[99] Senti M, Tomas M, Vila J, Marrugat J, Elosua R, Sala J, Masia R. Relationship of age related myocardial infarction risk and Gln/Arg 192 variants of the human paraoxonase 1 gene. The REGICOR study. *Atherosclerosis* 2001; 156: 443-9.

[100] Belogh Z, Seres I, Harangi M, Kovacs P, Kakuk G, Paragh G. Gemfibrozil increases paraoxonase activity in type 2 diabetic patients: a new hypothesis of the beneficial action of fibrates? *Diabetes Metab* 2001; 27: 604-10.

[101] Meroni PL, Raschi E, Testoni C, Tincani A, Balestrieri G, Molteni R, Khamashta MA, Tremoli E, Camera M. Statins prevent endothelial cell activation induced by antiphospholipid (anti-β□-glycoprotein I) antibodies. Effect on the proadhesive and proinflammatory phenotype. *Arthritis Rheum* 2001; 44: 2870-8.

[102] Ferro D, Iuliano L, Violi F, Valesini G, Conti F. Antioxidant treatment decreases the titer of circulating anticardiolipin antibodies. *Arthritis Rheum* 2002; 46: 3110-2.

In: Focus on Atherosclerosis Research
Editor: Leon V.Clark, pp. 115-131
ISBN 1-59454-044-6

Chapter VI

Impact of Lipoprotein Metabolism on Extreme Longevity

Francesco Panza*[*], *Anna M. Colacicco, Alessia D'Introno, Cristiano Capurso, Sabrina Capurso, Antonio Capurso and Vincenzo Solfrizzi
Department of Geriatrics, Center for Lipoprotein Metabolism
University of Bari, Italy

Abstract

The relationships among lipoprotein metabolism, genetic vascular factors, vascular disease and Alzheimer's disease (AD) suggest that the examination of centenarian populations in relation to certain genes or lipoprotein metabolism provide insights into the human longevity. The findings on the higher frequency of the apolipoprotein-E (APOE) ε4 allele in middle-aged subjects than in centenarians were substantially confirmed. On the contrary, our findings did not confirm previous data on increased prevalence of the high-risk angiotensin I converting enzyme 1 (ACE1) D allele in French centenarians. We hypothesized that the variability in the strength of association between ACE1 polymorphism and longevity could be related to regional differences in ACE1*D frequency in Europe showed in our study, as recently reported for apoE ε2 and ε4 allele in centenarians. Indeed some studies of cholesterol profiles in centenarians have also had contradictory outcomes. Notable amongst these is evidence that lower serum levels of high-density lipoprotein cholesterol (HDL-C), with higher HDL2-C subfraction, and higher lipoprotein(a) [Lp(a)] concentration in centenarians which is an apparently disadvantageous lipoprotein profile. Indeed, lower HDL-C concentration was linked with decreased serum albumin, elevated C-reactive protein, and interleukin-6 levels, suggesting that HDL-C could be a sensitive marker for frailty and comorbidity in centenarians. Our own findings suggest that elevated Lp(a) serum levels, increasing the risk for cerebrovascular disease, may play a role in determining clinical AD but that

* Address correspondence to: Francesco Panza, M.D., Department of Geriatrics, Center for Lipoprotein Metabolism, University of Bari - Policlinico, Piazza Giulio Cesare, 11 - 70124 Bari – Italy, Phone: + 39-80-5473685; FAX : + 39-80-5478860; E-mail: geriat.dot@geriatria.uniba.it

Lp(a) elevation in centenarians, in the absence of other coronary artery disease risk factors, appears as a positive survival factor. We have also shown how in different populations, there are significant trends in the reduction of serum APOE levels from APOE $\varepsilon 2$- to $\varepsilon 4$-carriers and significant differences in serum APOE levels with respect to age in $\varepsilon 4$-carriers but only after adjustment for HDL-C. While further studies are needed to confirm the possible role of APOE concentration as putative longevity factor this paper provides an overview of many of the investigated vascular factors that have investigated with respect to longevity.

Introduction

The search for factors involved in aging and longevity has gone forward extensively in the last years because of prolonged human life expectancy and increase in the number of elderly people. Different genetic and non genetic factors have been examined in the quest to understand the biological basis of human longevity. In fact, it can hypothesize that centenarians have environmental and genetic factors that confer unexpected survival advantage. Since centenarians are characterized by marked delay or escape from age-related diseases, such as coronary artery disease (CAD), cerebrovascular disease (CVD), and Alzheimer's disease (AD), the first, the third, and the fourth cause of death, respectively, in the western population, it can suppose that genes and biochemical factors likely to be implied in these disorders may have a role in human longevity.

Among the genetic markers, apolipoprotein E (APOE) gene has been widely examined in centenarian populations in different studies [1-3], due to its well documented role in AD [4] as well as vascular diseases [5]. In fact, the APOE $\varepsilon 4$ allele is associated with high serum total cholesterol (TC), low-density lipoprotein cholesterol (LDL-C), and apolipoprotein B (*APOB*) levels in many populations [6], and has been found to increase risk for CAD, myocardial infarction [5], and AD [4]. Another gene, the angiotensin I converting enzyme (ACE1), note to increase the risk for CAD [7], and also linked in some studies to late onset AD [8], has been associated with longevity [1]. However, the findings have not been replicated in independent populations [9-11]. Other few genes linked to lipoprotein metabolism and vascular disease have been investigated as putative markers for longevity, but, generally, with negative or inconsistent association results.

Evidence from epidemiological and biochemical studies linking lipid and lipoprotein factors, such as TC, LDL-C, high-density lipoprotein cholesterol (HDL-C), *APOB*, lipoprotein(a) (Lp(a)) [11, 12], and serum *APOE* [13, 14], to CAD and CVD suggest a role of these factors in human longevity. Further, proteins involved in lipid metabolism have also been associated, although with conflicting results, with AD. However, it cannot state that these genetic and non genetic factors are actually related to successful aging and longevity. Some more studies in different centenarian populations are needed to unravel clues in human longevity that appears as a complex network of various factors.

In this chapter, we will discuss the evidence that genetic factors linked to lipoprotein metabolism, lipids, and lipoproteins, likely to be associated to both vascular disease and AD, may have a role in determining human longevity.

Apolipoprotein E (*APOE*) Gene Polymorphism and Longevity

Apolipoprotein E *(APOE)* has three common isoforms, namely E2, E3, and E4, which are coded by the alleles ε2, ε3, and ε4 at a single locus on chromosome 19. The APOE gene has been extensively examined in populations worldwide for its association with an increased risk for CAD and AD [4, 16], and has also been evaluated in centenarian populations in different studies (Table 1). Particularly, in a recent study we showed that the frequency of the APOEε4 allele was significantly higher in middle-aged subjects than in centenarians, while the ε2 allele did tend to increase in centenarians, though not significantly [3]. Our study confirmed previous results showing a decrease of the APOEε4 allele frequency in elderly people (octogenarians and nonagenarians) compared with younger age groups [17, 18]. In contrast, in another Italian study, there was no difference in the ε4 allele frequency among healthy and disabled octo- and nonagenarians, and healthy adult controls [19]. In a study carried out on 1562 Han Chinese subjects (20-108 years), the APOEε4 allele frequency in the very old age group ($\geq$ 85 years) was significantly different from that in the youth (20-39 years), middle (40-59 years), and old (60-84 years) age group [20]. In nonagenarian subjects from Northern Ireland [21], where there is a high intrinsic incidence of cardiovascular disease, the ε4 allele frequency was reduced, the ε3 unchanged and the ε2 increased. Similarly, in French centenarians, compared to a control group of healthy adults, the ε4 allele frequency was significantly decreased while the ε2 allele frequency was significantly increased [1].

In a Finland study, the APOEε2 allele frequency was significantly higher (7.0%) and the ε4 allele frequency lower (8.4%) in centenarians than in younger populations [2]. The differences in the frequencies of the APOEε2 (4.5% vs 3.5%) and ε4 (4.5% vs 11.82%) alleles between Japanese centenarians and younger adults did not attain statistical significance [22-24]. In our study we also reported a significant increasing geographic trend for ε2 allele and a decreasing trend for ε4 allele in centenarian from Northern to Southern regions of Europe group [3]. We suggest that this geographical trend could affect the strenght of association between ε2 allele and longevity, as recently confirmed in Finnish centenarians with a significant trend for ε2 allele in persons aged 100 to 101 (9%), 102 to 103 (21%), and 104 years and older (25%) [25]. Nevertheless, in the same population, i.e. 224 Finnish centenarians identified from the Finnish National Population Register in 1991, previous findings showed no significant differences at 3-year survival among different APOE allele carriers, suggesting that in this very select group the protection/risk effect associated with the APOE polymorphism might be attenuated [26]. In another recent study conducted on 177 Danish centenarians and using data from a study of 40-year-old Danish men, 21% of the centenarians were APOE ε2-carriers and 15% were ε4-carriers compared to 13 and 29%, respectively, of the young men. The relative mortality risk-values were 0.95 (95% CI 0.88 to 1.02) for ε2-carriers and 1.13 (95% CI 1.05 to 1.22) for ε4-carriers, using ε3/ε3 and ε4/ε2 genotypes as reference, thus suggesting the view that the APOE gene is a "frailty gene" and not a "longevity gene" [27].

Table 1. Principal Studies on Apolipoprotein E (APOE) Polymorphism in Centenarians

Reference	Subjects	APOε4 allele Frequency (95% CI)	APOε2 allele Frequency (95% CI)	Results and conclusions
Shachter et al. 1994 [1]	325 French centenarians and 161 French middle aged subjects (20-70 years)	Centenarians: 0.05 (0.07-0.04) Middle aged subjects: 0.11 (0.15-0.08)	Centenarians: 0.13 (0.16-0.10) Middle aged subjects: 0.07 (0.10-0.04)	In French centenarians, compared to a control group of healthy adults, the ε4 allele frequency was significantly decreased (5.2% vs 11.2%) while the ε2 allele frequency was significantly increased (12.8% vs 6.8%)
Louhija et al. 1994 [2]	179 Finnish centenarians, 615 adult Finns (20-55 years), and 1577 young Finns (3-18)	Centenarians: 0.09 (0.12-0.06) Adult Finns: 0.23 (0.20-0.25) Young Finns: 0.19 (0.18-0.21)	Centenarians: 0.07 (0.10-0.05) Adult Finns: 0.04 (0.03-0.05) Young Finns: 0.04 (0.03-0.46)	In Finland, the apo ε2 allele was significantly higher (7.0%) and the ε4 allele lower (8.4%) in centenarians than in younger populations
Asada et al. 1996 [22]	33 Japanese centenarians (mean age: 102 years) and 880 younger historical Japanese controls (aged 16-78 years)	Centenarians: 0.05 Younger subjects: 0.04	Centenarians: 0.05 Younger subjects: 0.11	The differences in the frequencies of the ε2 (4.5% vs 3.5%) and ε4 (4.5% vs 11.82%) alleles between Japanese centenarians and younger adults did not attain statistical significance
Panza et al. 1999 [3]	67 centenarians from Apulia, Southern Italy and125 unrelated caregivers or volunteers (19-80 years)	Centenarians: 0.02 (0.07-0.01) Middle aged subjects: 0.08 (0.12-0.05)	Centenarians: 0.12 (0.19-0.07) Middle aged subjects: 0.08 (0.12-0.05)	In Italian centenarians, the frequency of the ε4 allele was significantly higher in middle-aged subjects than in centenarians, while ε2 allele frequency does tend to increase in centenarians, though not significantly
Choi et al. 2003 [28]	103 Korean centenarians (mean age 102.4 ± 2.6 years) and 6435 adults of mean age 50.7 ± 7.9 years	Centenarians: 0.07 Middle aged subjects: 0.09 (0.12-0.05)	Centenarians: 0.06 Middle aged subjects: 0.06	The frequencies of genotypes and alleles of the APOE gene of the centenarians were not significantly different from those of the control groups. However, the frequency of the ε4 allele of the APOE gene was significantly higher in centenarians with dementia than in centenarians without definitive dementia (9.1% vs 1.5%)

CI: confidence interval.

Finally, a very recent cross-sectional study has been investigated the distributions of the APOE and angiotensin I converting enzyme 1 (ACE1) genotypes and their relations with dementia status in 103 Korean centenarians and two control groups: 7232 apparently healthy adults of mean age 48.5 (9.6 SD) years for the ACE1 genotyping, and 6435 adults of mean age 50.7 (7.9 SD) years for the APOE genotyping. In this study, it was reported that the frequencies of the ACE1 and APOE genotypes and alleles were not significantly different in centenarians compared to the control groups, suggesting that neither gene is significantly associated with longevity in the Korean population. The frequencies of APOE genotypes and alleles in centenarians were not significantly different from those of the control groups, but the frequency of the APOE ε4 allele was significantly higher in centenarians with dementia than in centenarians without definitive dementia (9.1% vs 1.5%) [28].

Other Apolipoprotein Genes Related to Longevity

Different vascular susceptibility genes which have been independently associated with both vascular disease or AD have been investigated for playing a role in longevity.

Besides APOE, genetic polymorphisms in other apolipoproteins have been shown to modulate lipid metabolism and risk for CAD [29, 30], and therefore examined for its potential role in human longevity. In studies carried out in Italian centenarians, the 3'APOB-VNTR polymorphism, a marker located less than 100 bp downstream of the second transcription signal of the APOB gene, has been associated with longevity [31, 32]. *APOB* is the main protein in LDL-C and plays a major role in cholesterol homeostasis [33], and in both northern and southern Italian populations the frequency of the 3'APOB-VNTR alleles with less than 35 repeats (Small alleles) was significantly lower in centenarians than in 20-60-year-old subjects [30, 31]. In Danish population, neither allele nor genotype frequencies differed between centenarians and 20-64-year-old subjects, but a significant gene-sex interaction relative to alleles having more than 37 repeats (Long alleles) was found. Furthermore, a significant difference was observed in the frequencies of Small alleles in younger controls between Denmark and Italy, which disappeared in centenarians [34]. These findings confirmed the regional differences from Northern to Southern Europe in APOE and ACE1 allele and genotype frequencies reported in our studies in centenarians and younger controls [3, 11].

Genes encoding for apolipoprotein C-I (APOC-I), apolipoprotein A1 (APOA-I), apolipoprotein C3 (APOCIII), and apolipoprotein A4 (APOA-IV) have been also investigated. In a British study conducted on subjects aged 84 years and older, the authors analyzed APOE and APOC-I genes as these two genes are less than 10 kb apart, and APOC-I genotype and allele frequencies can be used to confirm associations at the APOE locus. This study reported a significant difference in APOC-I allele and genotype frequencies in elderly women compared to the younger sample, while no difference was observed in the elderly men. Differences in APOE genotype distribution were also reported between the elderly women and the young samples, but the APOE ε2 allele was not over-represented in the elderly women or men [35]. The *APOA-I*, *APOCIII* and *APOA-IV* genes are tandemly organized in a cluster on the long arm of the human chromosome 11 and have been extensively investigated in lipoprotein disorders and vascular disease [36]. An analysis of distribution of APOA1-MspI (-75 nt from the transcription starting site) A/P allele (Absence/Presence of the restriction site) frequency in three age classes (18-45 years, 46-80 years, 81-109 years) revealed that the frequency of P allele tends to increase in the oldest old males (81-109 years), but not in the female group [36]. Thus APOA-1 P allele seems to be linked to longevity in a sex specific way. Because the P allele is associated with higher serum LDL-C at middle-age and is over-represented in patients affected by cardiovascular disease with high LDL-C serum levels, these results give evidence of an intriguing genetic paradox in centenarians. These apparently contradictory findings confirm that genetic risk factors acting at middle-age can be present or also over-represented in centenarians [1, 37], or alleles which were protective at a young age, and therefore frequent in this age-group according to its genetic and environmental features, may change and be disadvantageous in older people [37]. The analysis of APOCIII-SstI (3'UTR 3238 nt) and APOA-IV-HincII (Asp_{127}/Ser_{127}) polymorphisms in the same three age classes showed no age-related variation either in males or females. Conversely, determination of APOCIII-SstI polymorphism in almost all Finnish centenarians alive in 1991 showed that APOC-III S2 allele (SstI restriction site present) occurred more often in the centenarians than in the youngest controls (frequency, 12.9% and 8.8%, respectively) [2]. Another polymorphism of the APOC-III gene located in the 5'UTR (T-455C) within a functional insulin element has been examined in Russian population and found to be associated with longevity. The greater frequency of the APOC-III –4555C allele was observed in subjects older than 80 years [38]. Analysis of common APOA-IV genetic variant at codon 360 (Gln-->His) revealed an over-representation of the 360/His allele in the aged healthy subjects compared with the younger group, suggesting a role for this polymorphism in successful aging and longevity [39]. However, these studies are unique and a role of the APOAI-CIII-AIV gene cluster in human longevity may be just suggested. Replication of these data is demanded, although it can be difficult because of the varying gene-gene and gene-environment interactions in different populations.

The Role of Total Cholesterol in Human Longevity

The strict relation between genetic apolipoprotein polymorphisms and serum levels of apolipoproteins, lipoproteins, and lipids has drawn attention on the role of lipid risk factors, such as TC, LDL-C, HDL-C, Lp(a), and serum *APOE,* in older people. Particularly, it has been suggested that cholesterol metabolism may differ between younger and older subjects [40], as showed by prospective studies where high serum TC levels in the oldest old were associated with increased survival [41]. Often a J or U-shaped association has been reported [42, 43], suggesting that subjects with very low or very high concentrations of cholesterol have an increased risk of death than subjects with moderate concentrations. Several studies have reported a positive association between low serum TC and the risk of dying in Japanese and Italian older people [40, 44]. Some studies have demonstrated a negative association between TC and mortality from cancer [45-47], while one study has shown that, in men and women aged 85 and older, high cholesterol is associated with lower mortality from cancer and infection [41]. One possible explanation for these contrasting findings could be selective mortality, that means that many individuals with high serum levels of TC were not included in the sample because they have already died. Subjects with higher TC might experience an earlier onset of severe CAD, leading to disproportionately greater mortality before study enrolment.

Recent studies have investigated the role of vascular risk factors in incident Mild Cognitive Impairment (MCI) [48, 49], the most widely used term to classify nondemented aged persons with a mild memory- or cognitive impairment that cannot be accounted for by any recognized medical or psychiatric condition [50]. MCI may be a prodromal phase of dementia and therefore of increased mortality from dementia, with estimates of 12% of MCI patients developing dementia in 1 year, 20% over 3 years, and rates of progression to dementia of 19.9% at 5 years in subjects classified as affected by uncertain dementia with the Clinical Dementia Rating scale [51]. In our recent study, age, CAD, and hypertension were risk factors for MCI, while higher levels of education, serum TC, and pack-years cigarettes appeared to have a protective effect [52]. On the contrary, Kivipelto et al. found that midlife elevated TC serum levels ($\geq$ 6,5 mmol/L) increased the risk for MCI (OR, 1.9), with systolic blood pressure showing a similar trend [48]. In addition, De Carli et al. showed that elevated midlife blood pressure increased the risk for MCI [48]. The same authors found that midlife diastolic blood pressure increased the risk for MCI only when a history of symptomatic cerebrovascular disease was excluded [49]. The discrepancy of findings relative to the association of TC serum levels with MCI can be explained by the fact that TC is a midlife determination (with average follow-up of 21 and 25 years). Furthermore, in a recent study we found lower TC serum levels in AD [53], confirming the data of cross-sectional and prospective studies [54, 55].

Lipoprotein Metabolism and Longevity

The Framingham and National Health and Nutrition Examination Survey (NHANES) III cross-sectional studies found that plasma HDL-C levels were comparable at different age-groups both in males and females [56, 57]. However, looking prospectively, plasma HDL-C levels seemed to decrease by 5 mg/dL per decade. These seemingly contradictory results may be explained by an HDL-C survival effect: a decrease in HDL-C levels below a certain range may result in the loss of cardiovascular protection and hence in increased mortality. Several studies examined the influence of HDL-C on longevity, with contrasting results [58-66]. In a study conducted on 42 octogenarians neither the total level of HDL particles nor the distribution of lipoprotein components among the subfractions accounted for the longevity of the majority of the study population [58]. On the contrary, HDL-C and the ratio of HDL/TC were significantly higher in healthy subjects aged 85-89 years than in patients without CAD aged 38-62 years and with triple-vessel disease aged 32-69 years [59]. Furthermore, during a 5-year follow-up in 535 people aged 85 years mortality was lowest at serum cholesterol level of 5.0-5.9 mmol/l for men and 7.0-7.9 mmol/l for women. The greatest mortality was observed with TC greater than or equal to 6.0 mmol/l and 8.0 mmol/l respectively in men and in women. Highest mortality was found (53.3%) in men with serum HDL-C less than 0.80 mmol/l, and lowest (16.5%) in women with serum HDL-C greater than 1.8 mmol/l [60]. In another population-based study, subjects of 80 years and older who were free of clinical CAD were characterized by differences in lipid and anthropometric values, particularly higher HDL-C and *APOA-I* levels [61]. HDL-C subfractions were measured in 1127 females and 825 males 65 years old and over who participated in the Cardiovascular Health Study. The slight increase in HDL-C with age appeared to be determined by an increase in the HDL2-C subfraction. HDL-C subfractions in older people were independently related to triglyceride levels, body weight, and insulin concentrations, all potentially modifiable risk factors. Both HDL2-C and HDL3-C were lower in older people with prevalent cardiovascular disease, although only HDL3-C was correlated with carotid atherosclerosis [62]. These findings are consistent with those of an Italian study in which centenarians had lower levels of HDL3a-C and higher levels of HDL2b-C, which were found to be inversely correlated with CAD risk, suggesting a role of HDL subfractions in extreme longevity [63]. Precedent data on the same sample demonstrated that, in spite of an independent and inverse association of total HDL-C levels with CAD in older women, no relation exists between any lipoprotein parameter and longevity in Italian centenarians [64]. In a collaborative study with our group, Paolisso et al. found that healthy Italian centenarians have LDL-C significantly lower and HDL-C with an opposite trend compared with aged subjects, suggesting a less atherogenic plasma lipid and lipoprotein profile in centenarians than aged subjects [65]. In a recent study, low concentrations of TC and LDL-C, and a relative predominance of HDL2-C were found in Japanese centenarians. In this study, only the number of APOE ε2 alleles was a significant determinant of LDL-C and *APOB*, suggesting that low levels of *APOB*-containing lipoproteins in centenarians may be attributable to a genetic cause. Lower levels of HDL-C in centenarians were associated with decreased serum albumin, elevated C-reactive protein (CRP) and interleukin 6 levels, and cognitive impairment, suggesting that HDL-C may be a marker for frailty and comorbidity in very advanced age [66]. Finally, Barzilai et al. studied plasma levels of lipoproteins in offspring of centenarians and compared them with offsprings' spouses and age- and gender-matched controls from the NHANES III Study. The authors demonstrated that offsprings of centenarians have a favourable lipid profile compared with matched controls, with higher HDL-C and lower LDL-C plasma levels in males [67].

Several studies have suggested a role of Lp(a) in extreme longevity [64, 66, 68-73]. Lp(a) is a unique lipoprotein particle with atherogenic and thrombotic properties in which the protein component consists of apolipoprotein B100 bound by a disulfide bond to apolipoprotein (a) (*APO(a)*) [74]. Lp(a) concentration in human plasma is genetically determined, with at least 90% of the variation determined by variations within the APO(a) gene [75]. High plasma concentrations of Lp(a) have been associated with increased risk for CAD in numerous retrospective case-control and prospective studies, with few notable exceptions [76]. This association was apparently not confirmed among the elderly [77-79]. The results of a recent study by our group carried out in 400 subjects enrolled in the prevalence survey (1992-1993) of the Italian Longitudinal Study on Aging (ILSA) [80] suggest that elevated serum concentration of Lp(a) may be associated with CAD, but only in interaction with high serum LDL-C and type 2 diabetes [73]. In older subjects Lp(a) seems to play a role as a dependent risk factor, according to recent reports that did not have a focus on the elderly. Actually, several studies on elderly subjects showed that high Lp(a) concentrations are stable with advancing age and this lipoprotein does not appear to be a risk indicator for CAD [77-79, 81]. These findings led to the hypothesis that Lp(a) is a CAD risk factor dependent on TC or LDL-C, suggesting a direct atherogenic action of Lp(a) for its high affinity binding to glycosaminoglycans and fibronectin [82]. Furthermore, increased plasma concentrations of Lp(a) have been associated with CVD and VaD [83, 84], and the recent vascular hypothesis of AD linking clinical expression of AD with cerebral infarct has suggested that Lp(a) could be a risk factor for AD, as well [85]. Mooser et al. showed that Lp(a) is an additional risk factor for late onset AD in APOE ε4-carriers, and a protective factor against the disease in ε4-non carriers older than 80 years [86]. We also found that Lp(a) serum levels were significantly associated with an increased risk for AD according to a nonlinear relationship, independently of APOE genotypes, and dependent on age [53]. Moreover, in another study we found that elevated Lp(a) serum levels were significantly associated with an increased risk for age-related cognitive decline, dependently of high serum levels of *APO*A-I [87], suggesting that elevated serum levels of Lp(a), while increasing the risk for subclinical atherosclerosis and silent CVD, may increase the risk for decline in cognitive functioning in the elderly.

It is generally admitted that Lp(a) serum levels are not age-dependent, but several papers have reported the effect of age, sex, and menopausal status on plasma Lp(a) levels [88-90]. A recent study showed that the mean plasma Lp(a) concentration and the frequency of Lp(a) $\geq$ 250 mg/L increased with age in all *APO(a)* phenotypes (low, intermediate, and high molecular-weight) [72]. In another study conducted in Japan, a significantly lower frequencies of low molecular-weight *APO(a)* phenotypes and alleles were found in men > 80 years old compared with younger subjects [85]. Two independent studies by our and French group, respectively, have clearly shown that Lp(a) levels were significantly elevated in healthy centenarians [68, 69]. This increase may be only partly explained by a distribution of *APO(a)* isoforms which are displaced towards isoforms of low molecular mass [69]. Data of our study lead to the hypothesis of a longevity-associated quantitative modulation of Lp(a) by APOA-IV gene polymorphism [68]. Furthermore, high plasma concentrations of the proinflammatory interleukin-6 in centenarians suggest that genetic control of the Lp(a) serum levels may be attenuated with increasing age and that environmental factors, such as chronic subclinical inflammatory processes, may play a role [70]. In Japanese centenarians, a significantly elevated concentration of Lp(a) was found [66], consistently with the results in Italian [68, 70] and French centenarians [69]. Arai et al. found positive correlation between the log-transformed Lp(a) and the log-transformed CRP [66], confirming the hypothesis that inflammatory activation may contribute to elevated Lp(a) concentrations in centenarians. Anyway, other hypotheses on the causes of increase in Lp(a) concentration in centenarians are worthwhile. Finally, the findings of all studies above examined suggest that an atherogenic marker as Lp(a), in the absence of other risk factors for CAD, appears as a positive survival factor.

Serum *APOE* levels in centenarian population have also been investigated in a very recent study by our group. Firstly, we demonstrated a significant trend in reduction of serum *APOE* levels from APOE ε2- to ε4-carriers in all study populations (healthy centenarians, AD patients, age-matched controls, and young healthy adults). Further, significant differences in serum *APOE* levels respect to age were found in APOE ε4-carriers between young healthy adults and age-matched controls, and between young healthy adults and centenarians, but only after adjustment for serum HDL-C levels [91, 92]. In fact, in normolipidemic subjects, the majority of *APOE* (>60%) in plasma is associated with HDL [93]. Recently, Reilly et al. determined serum *APOE* concentrations in 4284 subjects from 4 to 71 years, with females exhibiting higher *APOE* values than males until the age of 17-26 years. Conversely, after the age of 26 years, serum *APOE* concentrations were higher in men than in women [94]. The decrease of serum *APOE* levels with age in children, with a synergistic effect of age and puberty, and a lack of age variation in adults were in agreement with other findings [95]. The role of serum *APOE* concentration in extreme longevity may be explained by the relevance of this factor in CVD [96]. Furthermore, nonetheless the uncertain relation to AD, serum *APOE* concentration has been shown to modulate lipid metabolism [97] and in addition to polymorphism has been suggested as a possible risk factor for CAD [98].

Conclusions

Complex inter-relationships between age-associated illnesses, such as vascular disease and dementia, and lipoprotein metabolism and different vascular genetic risk factors suggest that biological and genetic pathways may be worthy of examination in centenarian populations to provide insights into the human longevity. Various genetic markers likely to be related with CAD, CVD, or AD have been therefore investigated in centenarians in numerous studies. The most attention has been drawn to APOE gene polymorphism which has been widely examined in different populations. Particularly, in study on Italian centenarians, we found that the frequency of the APOE ε4 allele was significantly higher in middle-aged subjects than in centenarians, while ε2 allele did tend to increase in centenarians. Other genetic polymorphisms involved in lipoprotein regulation (APOA-I, APOCIII, APOA-IV) have been examined as potential markers for longevity. However, the results were contrasting, or, in same cases, there was just one study. Apart from these genetic polymorphisms, lipid and lipoprotein factors, such as TC, LDL-C, HDL-C, APOB, Lp(a), and serum *APOE*, linked to CAD and CVD, have been suggested to play a role in human longevity. In fact, it has been shown that centenarians have lower serum levels of HDL-C (with higher HDL2-C subfraction) and higher Lp(a) concentration. Lower concentration of HDL-C was associated with decreased serum albumin, and elevated CRP and IL-6 levels, suggesting that HDL-C could be a sensitive marker for frailty and comorbidity in centenarians. When we evaluated interaction of Lp(a) with lipid and nonlipid CAD risk factors in free-living elderly subjects, high serum Lp(a) in the elderly population was not found as independent predictor factor for CAD, but dependent on type 2 diabetes mellitus and elevated LDL-C levels. We suggest that Lp(a) elevation in centenarians, in the absence of other coronary artery disease risk factors, appears as a positive survival factor. Finally, in different populations we demonstrated a significant trend in reduction of serum *APOE* levels from APOE ε2- to ε4-carriers and significant differences in serum APOE levels respect to age in ε4-carriers but only after adjustment for HDL cholesterol. However, further studies are needed to confirm the possible role of APOE concentration as putative longevity factor.

Acknowledgements

This group is supported by Italian Longitudinal Study on Ageing (ILSA) (Italian National Research Council - CNR-Targeted Project on Ageing - Grants 9400419PF40 and 95973PF40), by AFORIGE ("Associazione per la FOrmazione e la RIcerca in Geriatria"). FP, AMC, and CC participate in the Ph.D.: "Carcinogenesis, Ageing, and Immunoregulation", supported by European Union. The authors wish to thank Giuseppe Giorgio, Dr Giovanni Castellaneta, and Dr Amalia Terralavoro, Pfizer Inc., for skilful assistance.

References

[1] Shachter F, Faure-Delanef L, Guenot F, et al. Genetic associations with human longevity at the APOE and ACE loci. *Nat Genet* 1994;16: 29-32.

[2] Louhija J, Miettinen HE, Kontula K, Tikkanen MJ, Miettinen TA, Tilvis RS. Aging and genetic variation of plasma apolipoproteins. Relative loss of the apolipoprotein E4 phenotype in centenarians. *Arterioscl Thromb* 1994;14:1084-1089.

[3] Panza F, Solfrizzi V, Torres F, et al. Decreased frequency of apolipoprotein E epsilon4 allele from Northern to Southern Europe in Alzheimer's disease patients and centenarians. *Neurosci Lett* 1999;277:53-56.

[4] Saunders AM, Strittmatter WJ, Schmechel D, et al. Association of apolipoprotein E allele e4 with late-onset familial and sporadic Alzheimer's disease. *Neurology* 1993;43:1467-1472.

[5] Eichner JE, Kuller LH, Orchard TJ, et al. Relation of apolipoprotein E phenotype to myocardial infarction and mortality from coronary artery disease. *Am J Cardiol* 1993;71:160-165.

[6] Boerwinkle E, Utermann G. Simultaneous effects of the apolipoprotein E polymorphism on apolipoprotein E, apolipoprotein B, and cholesterol metabolism. *Am J Hum Genet* 1988;42:104-112.

[7] Cambien F, Poirier O, Lecerf L, et al.. Deletion polymorphism in the gene for angiotensin-converting enzyme is a potent risk factor for myocardial infarction. *Nature* 1992:359;641-644.

[8] Kehoe PG, Russ C, McIlory S, et al. Variation in DCP1, encoding ACE, is associated with susceptibility to Alzheimer disease. *Nat Genet* 1999;21:71-72.

[9] Bladbjerg EM, Andersen-Ranberg K, de Maat MP, et al. Longevity is independent of common variations in genes associated with cardiovascular risk. *Thromb Haemos* 1999;82:1100-1105.

[10] Blanchè H, Cabanne L, Sahbatou M, Thomas G. A study of French centenarians: are ACE and APOE associated with longevity? *C. R. Acad. Sci. III* 2001;324:129-135.

[11] Panza F, Solfrizzi V, D'Introno A, et al. Angiotensin I converting enzyme (ACE1) gene polymorphism in centenarians: different allele frequencies between the North and South of Europe. *Exp Gerontol* 2003 (in press)

[12] Danesh J, Collins R, Peto R. Lipoprotein(a) and coronary heart disease. Meta-analysis of prospective studies. *Circulation* 2000;102:1082-1085.

[13] Zenker G, Koltringer P, Bone G, et al. Lipoprotein(a) as a strong indicator for cerebrovascular disease. *Stroke* 1986;17:942-945.

[14] Siest G, Pillot T, Regis-Bailly A et al. Apolipoprotein E: an important gene and protein to follow in laboratory medicine. *Clin Chem* 1995;41:1068-1086.

[15] Couderc R, Mahieux F, Bailleul S, et al. Prevalence of apolipoprotein E phenotypes in ischemic cerebrovascular disease. A case-control study. *Stroke* 1993;24:661-664.

[16] Davignon J, Gregg RE, Sing CF,. Apolipoprotein E polymorphism and atherosclerosis. *Arteriosclerosis* 1988;8:1-21.

[17] Davignon J, Bouthillier D, Nestruck AC, Sing CF. Apolipoprotein E polymorphism and atherosclerosis: insight from a study in octagenarians. *Trans Am Clin Climatol Assoc* 1987;99:100-110.

[18] Kervinen K, Savolainen MJ, Salokannel J, et al. Apolipoprotein E and B polymorphisms - longevity factors assessed in nonagenarians. *Atherosclerosis* 1994;105, 89-95.

[19] Bader G, Zuliani G, Kostner GM, Fellin R. Apolipoprotein E polymorphism is not associated with longevity or disability in a sample of Italian octo- and nonagenarians. *Gerontology* 1998;44:293-239.

[20] Jian-Gang Z, Yong-Xing M, Chuan-Fu W, et al. Apolipoprotein E and longevity among Han Chinese population. *Mech Ageing Dev* 1998;104:159-167.

[21] Rea IM, Mc Dowell I, McMaster D, et al. Apolipoprotein E alleles in nonagenarian subjects in the Belfast Elderly Longitudinal Free-living Ageing Study (BELFAST). *Mech Ageing Dev* 2001;122:1367-1372.

[22] Asada T, Kariya T, Yamagata Z, et al. Apolipoprotein E allele in centenarians. *Neurology* 1996;46:1484.

[23] Asada T, Yamagata Z, Kinoshita T, et al. Prevalence of dementia and distribution of ApoE alleles in Japanese centenarians: an almost-complete survey in Yamanashi Prefecture, Japan. *J Am Geriatr Soc* 1996;44:151-155.

[24] Yamagata Z, Asada T, Kinoshita A, et al. Distribution of apolipoprotein E gene polymorphisms in Japanese patients with Alzheimer's disease and in Japanese centenarians. *Hum Hered* 1997;47:22-26.

[25] Frisoni GB, Louhija J, Geroldi C, Trabucchi M. Longevity and the epsilon2 allele of apolipoprotein E: the Finnish Centenarians Study. *J Gerontol A Biol Sci Med Sci* 2001;56:M75-78.

[26] Louhija J, Viitanen M, Aguero-Torres H, Tilvis R. Survival in Finnish centenarians in relation to apolipoprotein E polymorphism. *J Am Geriatr Soc* 2001;49:1007-1008.

[27] Gerdes LU, Jeune B, Ranberg KA, et al. Estimation of apolipoprotein E genotype-specific relative mortality risks from the distribution of genotypes in centenarians and middle-aged men: apolipoprotein E gene is a "frailty gene," not a "longevity gene". *Genet Epidemiol* 2000;19:202-210.

[28] Choi YH, Kim JH, Kim DK, et al. Distributions of ACE and APOE polymorphisms and their relations with dementia status in Korean centenarians. J Gerontol A Biol Sci Med Sci 2003;58:227-231.

[29] Menzel HJ, Boerwinkle E, Schrangl-Will S, et al. Human apolipoprotein A-IV polymorphism: frequency and effect on lipid and lipoprotein levels. *Hum Genet* 1988;79:368-372.

[30] Groenendijk M, Cantor RM, de Bruin TW, et al. The apoAI-CIII-AIV gene cluster. *Atherosclerosis* 2001;157:1-11.

[31] De Benedictis G, Falcone E, Rose G, et al. DNA multiallelic systems reveal gene/longevity associations not detected by diallelic systems. The APOB locus. *Hum Genet* 1997;99:312-318.

[32] De Benedictis G, Carotenuto L, Carrieri G, et al. Age-related changes of the 3'APOB-VNTR genotype pool in ageing cohorts. *Ann Hum Genet* 1998;62 (Pt 2):115-122.

[33] Havel JH & Kane JP. Introduction: structure and metabolism of plasma lipoproteins. In: *The metabolic basis of inherited disease* (eds. C.R. Scriver, AL Blaudet, W S. Sly & D. Valle), 8th edn., pp. 2705-2716. New York: McGraw-Hill, 2001.

[34] Varcasia O, Garasto S, Rizza T, et al. Replication studies in longevity: puzzling findings in Danish centenarians at the 3'APOB-VNTR locus. *Ann Hum Genet* 2001;65(Pt 4):371-376.

[35] Galinsky D, Tysoe C, Brayne CE, et al.. Analysis of the apo E/apo C-I, angiotensin converting enzyme and methylenetetrahydrofolate reductase genes as candidates affecting human longevity. *Atherosclerosis* 1997;129:177-183.

[36] Garasto S, Rose G, Derango F, et al. The study of APOA1, APOC3 and APOA4 variability in healthy ageing people reveals another paradox in the oldest old subjects. *Ann Hum Genet* 2003;67(Pt 1):54-62.

[37] Yashin AI, De Benedictis G, Vaupel JW, et al. Genes and longevity: lessons from studies of centenarians. *J Gerontol A Biol Sci Med Sci* 2000;55:B319-328.

[38] Anisimov SV, Volkova MV, Lenskaya LV, et al. Age-associated accumulation of the apolipoprotein C-III gene T-455C polymorphism C allele in a Russian population. *J Gerontol A Biol Sci Med Sci* 2001;56:B27-32.

[39] Merched A, Xia Y, Papadopoulou A, et al. Apolipoprotein AIV codon 360 mutation increases with human aging and is not associated with Alzheimer's disease. *Neurosci Lett* 1998;242:117-119.

[40] Schatz IJ, Masaki K, Yano K, et al. Cholesterol and all-cause mortality in elderly people from the Honolulu Heart Program: a cohort study. *Lancet* 2001;358:351-355.

[41] Weverling-Rijnsburger AW, Blauw GJ, Lagaay AM, et al. Total cholesterol and risk of mortality in the oldest old. *Lancet* 1997;350:1119-1123.

[42] Kagan A, McGee DL, Yano K, et al. Serum cholesterol and mortality in a Japanese-American population: the Honolulu Heart program. *Am J Epidemiol* 1981;114:11-20.

[43] Stemmermann GN, Chyou PH, Kagan A, et al. Serum cholesterol and mortality among Japanese-American men. The Honolulu (Hawaii) Heart Program. *Arch Intern Med* 1991;151:969-972.

[44] Brescianini S, Maggi S, Farchi G, et al. Low total cholesterol and increased risk of dying: are low levels clinical warning signs in the elderly? Results from the Italian Longitudinal Study on Aging. *J Am Geriatr Soc* 2003;51:991-996.

[45] Schuit AJ, Van Dijk CE, Dekker JM, et al. Inverse association between serum total cholesterol and cancer mortality in Dutch civil servants. *Am J Epidemiol* 1993;137:966-976.

[46] International Collaborative Group. Circulating cholesterol level and risk of death from cancer in men aged 40 to 69 years. Experience of an International Collaborative Group. *JAMA* 1982;248:2853-2859.

[47] Tornberg SA, Holm LE, Carstensen JM, et al. Cancer incidence and cancer mortality in relation to serum cholesterol. *J Natl Cancer Inst* 1989;81:1917-1921.

[48] Kivipelto M, Helkala E-L, Hanninen T, et al. Midlife vascular risk factors and late-life mild cognitive impairment. A population-based study. *Neurology* 2001;56:1683-1689.

[49] DeCarli C, Miller BL, Swan GE, et al. Cerebrovascular and brain morphologic correlates of mild cognitive impairment in the National Heart, Lung, and Blood Institute Twin Study. *Arch Neurol* 2001;58:643-647.

[50] Petersen RC, Smith GE, Waring SC, et al. Mild cognitive impairment: clinical characterization and outcome. *Arch Neurol* 1999;56:303-308

[51] Solfrizzi V, Panza F, Capurso A. The role of diet in cognitive decline. *J Neural Transm* 2003;110:95-110

[52] Solfrizzi V, Panza F, Colacicco AM, et al., for the Italian Longitudinal Study on Aging Working Group. Vascular risk factors, incidence of MCI, and rates of progression to dementia. *Neurology* 2004 [in press].

[53] Solfrizzi V, Panza F, D'Introno A, et al. Lipoprotein(a), apolipoprotein E genotype, and risk of Alzheimer's disease. *J Neurol Neurosurg Psychiatry* 2002;72:732-736.

[54] Kuusisto J, Koivisto K, Mykkanen L, et al. Association between features of the insulin resistance syndrome and Alzheimer's disease independently of apolipoprotein E4 phenotype: cross sectional population based study. *BMJ* 1997;315:1045-1049.

[55] Romas SN, Tang MX, Berglund L, Mayeux R. APOE genotype, plasma lipids, lipoproteins, and AD in community elderly. *Neurology* 1999;53:517-521.

[56] Wilson PW, Anderson KM, Harris T, et al. Determinants of change in total cholesterol and HDL-C with age: the Framingham Study. *J Gerontol* 1994;49:M252-257.

[57] U.S. Department of Health and Human Services (DHHS). National Center for Health Statistics: Third National Health and Nutrition Survey, 1988-1994, *NHANES III Second Laboratory Data File* (CD-ROM series 11, No. 2A). Hyattsville, MD: Centers for Disease Control and Prevention, 1998.

[58] Ron D, Aviram M, Brook JG. High density lipoprotein in octogenarians. *Biochem Med* 1983;30:253-260.

[59] Nikkila M, Heikkinen J. High-density lipoprotein cholesterol and longevity. *Age Ageing* 1990;19:119-124.

[60] Nikkila M, Heikkinen J. Serum cholesterol, high-density lipoprotein cholesterol and five-year survival in elderly people. *Age Ageing* 1990;19:403-408.

[61] Campbell AJ, Busby WJ, Robertson MC. Over 80 years and no evidence of coronary heart disease: characteristics of a survivor group. *J Am Geriatr Soc* 1993;41:1333-1338.

[62] Ettinger WH Jr, Verdery RB, Wahl PW, et al. High density lipoprotein cholesterol subfractions in older people. *J Gerontol* 1994;49:M116-122.

[63] Barbagallo CM, Averna MR, Frada G, et al. Lipoprotein profile and high-density lipoproteins: subfractions distribution in centenarians. *Gerontology* 1998;44:106-110.

[64] Barbagallo CM, Averna MR, Frada G, et al. Plasma lipid, apolipoprotein and Lp(a) levels in elderly normolipidemic women: relationships with coronary heart disease and longevity. *Gerontology* 1995;41:260-266.

[65] Paolisso G, Gambardella A, Ammendola S, et al.Preserved antilipolytic insulin action is associated with a less atherogenic plasma lipid profile in healthy centenarians. *J Am Geriatr Soc* 1997;45:1504-1509.

[66] Arai Y, Hirose N, Nakazawa S, et al. Lipoprotein metabolism in Japanese centenarians: effects of apolipoprotein E polymorphism and nutritional status. *J Am Geriatr Soc* 2001;49:1434-1341.

[67] Barzilai N, Gabriely I, Gabriely M, et al. Offspring of centenarians have a favorable lipid profile. *J Am Geriatr Soc.* 2001;49:76-9.

[68] Pepe G, Di Perna V, Resta F, et al. In search of a biological pattern for human longevity: impact of apo A-IV genetic polymorphism on lipoproteins and the hyper-Lp(a) in centenarians. *Atherosclerosis* 1998;137:407-417.

[69] Thillet J, Doucet C, Chapman J, et al.. Elevated lipoprotein(a) levels and small apo(a) isoforms are compatible with longevity: evidence from a large population of French centenarians. *Atherosclerosis* 1998;136:389-394.

[70] Baggio G, Donazzan S., Monti D, et al. Lipoprotein(a) and lipoprotein profile in healthy centenarians: a reappraisal of vascular risk factors. *FASEB J* 199812,433-437.

[71] Matsubara M, Akita H, Shibuya H, Chiba H. Decreased longevity in Japanese men, associated with low-molecular-weight apolipoprotein(a) phenotypes. *Ann Clin Biochem* 2001;38(Pt 2):120-123.

[72] Akita H, Matsubara M, Shibuya H, et al. Effect of ageing on plasma lipoprotein(a) levels. *Ann Clin Biochem* 2002;39(Pt 3):237-240.

[73] Solfrizzi V, Panza F, Colacicco AM, et al. Relation of lipoprotein(a) as coronary risk factor to type 2 diabetes mellitus and low-density lipoprotein cholesterol in patients >65 years of age (The Italian Longitudinal Study on Aging). *Am J Cardiol* 2002;89,825-829.

[74] Eaton DL, Fless GM, Kohr WJ, et al. Partial amino acids sequence of apolipoprotein(a) shows that it is homologous to plasminogen. *Proc Natl Acad Sci USA* 1987;84,3224-3228.

[75] Boerwinkle E, Leffert CC, Lin J, et al. Apolipoprotein(a) gene accounts for greater than 90% of the variation in plasma lipoprotein(a) concentrations. *J Clin Invest* 1992;90:52-60.

[76] Danesh J, Collins R, Peto R. Lipoprotein(a) and coronary heart disease. Meta-analysis of prospective studies. *Circulation* 2000;102:1082-1085.

[77] Rhoads GG, Dahlen G, Berg K, et al. Lp(a) lipoprotein as a risk factor for myocardial infarction. *JAMA* 1986;256:2540-2544.

[78] Simons LA, Friedlander Y, McCallum J, et al. Lipoprotein(a) is not associated with coronary heart disease in the elderly: cross-sectional data from the Dubbo study. *Atherosclerosis* 1993;99:87-95.

[79] Sunayama S, Watanabe Y, Ohmura H, et al. Lack of increased coronary atherosclerotic risk due to elevated lipoprotein(a) in women > or = 55 years of age. *Circulation* 1996;94:1263-1268.

[80] Solfrizzi, V., Panza, F., Torres, F., Mastroianni, F., Del Parigi, A., Venezia, A., Capurso, A., 1999. High monounsaturated fatty acids intake protects against age-related cognitive decline. *Neurology* 52, 1563-1569

[81] Solfrizzi V, Panza F, D'Introno A, et al. Lipoprotein(a) in the elderly: beyond atherosclerosis. *Arch Gerontol Geriatr* 2002;Suppl 8:337-343.

[82] Chapman MJ, Huby T, Nigon F, et al. Lipoprotein(a): implication in atherothrombosis. *Atherosclerosis* 1994;110 Suppl,S69-S75.

[83] Zenker G, Koltringer P, Bone G, et al. Lipoprotein(a) as a strong indicator for cerebrovascular disease. *Stroke* 1986;17:942-945.

[84] Urakami K, Wada-Isoe K, Wakutani Y, et al. Lipoprotein(a) phenotypes in patients with vascular dementia. *Dement Geriatr Cogn Disord* 2000;11:135-138.

[85] Snowdon DA, Greiner LH, Mortimer JA, et al. Brain infarction and the clinical expression of Alzheimer disease. The Nun Study. *JAMA* 1997;277:813-817.

[86] Mooser V, Helbecque N, Miklossy J, et al. Interactions between apolipoprotein E and apolipoprotein(a) in patients with late-onset Alzheimer disease. *Ann Intern Med* 2000;132:533-537.

[87] Panza F, Solfrizzi V, Torres F, et al. Lipoprotein(a) affects cognitive function in the elderly. The Italian Longitudinal Study on Aging. *Neurobiol Aging* 2000;21(Supplement 1):S34 (abstract).

[88] Nago N, Kayaba K, Hiraoka J, et al. Lipoprotein(a) levels in the Japanese population: influence of age and sex, and relation to atherosclerotic risk factors. The Jichi Medical School Cohort Study. *Am J Epidemiol* 1995;141:815-821.

[89] Jenner JL, Ordovas JM, Lamon-Fava S, et al. Effects of age, sex, and menopausal status on plasma lipoprotein(a) levels. The Framingham Offspring Study. *Circulation* 1993;87:1135-1141.

[90] Soma MR, Osnago-Gadda I, Paoletti R, et al. The lowering of lipoprotein(a) induced by estrogen plus progesterone replacement therapy in postmenopausal women. *Arch Intern Med* 1993;153:1462-1468.

[91] Solfrizzi V, Panza F, Colacicco AM, et al. Serum apolipoprotein E levels in Alzheimer's disease and extreme longevity. *Arch Gerontol Geriatr* 2002; Suppl 8:345-352.

[92] Panza F, Solfrizzi V, Colacicco AM, et al. Apolipoprotein E (APOE) polymorphism influences serum APOE levels in Alzheimer's disease patients and centenarians. *Neuroreport* 2003;14:605-608.

[93] Cohn JS, Tremblay M, Amiot M, et al. Plasma concentration of apolipoprotein E in intermediate-sized remnant-like lipoproteins in normolipidemic and hyperlipidemic subjects. *Arterioscler Thromb Vasc Biol* 1996;16:149-159.

[94] Reilly SL, Ferrell RE, Kottke BA, et al. The gender-specific apolipoprotein E genotype influence on the distribution of plasma lipids and apolipoproteins in the population of Rochester, MN. II. Regression relationships with concomitants. *Am J Hum Genet* 1992;51:1311-1324.

[95] Gracia V, Fiol C, Hurtado I, et al. An enzyme-linked immunosorbent assay method to measure human apolipoprotein E levels using commercially available reagents: effect of apolipoprotein E polymorphism on serum apolipoprotein E concentration. *Anal Biochem* 1994;223:212-217.

[96] Couderc R, Mahieux F, Bailleul S, et al. Prevalence of apolipoprotein E phenotypes in ischemic cerebrovascular disease. A case-control study. *Stroke* 1993;24:661-664.

[97] Bohnet K, Pillot T, Visvikis S, et al. Apolipoprotein (apo) E genotype and apoE concentration determine binding of normal very low density lipoproteins to HepG2 cell surface receptors. *J Lipid Res* 1996;37:1316-1324.

[98] Siest G, Pillot T, Regis-Bailly A, et al. Apolipoprotein E: an important gene and protein to follow in laboratory medicine. *Clin Chem* 1995;41:1068-1086.

In: Focus on Atherosclerosis Research
Editor: Leon V.Clark, pp. 133-164

ISBN 1-59454-044-6

Chapter VII

Fatty Acids: Determinants of Endothelial Function, Inflammation, Insulin Resistance and Atherosclerosis

Antonie J.H.H.M. van Oostrom and Manuel Castro Cabezas
Department of Vascular Medicine, University Medical Center Utrecht, the Netherlands

Abstract

The most common metabolic disorders in Western countries, obesity and the insulin resistance syndrome (IRS) are accompanied by an increased risk of coronary heart disease (CHD) and show an alarming increase in prevalence. Both disorders are characterized by increased plasma levels of free fatty acids (FFA) which are thought to play a critical role in the pathogenesis. FFA elevations are associated to obesity and a hypercaloric diet. Furthermore, FFA decrease insulin sensitivity in the adipose tissue, in the liver and in the muscle and aggravate pancreatic dysfunction, ultimately leading to insulin resistance. Insulin resistance at the level of the adipose tissue will further increase plasma FFA, since insulin is a key player in lipid metabolism with in general a lipogenic effect. Impaired adipose tissue FFA trapping, in particular in the postprandial phase, due to insulin resistance or dysfunction of the C3/ASP system (complement component 3/acylation stimulating protein) maintains this vicious circle. Under these conditions there is enhanced hepatic FFA flux, leading to the characteristic IRS lipid phenotype with high triglycerides (TG), low levels of high density lipoprotein (HDL) cholesterol and high levels of small-dense highly atherogenic low density lipoprotein (LDL) cholesterol. Furthermore, FFA may impair endothelial function via activation of proinflammatory pathways and production of reactive oxygen species by leukocytes and endothelial cells, processes that eventually may contribute to atherogenesis. Certain fatty acids, the polyunsaturated omega-3 fatty acids in particular, exert beneficial effects on lipid metabolism, endothelial function and inflammation. These effects are most certainly PPAR (peroxisome proliferator activating receptor) mediated since FFA are the natural ligands of these receptors. Treatment of FFA-related metabolic disorders first of all should include life-style intervention consisting of weight reduction and diet modification in favor of polyunsaturated fatty acids. In addition, drugs with PPAR activating properties, like fibrates and the novel class of thiazolidinediones, may be helpful. Furthermore, statins may reduce FFA as well, either via TG lowering or possibly also via PPAR activation.

Introduction

Coronary Heart Disease (CHD) is the major contributor to morbidity and mortality in Western societies [1] and is nowadays regarded as a low-grade chronic inflammatory disease in response to endothelial injury [2,3]. Important CHD risk factors include decreased insulin sensitivity, smoking, hypertension, dyslipidemia, increased body fat mass, unfavorable body fat distribution and a prothrombotic state [4-8]. Most of these risk factors are strongly interrelated and are part of the insulin resistance syndrome (IRS) as was elegantly described by Reaven in 1988 [9]. Type 2 diabetes is the ultimate consequence of the IRS. Among all riskfactors, obesity is most strongly related to the prevalence of type 2 diabetes ($r = 0.64$, $p < 0.001$, in a general population of 150.000 subjects) [10]. Type 2 diabetes is associated with a two- to fourfold increased risk of CHD mortality and it is accompanied with a similar risk for primary coronary events as the risk for a second event in non-diabetics [11-14]. The incidences of obesity, the IRS and type 2 diabetes are rapidly increasing in Western societies and therefore a dramatic rise in CHD has to be expected [15,16]. This increase is most likely a result of the changing Western lifestyle that is more and more sedentary and characterized by a hypercaloric diet rich in saturated fatty acids and carbohydrates [16].

In type 2 diabetes, or in general in the IRS, the major part of morbidity and mortality is attributed to microvascular (retinopathy, nephropathy and neuropathy) and macrovascular disease (myocardial infarction, stroke and peripheral vascular disease). When glucose levels are strictly regulated, microvascular complications can be significantly reduced [13,17], but to date macrovascular complications have only been decreased by treating other modifiable components of the IRS. In this respect, the United Kingdom Prospective Diabetes Study (UKPDS) has shown benefit from tight bloodpressure control [18]. In addition, as shown in subgroup analyses of the Heart Protection Study (HPS), the Scandinavian Simvastatin Survival Study (4S) and the Cholesterol And Recurrent Events (CARE) trial, improvement of dyslipidemia resulted in a marked reduction of macrovascular disease in type 2 diabetes patients as well [19-21]. Furthermore, multifactorial treatment targeted on hyperglycaemia, hyperlipidemia and hypertension using very strict treatment goals, with standard addition of aspirin and angiotensin converting enzyme (ACE) inhibitors resulted in 50% reduction of total cardiovascular and microvascular events in type 2 diabetes patients [22].

Impressive reductions of CHD mortality and morbidity have also been achieved by strict lifestyle modifications. The Lyon Diet Heart study is the best example by showing that a 46 months Mediterranean diet rich in unsaturated fatty acids, anti-oxidants and folic acid, reduced the rate of secondary cardiovascular events by almost 50% [23]. This result was achieved without any significant changes of low-density lipoprotein (LDL) cholesterol, suggesting that other anti-atherosclerotic mechanisms may have been implicated [23].

From all lipid parameters, triglycerides (TG) and high-density lipoprotein (HDL) cholesterol are the most important predictors of CHD in type 2 diabetes and the IRS, whereas total cholesterol is not a strong predictor [24-26]. The fact that besides TG, also HDL-cholesterol predicts CHD is not unexpected since those parameters are closely inversely correlated. This is due to transfer of cholesteryl esters from HDL to TG-rich particles (TRP's), and TG from TRP's to HDL, by cholesteryl ester transfer protein (CETP). This transfer, in concerted action with hepatic lipase (HL) and phospholipid transfer protein (PLTP), leads to small dense, relatively cholesterol depleted and TG-enriched, HDL particles, a process that mainly occurs under hyper-TG conditions [27]. In addition, when hypertriglyceridemia is present, via similar mechanisms, small dense LDL cholesterol will be generated. These particles are highly atherogenic after oxidative or glycaemic modification, due to the processing by upregulated scavenger receptors on activated monocytes, transforming them into macrophages and eventually leading to foam cell formation [28]. A high prevalence of small-dense LDL is very common in patients with the IRS [29,30].

Regarding TRP's, it is important to acknowledge that these lipoproteins are mainly produced postprandially and people are non-fasting most part of the day [31-35]. Of all patients with premature CHD, 40 % has normal fasting plasma lipids [36,37] whereas many of these patients have impaired clearance of postprandial lipoproteins, e.g. chylomicrons and their remnants [38-42]. It is therefore that atherosclerosis has been considered a postprandial phenomenon [43,44]. In addition, postprandial glucose may play a more important role in the IRS than fasting glucose [45-47].

During lipolysis of TG-rich lipoproteins by lipoprotein lipase (LPL) and HL, free fatty acids (FFA) are released in the circulation. Increased levels of FFA as a result of obesity and more important due to a hypercaloric diet, are regarded as one of the key etiologic components of the IRS [48]. FFA have detrimental effects on glucose and lipid metabolism, endothelial function, inflammatory pathways, pancreatic function [48,49], eventually result in obesity and are related to the development of atherosclerosis.

This paper deals with the effects of FFA on glucose and lipid metabolism, endothelial function and inflammatory pathways in the IRS and emphasizes their relevance in clinical practice. In addition, therapeutical interventions modifying FFA metabolism will be discussed. Among these, a novel class of drugs, peroxisome proliferator activating receptor (PPAR)-γ-agonists (thiazolidinediones) that modifies insulin sensitivity and lipid metabolism will be discussed.

The Effects of FFA on Glucose Metabolism

The "glucose-fatty acid cycle" was elegantly described by Randle et al as early as 1963, showing that FFA largely influence the metabolism of glucose [50]. The central concept of this cycle, which is also known as the "Randle cycle", is competition between glucose and FFA for uptake in the muscle cell [51]. In the case of increased availability of one of those two substrates, the metabolism of the other will decrease [48,52,53]. Randle postulated that increased FFA levels result in elevation of mitochondrial Acetyl-Coenzyme A (Acetyl-CoA), indirectly leading to inhibition of Pyruvatedehydrogenase, the key enzyme in glucose oxidation. Eventually this would result in increased cellular glucose concentrations, leading to reduced glucose uptake by the muscle cell and therefore rising plasma glucose levels [51].

Others postulated that FFA activate an isoform of Pyruvatekinase C leading to downregulation of the glucose transporter (GLUT-4) on the muscle cell membrane [54].

Besides the effects on muscle cells, it has been demonstrated that FFA partly antagonize the inhibitory effect of insulin on hepatic glucose output [52,55] and may reduce hepatic insulin clearance [56], thereby further increasing plasma glucose and insulin which may lead to decreased insulin sensitivity by unknown mechanisms.

Postprandial Lipemia: Crossroad of Endogenous and Exogenous Free Fatty Acids

Triglycerides contain three fatty acid chains esterified to a glycerol backbone. Fatty acids differ in carbon chain length and the degree of saturation, depending on the type of dietary fat. TG are transported in the blood in lipoprotein particles. These particles contain a lipid core of TG and cholesterol esterified to fatty acids, surrounded by an amphipatic monolayer of phospholipids, free cholesterol and apolipoproteins with protruding hydroxide groups. Lipoproteins are classified on the base of density and apolipoprotein (apo) constitution. Apolipoproteins are essential for receptor recognition and the action of enzymes involved in lipid metabolism.

Two different lipoprotein pathways are known: the endogenous and the exogenous pathways (Figure 1). The former starts in the liver were TG and cholesterol are incorporated into one apoB100-containing particle, the resulting lipoprotein is called very low-density lipoprotein (VLDL). The exogenous pathway finds its origin in the small intestine were dietary TG and cholesteryl esters are packed in one apoB48-containing particle, called chylomicron. Both VLDL and chylomicrons compete for the same lipolytic pathway, e.g. endothelium bound lipoprotein lipase (LPL) that hydrolyzes TG into glycerol and three FFA. Depending on density, the remaining lipoproteins are called IDL and chylomicron remnants, respectively. Both these particles are still substrate for LPL and HL, resulting in LDL and smaller chylomicron remnants, respectively. A prerequisite for the action of LPL is availability of apoCII from lipoproteins as a cofactor. Since chylomicrons are mainly produced in the postprandial phase, in this period there is accumulation of TRP's due to limited LPL availability.

In the IRS, fasting hyperlipidemia, characterized by VLDL overproduction, is present due to enhanced hepatic FFA fluxes (Figure 1). One of the contributing mechanisms is decreased inhibition of adipose tissue hormone-sensitive lipase [57-59]. In addition, inhibition of hepatic VLDL secretion by insulin is reduced in the IRS [29,60]. Postprandially when additional TRP's are produced, hypertriglyceridemia is enhanced in the IRS due to competition for LPL [61]. Furthermore, in the IRS the activity of LPL is slightly reduced [62].

Postprandial dyslipidemia is one of the most striking lipid abnormalities in the IRS, however, while fasting TG is an independent CHD riskfactor [63], postprandial hypertriglyceridemia has not yet been investigated in a prospective study as a CHD predictor [64,65].

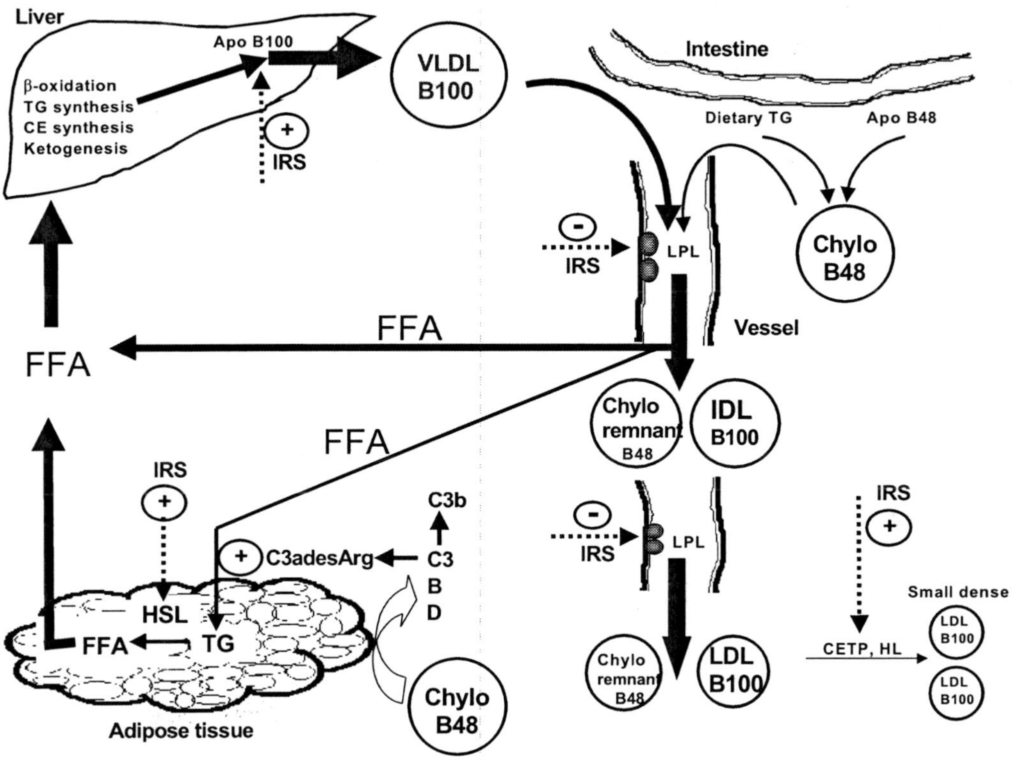

Figure 1. Schematic Representation of Free Fatty Acids (FFA) Metabolism. Specific abnormalities that occur due to insulin resistance (IR) are indicated separately. The thickness of an arrow indicates the magnitude of the flux. In IR, increased hepatic flux of free fatty acids (FFA) occurs due to a reduced inhibition of adipose tissue hormone sensitive lipase (HSL) mediated lipolysis. Reduced inhibition of very low density lipoprotein (VLDL) secretion in IR and increased substrate availability results in increased plasma VLDL concentrations, represented by fasting dyslipidemia. Rather than an increased VLDL particle number, IR is characterised by a relative triglyceride (TG) enrichment of VLDL, resulting in large VLDL particles. Furthermore, hepatic FFA are substrate for β-oxidation, cholesterol ester (CE) synthesis and ketone body formation. In the postprandial state additional TG-rich lipoproteins, e.g. chylomicrons, are produced in the small intestine. Both VLDL and chylomicrons compete for the same clearance pathway, the endothelium bound lipoprotein lipase (LPL). Hydrolysis of TG from TG-rich particles, mediated by LPL, results in FFA, smaller intermediate density lipoproteins (IDL) and chylomicron remnants. Due to increased VLDL and a slightly reduced LPL, postprandial lipemia is exaggerated in IR. FFA produced by lipolysis of TG-rich lipoproteins can be stored as TG in the adipose tissue. This process is in part mediated by acylation stimulating protein (ASP), which is identical to C3adesArg. Chylomicrons induce adipocyte production of complement component 3 (C3) together with cofactors B and D. By activation of the complement system, C3 will then be split in C3b and C3a, the latter becomes ASP after desargination. Cholesterolester transfer protein (CETP), exchanges cholesterol from low density lipoprotein (LDL) for TG from chylomicron remnants and IDL. Hepatic lipase (HL) hydrolyses part of the TG in LDL, resulting in small dense LDL. Due to a slight increase of HL activity in IR, these processes are exaggerated. Small dense LDL, in particular after oxidative and glycaemic modification, is a highly atherogenic particle. TG-rich particles and FFA are related to atherosclerotic disease as well.

FFA and the Liver

Circulating FFA are cytotoxic and therefore FFA are bound to proteins, albumin in particular. As previously mentioned, FFA can be transported to the liver to be re-esterified with a glycerol molecule to form TG and become incorporated in VLDL. In addition, FFA can also be transported into the hepatocyte mitochondria via the carnitine shuttle. β-Oxidation, of these fatty acids provides Acetyl-CoA and ATP. Acetyl-CoA can either enter the aerobic citric acid pathway to provide additional ATP, or the anaerobic pathway to provide lactate and a smaller amount of ATP [48,66]. In the postabsorptive period or when abundant FFA are available, for instance postprandially, Acetyl-CoA can enter the ketogenesis pathway [48,66]. In humans, 3-β-hydroxybutyrate (3HB) is the most abundant ketone body and like the other ketone bodies it is mainly produced in the liver. In insulin sensitive subjects, insulin inhibits ketogenesis by reducing hepatic FFA fluxes and by activation of Acetyl-CoA carboxylase resulting in decreased mitochondrial FFA flux. In situations when limited carbohydrates are available, for instance in the postabsorptive period, organs like the heart, the kidney cortex and skeletal muscle are fully dependent on FFA and ketone bodies to generate energy. Recently, we have provided for the first time *in vivo* evidence for increased postprandial hepatic FFA flux, represented by elevated HBA levels, in patients with the IRS and familial combined hyperlipidemia (FCHL) after an oral fat load [67]. It was hypothesized that in non-FCHL patients with the IRS, similar findings may occur. Furthermore, we have shown gender differences in postprandial HBA generation, with higher levels in healthy females in comparison to males [68].

FFA and the Adipose Tissue: A Link with Inflammation and the Complement System

Under normal circumstances, the adipose tissue is a repository for excess FFA. Since FFA fluxes to the muscle and liver aggravate insulin resistance and hyperlipidemia, peripheral FFA trapping in particular in the postprandial phase, is regarded as a determinant of a healthy lipoprotein phenotype [69,70]. Postprandial FFA trapping depends on LPL-mediated hydrolysis of TG from TRP's. Endothelium-bound LPL is mainly expressed in skeletal muscles and adipose tissue but exists also in a variety of other tissues including heart, liver, kidney, pancreas, nervous system, adrenals and macrophages [71]. While LPL activity in subcutaneous tissue is similar in healthy subjects and type 2 diabetes patients [72], the latter may express lower LPL levels in the visceral fat depot [73]. The visceral depot has a much smaller contribution to total body fat (approximately 10% versus 80% for subcutaneous fat), but it is regarded as more important in the pathogenesis of the IRS, since it is metabolically more active, more resistant to insulin's actions and its vascularization drains directly into the portal vein [74-77].

Albumin bound FFA or FFA locally released after the action of LPL can enter the adipocyte by different pathways, to form TG after re-esterification to glycerol. In general, FFA can cross a cell membrane by passive diffusion through the lipid bilayer [78] or by an active protein mediated transport system for which CD36 (fatty acid translocase) [79], plasma membrane fatty acid-binding protein [80] and fatty acid transport protein [81] have been proposed as important candidates. It is thought that these transport mechanisms may depend on the type of fatty acids [78]. It is unknown what the actual contribution of passive transport is, since in the adipocyte FFA uptake is thought to be largely hormonally controlled. Insulin stimulates FFA uptake most likely coupled to increased glycerol synthesis, which is necessary for FFA re-esterification into TG. The stimulation of glycerol synthesis is indirect via increased adipocyte glucose uptake [82] and possibly also via direct effects on key enzymes involved in the synthesis of glycerol [83].

Recently the C3/ASP (complement component 3/acylation stimulating protein) pathway, a novel pathway involved in adipocyte FFA trapping, has received major attention [84]. This was especially due to the identification of human models with a defect in this pathway [38,85,86]. It has been shown that chylomicrons are the strongest activators of adipocyte C3 *in vitro* [87]. The complement system is the most important and phylogenetically best conserved mechanism in human immunology and comprises three different pathways, e.g. the alternative, the classical and the mannose binding lectin pathways [88]. Activation of these pathways results in cleavage of C3 leading to formation of C3a and C3b (Figure 1). In inflammation, the terminal route of the complement cascade is active, starting with C3b and leading to formation of the membrane attack complex [88]. In addition, C3b attached to pathogens improves adherence to leukocytes, and other terminal complement components like C5a are strong chemoattractants [88]. Adipose tissue has been shown to produce C3/ASP most likely by activation of the alternative complement pathway. This comprises interaction of adipocyte produced C3 and the alternative pathway cofactors B and D (the latter also known as the adipocyte hormone adipsin), resulting in cleavage of C3 in C3a and C3b [85]. Under normal circumstances C3b will not enter the terminal complement activation routes leading to inflammation, since there are abundant inactivators of these routes. The early split-product C3a, in plasma exits only in the desarginated form (C3adesArg), due to the action of carboxypeptidase B on the carboxyl terminus of C3a [85]. C3adesArg is identical to ASP, the newly recognized adipocyte-derived hormone. It has been shown that ASP does not bind to the classic C3a receptor [89], but very likely to a specific receptor called the chemoattractant receptor-like protein, C5L2 [90]. Via mechanisms that still have to be fully elucidated, C3/ASP has been shown to improve TG synthesis in adipocytes. These mechanisms include improvement of adipocyte glucose transport and inhibition of adipocyte HSL and are thought to be additional to, but not depending on, the effects of insulin [85,91,92]. In addition, *in vitro* studies have shown that ASP promotes glucose uptake in muscle cells [85]. This effect is theoretically beneficial by in part resolving the resistance for the effects of insulin in this tissue.

Both ASP and C3 have been positively associated to obesity, type 2 diabetes, the IRS, dyslipidemia and hypertension [38,85,86]. It is therefore not unexpected that both parameters are elevated in CHD patients [38,85] and that C3 is a strong predictor of myocardial infarction [93,94]. It has also been shown that complement activation plays a role in the induction of tissue damage after myocardial infarction [95] and complement deposition has been observed in atherosclerotic plaques [96]. The strong relationship between C3/ASP and adipose tissue homeostasis is underlined in studies showing a decrease of C3 and ASP during weight loss [85].

Considering that chylomicrons are a major activator of the C3/ASP system, studies on postprandial C3 and ASP changes have been carried out. Using standardized oral fat loads it has been shown that plasma ASP does not change postprandially in healthy lean and obese subjects or in patients with FCHL [97-99]. The proposed reasons for these findings are the relatively low concentrations of ASP in plasma and the paracrine action of the hormone. Indeed when ASP was measured locally in adipose tissue by cannulation of a vein and estimation of arterio-venous gradients, the expected postprandial increase has been demonstrated [100]. In contrast, C3 that shows much higher concentrations in plasma (g/L), did show postprandial changes in plasma. However, the results of different studies are inconsistent, which might be due to the study meals that were used. Two studies showed unchanged postprandial plasma C3 after a mixed meal in healthy subjects [97,98], while in obese subjects a small but significant increase was observed [98]. When standardized oral fat loads were administered, C3 increments were observed in healthy subjects, in normolipidemic CHD patients and in FCHL patients [38,86,101]. We have recently shown that the discrepancy in the literature about postprandial C3 changes was due to the different study meals that were used. When healthy insulin sensitive subjects ingested a standardized oral fat load, a 10% postprandial C3 increase was observed that was abolished when glucose was added to the fat load [102]. We therefore hypothesized that the C3/ASP system is a backup system for postprandial FFA trapping that is active in low insulin or insulin resistant circumstances, explaining the strong positive relation between C3/ASP and the IRS [102]. Indeed a recent study has shown that upon a dramatic weight reduction insulin sensitivity improved with a concomitant fall in plasma ASP levels [103].

Besides insulin resistance, C3/ASP-resistance has been described in humans, both leading to failure of adipose tissue FFA trapping resulting in abnormal diversion of FFA to the liver and muscle, leading to the above described metabolic disturbances. Therefore, modulation of the action of ASP may have clinical implications. It has recently been shown that ASP levels and concomitantly concentrations of FFA depend on the type of fatty acids in the diet. Low ASP and high FFA levels were seen in response to a diet enriched in trans fatty acids while a diet enriched in unsaturated fatty acids resulted in the opposite [104].

The adipose tissue is far from an inert organ since there is growing evidence that this tissue is actually an endocrine gland. Besides the above described production of ASP, the adipocyte has been shown to secrete a variety of other hormones involved in homeostasis, amongst them leptin, resistin, adiponectin and sex steroids [105,106]. The exact actions of these hormones are not yet completely elucidated. Furthermore, adipose tissue is controlled by the sympathetic [107,108] and as recently has been shown also by the parasympathetic nervous system [109].

In addition, the adipose tissue has been shown to secrete a variety of chemokines (adipokines) that may link adiposity and the IRS to the inflammatory component of atherosclerosis. Amongst these, interleukin-6 (IL-6) and tumor necrosis factor α (TNFα) are the most important. One third of plasma IL-6 is produced by the adipose tissue, visceral fat in particular [110,111]. Endothelial cells and inflammatory cells like lymphocytes and monocytes are also known to secrete IL-6 [112,113]. IL-6 synthesis is under negative control of glucocorticoids and positively regulated by catecholamines [110,112]. Plasma IL-6 and adipose tissue arterial bed IL-6 concentrations are associated to CHD [114], obesity, FFA and insulin sensitivity [111,115,116]. On the other hand IL-6 is thought to limit adiposity since it reduces adipose LPL activity [111,116]. *In vivo* studies have shown an increase in plasma IL-6 concentrations during the day, however a circadian rhythm seems more important than feeding [111,117-119]. IL-6 is the major chemokine responsible for hepatic C-reactive protein (CRP) production. From all inflammatory markers, CRP is the strongest independent predictor of the development of CHD in healthy subjects and patients with established CHD [120]. The strongest correlate to CRP seems to be body composition, in particular abdominal fat deposition [121-123]. Data about relationships between CRP and other components of the metabolic syndrome are inconsistent after correction for body composition parameters. However, some studies show a correlation with insulin sensitivity parameters, triglyceride and HDL-cholesterol levels, the relation with total cholesterol and LDL-cholesterol remains equivocal [121-123]. The current theories about the relationship between CRP and atherosclerosis most often include endothelial cell activation by atherosclerotic disease leading to IL-6 production by these cells [124]. However, a relationship with adipose tissue also seems plausible since most of IL-6 is produced by this tissue and the adipose tissue directly drains to the portal system. There is growing evidence that CRP is not only a marker of atherosclerotic disease, but also directly proatherogenic and pro-inflammatory by inducing activation of endothelial cells [125]. The mechanism involved is thought to include activation of the complement system [125,126].

The other important adipokine, TNFα, is under normal conditions a key regulator of adipocyte mass by promoting adipocyte apoptosis and inhibiting adipogenesis [127]. TNFα is involved in lipid metabolism by reducing LPL activity and downregulation of the adipocyte FFA transporters [127,128]. There is also evidence for reduction of adipocyte lipogenesis and it is thought that adipocyte lipolysis is enhanced due to the action of TNFα [127]. Overall, the suggested effects of TNFα result in weight loss, however, the role of TNFα in obesity is unclear. Increased plasma levels of TNFα have been associated with the IRS and in animal models it has been shown that TNFα administration or neutralization regulates insulin sensitivity [128]. The mechanisms behind this could be increased hepatic FFA fluxes due to TNFα stimulated lipolysis, or downregulation of adipocyte GLUT4 and reduction of insulin signaling and action in the adipocyte and probably also the hepatocyte [128,129]. In addition, TNFα has also been shown to antagonize peroxisome proliferator activator receptor γ (PPARγ) signaling pathways [128,129]. PPARγ is a nuclear receptor involved in adipogenesis and adipocyte function and is also active in many other tissues. The role of the PPARγ receptor will be discussed later, since it is an important target of insulin sensitizing drugs.

Endothelial Function and Free Fatty Acids

Endothelial dysfunction is inextricably connected with atherogenesis, the IRS and with the development of CHD events [130,131]. The endothelium controls vasomotion, permeability, proliferation and migration of smooth muscle cells. Above all these effects, the endothelium plays a key role in the control of the balance between pro- and anti-inflammatory, thrombotic and atherogenic processes [130,132-135]. Different tools are available for assessment of endothelial function. An indirect method is measurement of plasma levels of soluble adhesion molecules and selectins, endothelin-1 (ET-1), plasminogen activator inhibitor (PAI-1) and von Willebrand factor (vWf), markers which have all been associated to cardiovascular disease [136-139]. Direct methods include the non-invasive measurement of the combined carotid intima-media thickness (IMT) and determination of nitric oxide (NO) dependent flow-mediated dilation (FMD) of the brachial artery. These techniques are a good representative of endothelial function of the coronary artery (the golden standard) and acceptable for implication in large clinical and epidemiological studies [140-142]. More invasive methods comprise Doppler velocimetry during infusion of agonists of the NO system in the brachial (forearm blood flow, FBF), femoral or coronary artery [143-146].

It is well known that triglycerides, but also glucose, can reversibly impair FMD, a phenomenon of particular importance in the postprandial phase [118,147-149]. In addition, FFA have been shown to impair FBF [145,150], however, the type of fatty acids seems to determine the response of the endothelium. In particular omega-3 or n-3 fatty acids abundant in fish oil, but also in plant oils, have shown beneficial effects [144,151,152], while in another *in vivo* study the chain length of fatty acids seemed less important [150]. The most important n-3 fatty acid is linolenic acid which is the origin of eicosapentaenoic acid (EPA) and docosahexaenoic acid (DHA), that together constitute 26% of fish oil fatty acids [153].

Clinical estimates of endothelial cell activation have shown reductions upon n-3 fatty acid supplementation *in vitro* and *in vivo*. Incubation of human endothelial cells with DHA and or EPA in some but not all studies, resulted in a lower expression of adhesion molecules on endothelial cells and reduced the production of pro-inflammatory cytokines like IL-6 and IL-8 [151,152,154]. In addition, *in vivo* diet intervention with n-3 fatty acids resulted in lower circulating levels of adhesion molecules and prothrombotic vWF [151]. These findings are of importance for the pathogenesis of atherosclerosis, since it has been shown that prevention of adherence of leukocytes to the endothelium by blocking specific endothelial cell adhesion molecules results in reduced plaque formation [72,155] and prevention of endothelial dysfunction [156]. Prerequisite for the interaction between leukocytes and endothelial cells is activation of both cell types resulting in a sequential upregulation of integrins and selectins [157]. Besides the above mentioned effects on endothelial cells, fatty acids exert multiple effects on leukocytes (Figure 2). It has been shown that fatty acids derived from lipolysis of an artificial TG-rich emulsion can activate leukocytes, neutrophils in particular [158,159]. These effects were most striking for medium chain fatty acids while they were not observed after stimulation with long chain fatty acids. The growing evidence of fatty acid mediated attenuation of leukocyte and endothelial cell activation implicates effects on the interaction between leukocytes and endothelial cells. Indeed studies with DHA have shown reduced monocyte and lymphocyte endothelium adherence *in vitro* [151]. Another endothelial cell site of action of FFA is NO. Incubation of endothelial cells with saturated fatty acids inhibited NO release [160], while EPA resulted in the contrary [152,161].

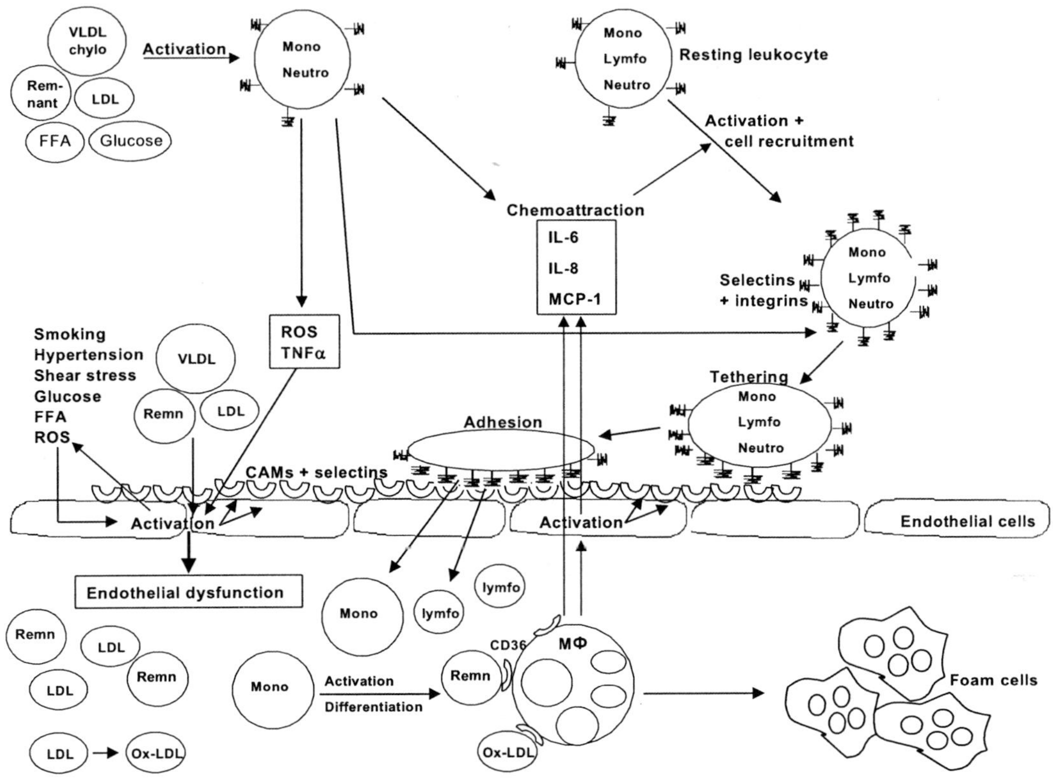

Figure 2. The Inflammatory Pathogenesis of Atherosclerosis. Endothelial cells become activated as a result of smoking, hypertension, shear stress, hyperglycaemia, reactive oxygen species (ROS) and dyslipidemia. The latter can be caused by very low density lipoprotein (VLDL) particles, remnant particles, low density lipoprotein (LDL) particles and free fatty acids (FFA). The results of endothelial cell activation are ROS production, dysfunction of the endothelium and expression of cellular adhesion molecules (CAMs) and selectins, facilitating adherence of all leukocytes and transmigration of monocytes and lymphocytes, but not neutrophils. Monocytes residing in the arterial wall become activated as a result of proinflammatory cytokines and differentiate into macrophages. LDL and remnant particles can enter the vessel wall as well. Oxidative modification of LDL results in a highly atherogenic particle that can easily be taken up by macrophages via the scavenger receptor (CD36). Activated monocytes and macrophages in the vessel wall can activate endothelial cells resulting in production of CAMs and cytokines like interleukin-6 (IL-6) and 8 and monocyte chemoattractant protein-1 (MCP-1). These effects combined will lead to recruitment and activation of leukocytes, as expressed by increased selectins and integrins on the outer membrane. Activated leukocytes are attracted to the activated endothelium and will eventually firmly adhere to the vessel wall. Monocytes and lymphocytes can then transmigrate through the endothelial barrier. Furthermore, monocytes and neutrophils can also become directly activated by lipoproteins, FFA and glucose. This will lead to adherence of these cells to the endothelium as well as production of chemoattractants leading recruitment and activation of additional leukocytes. Furthermore, these directly activated leukocytes can activate the endothelium by the production of ROS and tumor necrosis factor α (TNFα).

Besides the beneficial effects of n-3 fatty acids on endothelium function and activation, also HDL-cholesterol-increasing and TG-lowering properties, postprandial TG in particular, have been described [162-164]. In addition, n-3 fatty acids can influence positively the size and oxidizability of lipids (ox-LDL in particular) [162]. Besides effects on lipid metabolism, n-3 fatty acids also have anti-arrhythmic effects [165] which may explain the protective effect with regard to sudden death that was frequently observed in n-3 fatty acid prospective studies [166]. In higher doses, n-3 fatty acids have been shown to lower bloodpressure [167] and decrease thrombogenecity [162]. Also plaque stabilizing effects via reduction of migration of vascular smooth muscle cells and monocytes have been described [162]. The effects of n-3 fatty acids summarized are suggestive of an atheroprotective effect of these fatty acids. Indeed, CHD is less common in populations with a high dietary fish contribution like Alaskan natives, Greenland Eskimos and Japanese islanders [168-170]. In addition, people from Western European countries with a high fish intake have a lower CHD risk than subjects who rarely consume fish [171,172]. Intervention studies based on a diet enriched in n-3 fatty acids showed a reduced CHD mortality as well [171,173]. Recommendation of a n-3 fatty acids enriched diet also has been shown to improve survival after a primary CHD event [174,175].

The mechanism behind the effects of FFA on endothelial cells and probably also leukocytes, is thought to involve the production of reactive oxygen species (ROS) [152,162,176-180]. This theory is supported by studies showing that antioxidants can attenuate FFA induced alteration in cellular redox balance [181]. Besides harmful effects on the endothelium, FFA like glucose, are thought to exert at least part of the pancreatic β-cell toxicity via ROS production [180]. However, in contrast to glucose [182], the exact mechanisms for FFA induced ROS formation have not yet been elucidated.

The atheroprotective properties of n-3 fatty acids most certainly include lower production of ROS leading to a reduction of intracellular protein kinase C (PKC) and nuclear factor κB (NF-κB) activation resulting in reduced transcription and synthesis of proteins like adhesion molecules and pro-inflammatory cytokines by endothelial cells [152,162,183,184]. ROS have also been shown to induce uncoupling of NO synthesis, resulting in production of superoxide radicals instead of the vasodilatory and atheroprotective agent NO [185]. A reduction of ROS could therefore directly and probably also indirectly via lowering of intracellular messengers, lead to a higher NO availability [152]. In addition, FFA have been shown to bind to peroxisome proliferator activating receptors (PPAR), since fatty acids are one of the natural ligands of this receptor [186].

Therapeutical Strategies to Improve FFA Metabolism

Undoubtedly, non-pharmacological life-style interventions are the most important strategies to improve FFA metabolism and consequently enhancement of insulin sensitivity. Since obesity is the key component of the IRS, it is not unexpected that the main target of therapy should be weight reduction by lowering dietary energy intake. When the different macronutrients are normally balanced in the diet, reduction of total energy intake was more effective in improving FFA metabolism, than an isoenergetic low fat diet [187]. Via weight reduction plasma levels of FFA will fall, leading to improvement of insulin sensitivity and endothelial function via the above described mechanisms. The most striking observation is that weight loss per se has been shown to reduce the progression of atherosclerosis measured by IMT [188]. Besides a lower bodyweight, more specifically a redistribution of fat stores in favor of the subcutaneous area would be preferable with regard to the IRS [189]. Reduction of bodyweight together with improved insulin sensitivity is accompanied with a reduction of pro-inflammatory markers like CRP, IL-6 and TNFα [190,191] and a reduction of adipocyte derived ASP [103]. Unfortunately, the clinical experience is that long term maintenance of the lower body weight is very difficult.

Other lifestyle modification should include regular exercise, which is first of all important to achieve and maintain weight loss, but has also shown direct improvement of insulin sensitivity [192]. In addition, exercise has been shown to sensitize lipolytic pathways by increasing the activity of LPL and HL [192].

Besides a lower total dietary energy intake to reduce body weight, also a balanced diet has proven efficacy. A diet low in saturated fatty acids has been shown to reduce total cholesterol and LDL-cholesterol, while TG levels remained unchanged [193]. Furthermore, in another study substitution of saturated fatty acids by polyunsaturated fatty acids resulted in improvement of insulin sensitivity and a decrease of abdominal fat, while total bodyweight remained unchanged [194]. As mentioned in the previous section, n-3 polyunsaturated fatty acids are the most beneficial unsaturated fatty acids, via effects on lipid metabolism, endothelial function and inflammatory processes.

The effects of n-3 fatty acids are most reasonably mediated via PKC and NF-κB activation but also via binding to PPARs [195,196]. PPARs are nuclear receptors that after ligand-induced activation form a heterodimer with a ligand-activated retinoic acid receptor (RXR). This activated complex then binds to a nuclear PPAR responsive element, which subsequently leads to gene transcription and synthesis of proteins with a variety of effects. PPARs play a role in protein, glucose and lipid metabolism, cellular differentiation, proliferation and apoptosis, neoplastic proliferation, inflammatory processes and endothelial function [197-201]. Three subtypes, PPARα, PPARβ/δ and PPARγ with distinct functions and distribution have been identified. PPARα is expressed in liver, heart, muscle, and kidney and is involved in the catabolism of fatty acids by stimulating uptake and peroxisomal β-oxidation [195,199,201]. Herewith hepatic FFA fluxes can be diverted from the route leading to a VLDL overproduction, resulting in lower plasma TG, increased HDL-cholesterol and less small dense LDL-cholesterol. It has also been shown that PPARα downregulate ApoCIII synthesis, leading to reduced inhibition of LPL mediated lipolysis [199,201]. In contrast, the production of ApoAI and ApoAII is upregulated, which may contribute to the increase of HDL-cholesterol, since these apolipoproteins are the major constituent of this particle [201]. Further PPARα induced enhancement of reversed cholesterol transport is via augmenting the macrophage lipid transporter ABCA1 (ATP-binding cassette, subfamily A, member 1) [202,203]. It is via PPARα activation that certain fatty acids, among them the n-3 polyunsaturated fatty acids, exert the hypolipidemic effects [195]. In addition, PPARα activation has been shown to exert direct anti-atherogenic effects [201]. This may be due to PPARα expression in macrophages, dendritic cells, B and T lymphocytes and also in vascular cells [197,204]. It is thought that via PPARα activation, pro-inflammatory pathways like NF-κB and mitogen-activated protein kinase cascades are controlled [197,201]. It is most likely that via these mechanisms leukocyte-endothelial interaction, transmigration of inflammatory cells, proliferation of vascular smooth muscle cells, thrombus formation and apoptosis of cells in the necrotic core in the plaque can be repressed [201].

The second type of PPAR, PPARβ/δ is ubiquitously expressed, however this receptor is under-exposed since a lack of specific agonists hampered further investigation. Very recently this receptor has been implicated in wound healing [197,205]. In addition, PPARβ/δ activation has shown hypolipidemic effects by increasing lipid loading in macrophages, an effect that on the other hand may increase HDL formation [197,201]. So far, the relation between certain fatty acids and PPARβ/δ activation as well as the clinical relevance is unknown.

The third member of PPAR is PPARγ, that is highly expressed in adipocytes and macrophages. In lower levels, this receptor is also found in the pancreas, skeletal muscle, vasculature, T cells, neutrophils and smooth muscle cells [197,198,200]. PPARγ promotes adipogenesis via a variety of adipose specific genes, resulting in smaller more insulin sensitive adipocytes as well as a redistribution of fat stores in favor of the subcutaneous depots, whereas visceral fat is reduced [200,206]. As a consequence of adipogenesis a mild weight gain has been described upon PPARγ activation [198]. On the other hand, the adipogenetic effects result in a decrease of plasma TG and FFA and a subsequent increase in HDL-cholesterol. The mechanisms for this are not fully elucidated, but may involve an increase in LPL activity and improvement of FFA clearance by adipose tissue and in part also by the liver [199]. Increased clearance is thought to involve both enhanced uptake as well as storage of FFA via upregulation of FFA transporters and enzymes involved in lipogenesis [199,207]. Furthermore, like for PPARα, cholesterol efflux from macrophages is stimulated upon PPARγ activation [201,202]. More importantly, PPARγ activation leads to improvement of insulin sensitivity, however the mechanisms are a matter of debate [197-201]. This effect is most certainly indirect via reduction of plasma FFA and adipose tissue TNFα production [198-200]. In addition, it is thought that insulin sensitivity is directly enhanced due to reduced hepatic gluconeogenesis and enhanced muscle and adipose tissue glucose uptake [199,200]. Furthermore, a variety of anti-inflammatory and vascular effects have been ascribed to PPARγ activation, most of them via inhibition of NF-κB activation leading to a reduction of pro-inflammatory cytokine and adhesion molecule production. In addition, activation of PPARγ may lead to reductions of prothrombotic markers, decrease of vascular intimal hyperplasia and a lowering of bloodpressure albeit an increase in plasma volume [198,200,201]. Recently, it has also been suggested that PPARγ activation may play a protective role in carcinogenesis due to the beneficial effect on cell differentiation and apoptosis [208]. Among the emerging roles of PPARs it is also becoming evident that PPARα and PPARγ play an important role in innate and adaptive immunity [197]. A variety of fatty acids are among the natural ligands of PPARγ, however, the differences in response to these fatty acids still have to be elucidated [198].

Pharmacological options to improve FFA metabolism are first of all drugs inducing PPARs. For many years the working mechanism of fibrates was unknown, however upon the discovery of the PPARs, it became apparent that fibrates are strong activators of PPARα [195]. Via the above described mechanisms, these drugs have been shown to reduce plasma TG and increase HDL-cholesterol levels [201,209,210]. In addition, fibrate induced reduction of postprandial FFA levels by improvement of LPL action has been described [211,212]. On the other hand, fibrate treatment did not show a reduction of release of endogenous FFA by the adipose tissue, as may have been expected by the low expression of PPARα in this tissue

[213]. Furthermore, in some but not all studies a reduction of LDL-cholesterol was observed [201,209,210]. This effect may be due to lower CETP mediated TG and cholesteryl ester exchange between TRP's and LDL-cholesterol resulting in a decreased small dense LDL formation [209,210,213]. In addition as discussed above, fibrates most certainly have direct anti-atherogenic effects [201]. In high cholesterol fed animals fibrate treatment resulted in plaque regression [214] and very recently improvement of endothelial function in hyperTG patients upon fibrate treatment has been described [211]. Furthermore, in a series of large intervention studies, the combined effects of fibrates have shown benefit in the primary and secondary prevention of CHD events and mortality [201].

So far two different clinically applicable PPARγ agonists (thiazolidinediones) are available: rosiglitazon and pioglitazon. As described, these drugs have profound effects on insulin sensitivity and lipid metabolism and therewith they are very promising agents in the treatment of the IRS. Regarding the TG lowering and HDL-cholesterol increasing effects of both agents, a recent meta-analysis suggested benefit in favor of pioglitazon [215]. Unfortunately both drugs induce an initial increase in LDL-cholesterol, however this effect is limited and involves a shift from small-dense LDL particles to less atherogenic larger buoyant LDL particles [215]. Animal studies have shown plaque size reduction and in a human study with type 2 diabetes patients, inhibition of restenosis after angioplasty upon treatment with PPARγ agonists have been shown [216-218]. Furthermore, improvement of endothelial function has been described in animal studies and in humans [200,219,220]. Unfortunately, long-term studies on the effects on development of the IRS and cardiovascular outcome are not available yet. Present and future research on PPAR agonists will be focussed on the development of tissue selective PPAR agonists, like selective PPARγ agonists. These drugs have shown a more pronounced effect on insulin sensitivity than on adipogenesis and thus result in a lower weight gain [221,222]. In addition, dual PPARα/γ agonists have been developed that may have major effects on different components of the IRS [223,224].

HMG-CoA reductase inhibitors (statins) induce the partial inhibition of HMG-CoA reductase, the rate-limiting enzyme in cholesterol synthesis. The resulting decrease in hepatic intracellular cholesterol concentration leads to an upregulation of LDL-receptors, enabling enhanced clearance of apoB and apoE containing lipoproteins, LDL in particular, but also VLDL and IDL [225]. Furthermore, it is thought that statins to some extent inhibit hepatic VLDL synthesis, which may explain the TG lowering properties of these drugs [193,226]. Besides reduction of fasting TG, also improvement of postprandial triglyceridemia has been shown [227,228]. Statin induced improvement of peripheral FFA trapping has been suggested by two studies showing a lowering of fasting and postprandial C3 levels [38,101] with a concomitant reduction of hepatic FFA flux in one study [101]. However, direct FFA lowering effects of statins have not been observed yet. Other actions of statins may include partial agonism of PPARα and γ [229,230]. Further evidence for this theory is given by the overlap between the non-lipid lowering (pleiotrophic) effects of statins and the variety of effects of PPAR agonists. Pleiotrophic effects of statins include acute improvement of endothelial function, anti-inflammatory effects resulting in reduced adherence and transmigration of leukocytes, reduction of ROS generation, plaque stabilization and regression, re-endothelialization and antithrombotic effects [231-236]. Besides the pleiotrophic effects relevant for cardiovascular disease, statins may have anticancer properties, may reduce osteoporosis, and may decrease the development of dementia and multiple sclerosis [237-239]. In a series of large clinical studies, statins have shown reductions of CHD morbidity and mortality in various groups of patients and healthy subjects [19,101,240-243]. In addition, a reduction of cerebrovascular and peripheral vascular diseases upon statin treatment has been described [19,233].

Other drugs that in potency may affect FFA metabolism include orlistat, an inhibitor of pancreas lipase that decreases the intestinal uptake of fat [244] and significantly improves postprandial lipemia and LDL-cholesterol levels [245]. Secondly, nicotinic acid (niacin), a drug that reduces VLDL production and peripheral FFA release, and raises HDL-cholesterol, may be of importance [246]. Unfortunately, both agents have a variety of side effects that limit their clinical use [244,246].

In conclusion, improved understanding of the metabolism of FFA may help to develop targeted approaches to frequent diseases like obesity, insulin resistance, type 2 diabetes and atherosclerosis.

References

[1] Braunwald E. Shattuck lecture--cardiovascular medicine at the turn of the millennium: triumphs, concerns, and opportunities. *N Engl J Med* 1997;337:1360-1369.

[2] Lusis AJ. Atherosclerosis. *Nature* 2000;407:233-241.

[3] Ross R. Atherosclerosis--an inflammatory disease. *N Engl J Med* 1999;340:115-126.

[4] Calle EE, Thun MJ, Petrelli JM, Rodriguez C, Heath CW, Jr. Body-mass index and mortality in a prospective cohort of U.S. adults. *N Engl J Med* 1999;341:1097-1105.

[5] Despres JP, Moorjani S, Lupien PJ, Tremblay A, Nadeau A, Bouchard C. Regional distribution of body fat, plasma lipoproteins, and cardiovascular disease. *Arteriosclerosis* 1990;10:497-511.

[6] Despres JP, Lamarche B, Mauriege P, Cantin B, Dagenais GR, Moorjani S, Lupien PJ. Hyperinsulinemia as an independent risk factor for ischemic heart disease. *N Engl J Med* 1996;334:952-957.

[7] Kuulasmaa K, Tunstall-Pedoe H, Dobson A, Fortmann S, Sans S, Tolonen H, Evans A, Ferrario M, Tuomilehto J. Estimation of contribution of changes in classic risk factors to trends in coronary-event rates across the WHO MONICA Project populations. *Lancet* 2000;355:675-687.

[8] Willeit J, Kiechl S, Oberhollenzer F, Rungger G, Egger G, Bonora E, Mitterer M, Muggeo M. Distinct risk profiles of early and advanced atherosclerosis: prospective results from the Bruneck Study. *Arterioscler Thromb Vasc Biol* 2000;20:529-537.

[9] Reaven GM. Banting lecture 1988. Role of insulin resistance in human disease. *Diabetes* 1988;37:1595-1607.

[10] Mokdad AH, Ford ES, Bowman BA, Nelson DE, Engelgau MM, Vinicor F, Marks JS. Diabetes trends in the U.S.: 1990-1998. *Diabetes Care* 2000;23:1278-1283.

[11] Haffner SM, Lehto S, Ronnemaa T, Pyorala K, Laakso M. Mortality from coronary heart disease in subjects with type 2 diabetes and in nondiabetic subjects with and without prior myocardial infarction. *N Engl J Med* 1998;339:229-234.

[12] Kannel WB, McGee DL. Diabetes and glucose tolerance as risk factors for cardiovascular disease: the Framingham study. *Diabetes Care* 1979;2:120-126.

[13] Nathan DM, Meigs J, Singer DE. The epidemiology of cardiovascular disease in type 2 diabetes mellitus: how sweet it is ... or is it? *Lancet* 1997;350 Suppl 1:SI4-SI9.

[14] Turner RC, Millns H, Neil HA, Stratton IM, Manley SE, Matthews DR, Holman RR. Risk factors for coronary artery disease in non-insulin dependent diabetes mellitus: United Kingdom Prospective Diabetes Study (UKPDS: 23). *BMJ* 1998;316:823-828.

[15] Mokdad AH, Bowman BA, Ford ES, Vinicor F, Marks JS, Koplan JP. The continuing epidemics of obesity and diabetes in the United States. *JAMA* 2001;286:1195-1200.

[16] Zimmet P, Alberti KG, Shaw J. Global and societal implications of the diabetes epidemic. *Nature* 2001;414:782-787.

[17] Intensive blood-glucose control with sulphonylureas or insulin compared with conventional treatment and risk of complications in patients with type 2 diabetes (UKPDS 33). UK Prospective Diabetes Study (UKPDS) Group. *Lancet* 1998;352:837-853.

[18] Tight blood pressure control and risk of macrovascular and microvascular complications in type 2 diabetes: UKPDS 38. UK Prospective Diabetes Study Group. *BMJ* 1998;317:703-713.

[19] MRC/BHF Heart Protection Study of cholesterol lowering with simvastatin in 20,536 high-risk individuals: a randomised placebo-controlled trial. *Lancet* 2002;360:7-22.

[20] Goldberg RB, Mellies MJ, Sacks FM, Moye LA, Howard BV, Howard WJ, Davis BR, Cole TG, Pfeffer MA, Braunwald E. Cardiovascular events and their reduction with pravastatin in diabetic and glucose-intolerant myocardial infarction survivors with average cholesterol levels: subgroup analyses in the cholesterol and recurrent events (CARE) trial. The Care Investigators. *Circulation* 1998;98:2513-2519.

[21] Pyorala K, Pedersen TR, Kjekshus J, Faergeman O, Olsson AG, Thorgeirsson G. Cholesterol lowering with simvastatin improves prognosis of diabetic patients with coronary heart disease. A subgroup analysis of the Scandinavian Simvastatin Survival Study (4S). *Diabetes Care* 1997;20:614-620.

[22] Gaede P, Vedel P, Larsen N, Jensen GV, Parving HH, Pedersen O. Multifactorial intervention and cardiovascular disease in patients with type 2 diabetes. *N Engl J Med* 2003;348:383-393.

[23] de Lorgeril M, Salen P, Martin JL, Monjaud I, Delaye J, Mamelle N. Mediterranean diet, traditional risk factors, and the rate of cardiovascular complications after myocardial infarction: final report of the Lyon Diet Heart Study. *Circulation* 1999;99:779-785.

[24] Fontbonne A, Eschwege E, Cambien F, Richard JL, Ducimetiere P, Thibult N, Warnet JM, Claude JR, Rosselin GE. Hypertriglyceridaemia as a risk factor of coronary heart disease mortality in subjects with impaired glucose tolerance or diabetes. Results from the 11-year follow-up of the Paris Prospective Study. *Diabetologia* 1989;32:300-304.

[25] Laakso M, Lehto S, Penttila I, Pyorala K. Lipids and lipoproteins predicting coronary heart disease mortality and morbidity in patients with non-insulin-dependent diabetes. *Circulation* 1993;88:1421-1430.

[26] Syvanne M, Taskinen MR. Lipids and lipoproteins as coronary risk factors in non-insulin-dependent diabetes mellitus. *Lancet* 1997;350 Suppl 1:SI20-SI23.

[27] Barter PJ, Brewer HB, Jr., Chapman MJ, Hennekens CH, Rader DJ, Tall AR. Cholesteryl ester transfer protein: a novel target for raising HDL and inhibiting atherosclerosis. *Arterioscler Thromb Vasc Biol* 2003;23:160-167.

[28] Kawamura M, Heinecke JW, Chait A. Pathophysiological concentrations of glucose promote oxidative modification of low density lipoprotein by a superoxide-dependent pathway. *J Clin Invest* 1994;94:771-778.

[29] Taskinen MR. Diabetic dyslipidaemia: from basic research to clinical practice. *Diabetologia* 2003;46:733-749.

[30] Sniderman AD, Lamarche B, Tilley J, Seccombe D, Frohlich J. Hypertriglyceridemic hyperapoB in type 2 diabetes. *Diabetes Care* 2002;25:579-582.

[31] Castro Cabezas M, Halkes CJ, Meijssen S, van Oostrom AJ, Erkelens DW. Diurnal triglyceride profiles: a novel approach to study triglyceride changes. *Atherosclerosis* 2001;155:219-228.

[32] Delawi D, Meijssen S, Castro Cabezas M. Intra-individual variations of fasting plasma lipids, apolipoproteins and postprandial lipemia in familial combined hyperlipidemia compared to controls. *Clin Chim Acta* 2003;328:139-145.

[33] van Oostrom AJ, Castro Cabezas M, Ribalta J, Masana L, Twickler TB, Remijnse TA, Erkelens DW. Diurnal triglyceride profiles in healthy normolipidemic male subjects are associated to insulin sensitivity, body composition and diet. *Eur J Clin Invest* 2000;30:964-971.

[34] van Wijk JP, Castro Cabezas M, Halkes CJ, Erkelens DW. Effects of different nutrient intakes on daytime triacylglycerolemia in healthy, normolipemic, free-living men. *Am J Clin Nutr* 2001;74:171-178.

[35] van Wijk JP, Halkes CJ, Erkelens DW, Castro Cabezas M. Fasting and daylong triglycerides in obesity with and without type 2 diabetes. *Metabolism* 2003;52:1043-1049.

[36] Genest JJ, McNamara JR, Salem DN, Schaefer EJ. Prevalence of risk factors in men with premature coronary artery disease. *Am J Cardiol* 1991;67:1185-1189.

[37] Miller M, Seidler A, Moalemi A, Pearson TA. Normal triglyceride levels and coronary artery disease events: the Baltimore Coronary Observational Long-Term Study. *J Am Coll Cardiol* 1998;31:1252-1257.

[38] Halkes CJM, van Dijk H, de Jaegere PP, Plokker HW, van der Helm Y, Erkelens DW, Castro Cabezas M. Postprandial increase of complement component 3 in normolipidemic patients with coronary artery disease: effects of expanded-dose simvastatin. *Arterioscler Thromb Vasc Biol* 2001;21:1526-1530.

[39] Meyer E, Westerveld HT, Ruyter-Meijstek FC, van Greevenbroek MM, Rienks R, van Rijn HJ, Erkelens DW, de Bruin TW. Abnormal postprandial apolipoprotein B-48 and triglyceride responses in normolipidemic women with greater than 70% stenotic coronary artery disease: a case-control study. *Atherosclerosis* 1996;124:221-235.

[40] Patsch JR, Miesenbock G, Hopferwieser T, Muhlberger V, Knapp E, Dunn JK, Gotto AM, Jr., Patsch W. Relation of triglyceride metabolism and coronary artery disease. Studies in the postprandial state. *Arterioscler Thromb* 1992;12:1336-1345.

[41] van Wijk JP, Halkes CJ, de Jaegere PP, Plokker HW, Erkelens DW, Castro Cabezas M. Normalization of daytime triglyceridemia by simvastatin in fasting normotriglyceridemic patients with premture coronary sclerosis. *Atherosclerosis* 2003, in press.

[42] Weintraub MS, Grosskopf I, Rassin T, Miller H, Charach G, Rotmensch HH, Liron M, Rubinstein A, Iaina A. Clearance of chylomicron remnants in normolipidaemic patients with coronary artery disease: case control study over three years. *BMJ* 1996;312:936-939.

[43] Zilversmit DB. Atherogenesis: a postprandial phenomenon. *Circulation* 1979;60:473-485.

[44] Castro Cabezas M, Erkelens DW. The direct way from gut to vessel wall: atheroinitiation. *Eur J Clin Invest* 1998;28:504-505.

[45] Ceriello A. The possible role of postprandial hyperglycaemia in the pathogenesis of diabetic complications. *Diabetologia* 2003;46 Suppl 1:M9-M16.

[46] Lefebvre PJ, Scheen AJ. The postprandial state and risk of cardiovascular disease. *Diabet Med* 1998;15 Suppl 4:S63-S68.

[47] Temelkova-Kurktschiev TS, Koehler C, Henkel E, Leonhardt W, Fuecker K, Hanefeld M. Postchallenge plasma glucose and glycemic spikes are more strongly associated with atherosclerosis than fasting glucose or HbA1c level. *Diabetes Care* 2000;23:1830-1834.

[48] McGarry JD. Banting lecture 2001: dysregulation of Fatty Acid metabolism in the etiology of type 2 diabetes. *Diabetes* 2002;51:7-18.

[49] Boden G, Shulman GI. Free fatty acids in obesity and type 2 diabetes: defining their role in the development of insulin resistance and beta-cell dysfunction. *Eur J Clin Invest* 2002;32 Suppl 3:14-23.

[50] Randle PJ, Garland PB, Hales CN, Newsholme EA. The glucose fatty-acid cycle. Its role in insulin sensitivity and the metabolic disturbances of diabetes mellitus. *Lancet* 1963;1:785-789.

[51] Randle PJ. Regulatory interactions between lipids and carbohydrates: the glucose fatty acid cycle after 35 years. *Diabetes Metab Rev* 1998;14:263-283.

[52] Boden G, Chen X, Ruiz J, White JV, Rossetti L. Mechanisms of fatty acid-induced inhibition of glucose uptake. *J Clin Invest* 1994;93:2438-2446.

[53] Wolfe BM, Klein S, Peters EJ, Schmidt BF, Wolfe RR. Effect of elevated free fatty acids on glucose oxidation in normal humans. *Metabolism* 1988;37:323-329.

[54] Shulman GI. Cellular mechanisms of insulin resistance. *J Clin Invest* 2000;106:171-176.

[55] Ferrannini E, Barrett EJ, Bevilacqua S, DeFronzo RA. Effect of fatty acids on glucose production and utilization in man. *J Clin Invest* 1983;72:1737-1747.

[56] Svedberg J, Bjorntorp P, Smith U, Lonnroth P. Free-fatty acid inhibition of insulin binding, degradation, and action in isolated rat hepatocytes. *Diabetes* 1990;39:570-574.

[57] Swislocki AL, Chen YD, Golay A, Chang MO, Reaven GM. Insulin suppression of plasma-free fatty acid concentration in normal individuals and patients with type 2 (non-insulin-dependent) diabetes. *Diabetologia* 1987;30:622-626.

[58] Taskinen MR, Lahdenpera S, Syvanne M. New insights into lipid metabolism in non-insulin-dependent diabetes mellitus. *Ann Med* 1996;28:335-340.

[59] Basu A, Basu R, Shah P, Vella A, Rizza RA, Jensen MD. Systemic and regional free fatty acid metabolism in type 2 diabetes. *Am J Physiol Endocrinol Metab* 2001;280:E1000-E1006.

[60] Lewis GF, Uffelman KD, Szeto LW, Weller B, Steiner G. Interaction between free fatty acids and insulin in the acute control of very low density lipoprotein production in humans. *J Clin Invest* 1995;95:158-166.

[61] Coppack SW. Postprandial lipoproteins in non-insulin-dependent diabetes mellitus. *Diabet Med* 1997;14 Suppl 3:S67-S74.

[62] Taskinen MR. Lipoprotein lipase in diabetes. *Diabetes Metab Rev* 1987;3:551-570.

[63] Lewis GF, O'Meara NM, Soltys PA, Blackman JD, Iverius PH, Pugh WL, Getz GS, Polonsky KS. Fasting hypertriglyceridemia in noninsulin-dependent diabetes mellitus is an important predictor of postprandial lipid and lipoprotein abnormalities. *J Clin Endocrinol Metab* 1991;72:934-944.

[64] Lewis GF, Steiner G. Hypertriglyceridemia and its metabolic consequences as a risk factor for atherosclerotic cardiovascular disease in non-insulin-dependent diabetes mellitus. *Diabetes Metab Rev* 1996;12:37-56.

[65] Syvanne M, Hilden H, Taskinen MR. Abnormal metabolism of postprandial lipoproteins in patients with non-insulin-dependent diabetes mellitus is not related to coronary artery disease. *J Lipid Res* 1994;35:15-26.

[66] Laffel L. Ketone bodies: a review of physiology, pathophysiology and application of monitoring to diabetes. *Diabetes Metab Res Rev* 1999;15:412-426.

[67] Meijssen S, Castro Cabezas M, Twickler TB, Jansen H, Erkelens DW. In vivo evidence of defective postprandial and postabsorptive free fatty acid metabolism in familial combined hyperlipidemia. *J Lipid Res* 2000;41:1096-1102.

[68] Halkes CJ, van Dijk H, Verseyden C, de Jaegere PP, Plokker HW, Meijssen S, Erkelens DW, Castro Cabezas M. Gender Differences in Postprandial Ketone Bodies in Normolipidemic Subjects and in Untreated Patients With Familial Combined Hyperlipidemia. *Arterioscler Thromb Vasc Biol* 2003, in press.

[69] Frayn KN. Non-esterified fatty acid metabolism and postprandial lipaemia. *Atherosclerosis* 1998;141 Suppl 1:S41-S46.

[70] Sniderman AD, Cianflone K, Arner P, Summers LK, Frayn KN. The adipocyte, fatty acid trapping, and atherogenesis. *Arterioscler Thromb Vasc Biol* 1998;18:147-151.

[71] Merkel M, Eckel RH, Goldberg IJ. Lipoprotein lipase: genetics, lipid uptake, and regulation. *J Lipid Res* 2002;43:1997-2006.

[72] Eriksson JW, Buren J, Svensson M, Olivecrona T, Olivecrona G. Postprandial regulation of blood lipids and adipose tissue lipoprotein lipase in type 2 diabetes patients and healthy control subjects. *Atherosclerosis* 2003;166:359-367.

[73] Kobayashi J, Tashiro J, Murano S, Morisaki N, Saito Y. Lipoprotein lipase mass and activity in post-heparin plasma from subjects with intra-abdominal visceral fat accumulation. *Clin Endocrinol (Oxf)* 1998;48:515-520.

[74] Arner P. Free fatty acids – do they play a central role in type 2 diabetes? *Diabetes Obes Metab* 2001;3 Suppl 1:11-19.

[75] Jansson PA, Smith U, Lonnroth P. Interstitial glycerol concentration measured by microdialysis in two subcutaneous regions in humans. *Am J Physiol* 1990;258:E918-E922.
[76] Jensen MD. Lipolysis: contribution from regional fat. *Annu Rev Nutr* 1997;17:127-139.
[77] Meek SE, Nair KS, Jensen MD. Insulin regulation of regional free fatty acid metabolism. *Diabetes* 1999;48:10-14.
[78] Hamilton JA, Kamp F. How are free fatty acids transported in membranes? Is it by proteins or by free diffusion through the lipids? *Diabetes* 1999;48:2255-2269.
[79] Abumrad NA, el Maghrabi MR, Amri EZ, Lopez E, Grimaldi PA. Cloning of a rat adipocyte membrane protein implicated in binding or transport of long-chain fatty acids that is induced during preadipocyte differentiation. Homology with human CD36. *J Biol Chem* 1993;268:17665-17668.
[80] Berk PD, Zhou SL, Kiang CL, Stump D, Bradbury M, Isola LM. Uptake of long chain free fatty acids is selectively up-regulated in adipocytes of Zucker rats with genetic obesity and non-insulin-dependent diabetes mellitus. *J Biol Chem* 1997;272:8830-8835.
[81] Binnert C, Koistinen HA, Martin G, Andreelli F, Ebeling P, Koivisto VA, Laville M, Auwerx J, Vidal H. Fatty acid transport protein-1 mRNA expression in skeletal muscle and in adipose tissue in humans. *Am J Physiol Endocrinol Metab* 2000;279:E1072-E1079.
[82] Hainque B, Guerre-Millo M, Hainault I, Moustaid N, Wardzala LJ, Lavau M. Long term regulation of glucose transporters by insulin in mature 3T3-F442A adipose cells. Differential effects on two glucose transporter subtypes. *J Biol Chem* 1990;265:7982-7986.
[83] Frayn KN, Shadid S, Hamlani R, Humphreys SM, Clark ML, Fielding BA, Boland O, Coppack SW. Regulation of fatty acid movement in human adipose tissue in the postabsorptive-to-postprandial transition. *Am J Physiol* 1994;266:E308-E317.
[84] Cianflone KM, Sniderman AD, Walsh MJ, Vu HT, Gagnon J, Rodriguez MA. Purification and characterization of acylation stimulating protein. *J Biol Chem* 1989;264:426-430.
[85] Cianflone K, Xia Z, Chen LY. Critical review of acylation-stimulating protein physiology in humans and rodents. *Biochim Biophys Acta* 2003;1609:127-143.
[86] Meijssen S, van Dijk H, Verseyden C, Erkelens DW, Castro Cabezas M. Delayed and exaggerated postprandial complement component 3 response in familial combined hyperlipidemia. *Arterioscler Thromb Vasc Biol* 2002;22:811-816.
[87] Scantlebury T, Sniderman AD, Cianflone K. Regulation by retinoic acid of acylation-stimulating protein and complement C3 in human adipocytes. *Biochem J* 2001;356:445-452.
[88] Walport MJ. Complement. First of two parts. *N Engl J Med* 2001;344:1058-1066.
[89] Wilken HC, Gotze O, Werfel T, Zwirner J. C3a(desArg) does not bind to and signal through the human C3a receptor. *Immunol Lett* 1999;67:141-145.

[90] Kalant D, Cain SA, Maslowska M, Sniderman AD, Cianflone K, Monk PN. The chemoattractant receptor-like protein C5L2 binds the C3a des-Arg77/acylation-stimulating protein. *J Biol Chem* 2003;278:11123-11129.

[91] Germinario R, Sniderman AD, Manuel S, Lefebvre SP, Baldo A, Cianflone K. Coordinate regulation of triacylglycerol synthesis and glucose transport by acylation-stimulating protein. *Metabolism* 1993;42:574-580.

[92] Van Harmelen V, Reynisdottir S, Cianflone K, Degerman E, Hoffstedt J, Nilsell K, Sniderman A, Arner P. Mechanisms involved in the regulation of free fatty acid release from isolated human fat cells by acylation-stimulating protein and insulin. *J Biol Chem* 1999;274:18243-18251.

[93] Muscari A, Massarelli G, Bastagli L, Poggiopollini G, Tomassetti V, Volta U, Puddu GM, Puddu P. Relationship between serum C3 levels and traditional risk factors for myocardial infarction. *Acta Cardiol* 1998;53:345-354.

[94] Muscari A, Massarelli G, Bastagli L, Poggiopollini G, Tomassetti V, Drago G, Martignani C, Pacilli P, Boni P, Puddu P. Relationship of serum C3 to fasting insulin, risk factors and previous ischaemic events in middle-aged men. *Eur Heart J* 2000;21:1081-1090.

[95] Griselli M, Herbert J, Hutchinson WL, Taylor KM, Sohail M, Krausz T, Pepys MB. C-reactive protein and complement are important mediators of tissue damage in acute myocardial infarction. *J Exp Med* 1999;190:1733-1740.

[96] Torzewski J, Torzewski M, Bowyer DE, Frohlich M, Koenig W, Waltenberger J, Fitzsimmons C, Hombach V. C-reactive protein frequently colocalizes with the terminal complement complex in the intima of early atherosclerotic lesions of human coronary arteries. *Arterioscler Thromb Vasc Biol* 1998;18:1386-1392.

[97] Charlesworth JA, Peake PW, Campbell LV, Pussell BA, O'Grady S, Tzilopoulos T. The influence of oral lipid loads on acylation stimulating protein (ASP) in healthy volunteers. *Int J Obes* 1998;22:1096-1102.

[98] Koistinen HA, Vidal H, Karonen SL, Dusserre E, Vallier P, Koivisto VA, Ebeling P. Plasma acylation stimulating protein concentration and subcutaneous adipose tissue C3 mRNA expression in nondiabetic and type 2 diabetic men. *Arterioscler Thromb Vasc Biol* 2001;21:1034-1039.

[99] Ylitalo K, Pajukanta P, Meri S, Cantor RM, Mero-Matikainen N, Vakkilainen J, Nuotio I, Taskinen MR. Serum C3 but not plasma acylation-stimulating protein is elevated in Finnish patients with familial combined hyperlipidemia. *Arterioscler Thromb Vasc Biol* 2001;21:838-843.

[100] Saleh J, Summers LK, Cianflone K, Fielding BA, Sniderman AD, Frayn KN. Coordinated release of acylation stimulating protein (ASP) and triacylglycerol clearance by human adipose tissue in vivo in the postprandial period. *J Lipid Res* 1998;39:884-891.

[101] Verseyden C, Meijssen S, Dijk HH, Jansen H, Castro CM. Effects of atorvastatin on fasting and postprandial complement component 3 response in familial combined hyperlipidemia. *J Lipid Res* 2003;44:2100-8.

[102] van Oostrom AJ, van Dijk H, Verseyden C, Sniderman A, Cianflone K, Rabelink TJ, Castro Cabezas M. Addition of glucose to an oral fat load reduces postprandial free

fatty acids and prevents the postprandial rise of complement component 3. *Am J Clin Nutr* 2004;79:510-5.

[103] Faraj M, Havel PJ, Phelis S, Blank D, Sniderman AD, Cianflone K. Plasma acylation-stimulating protein, adiponectin, leptin, and ghrelin before and after weight loss induced by gastric bypass surgery in morbidly obese subjects. *J Clin Endocrinol Metab* 2003;88:1594-1602.

[104] Matthan NR, Cianflone K, Lichtenstein AH, Ausman LM, Jauhiainen M, Jones PJ. Hydrogenated fat consumption affects acylation-stimulating protein levels and cholesterol esterification rates in moderately hypercholesterolemic women. *J Lipid Res* 2001;42:1841-1848.

[105] Guerre-Millo M. Adipose tissue hormones. *J Endocrinol Invest* 2002;25:855-861.

[106] Mora S, Pessin JE. An adipocentric view of signaling and intracellular trafficking. *Diabetes Metab Res Rev* 2002;18:345-356.

[107] Meijssen S, Castro Cabezas M, Ballieux CG, Derksen RJ, Bilecen S, Erkelens DW. Insulin mediated inhibition of hormone sensitive lipase activity in vivo in relation to endogenous catecholamines in healthy subjects. *J Clin Endocrinol Metab* 2001;86:4193-4197.

[108] Lawrence VJ, Coppack SW. The endocrine function of the fat cell-regulation by the sympathetic nervous system. *Horm Metab Res* 2000;32:453-467.

[109] Kreier F, Fliers E, Voshol PJ, Van Eden CG, Havekes LM, Kalsbeek A, Van Heijningen CL, Sluiter AA, Mettenleiter TC, Romijn JA, Sauerwein HP, Buijs RM. Selective parasympathetic innervation of subcutaneous and intra-abdominal fat--functional implications. *J Clin Invest* 2002;110:1243-1250.

[110] Fried SK, Bunkin DA, Greenberg AS. Omental and subcutaneous adipose tissues of obese subjects release interleukin-6: depot difference and regulation by glucocorticoid. *J Clin Endocrinol Metab* 1998;83:847-850.

[111] Mohamed-Ali V, Goodrick S, Rawesh A, Katz DR, Miles JM, Yudkin JS, Klein S, Coppack SW. Subcutaneous adipose tissue releases interleukin-6, but not tumor necrosis factor-alpha, in vivo. *J Clin Endocrinol Metab* 1997;82:4196-4200.

[112] Papanicolaou DA, Wilder RL, Manolagas SC, Chrousos GP. The pathophysiologic roles of interleukin-6 in human disease. *Ann Intern Med* 1998;128:127-137.

[113] Papanicolaou DA, Vgontzas AN. Interleukin-6: the endocrine cytokine. *J Clin Endocrinol Metab* 2000;85:1331-1333.

[114] Ridker PM, Rifai N, Stampfer MJ, Hennekens CH. Plasma concentration of interleukin-6 and the risk of future myocardial infarction among apparently healthy men. *Circulation* 2000;101:1767-1772.

[115] Fernandez-Real JM, Vayreda M, Richart C, Gutierrez C, Broch M, Vendrell J, Ricart W. Circulating interleukin 6 levels, blood pressure, and insulin sensitivity in apparently healthy men and women. *J Clin Endocrinol Metab* 2001;86:1154-1159.

[116] Kern PA, Ranganathan S, Li C, Wood L, Ranganathan G. Adipose tissue tumor necrosis factor and interleukin-6 expression in human obesity and insulin resistance. *Am J Physiol Endocrinol Metab* 2001;280:E745-E751.

[117] Orban Z, Remaley AT, Sampson M, Trajanoski Z, Chrousos GP. The differential effect of food intake and beta-adrenergic stimulation on adipose-derived hormones and cytokines in man. *J Clin Endocrinol Metab* 1999;84:2126-2133.

[118] van Oostrom AJ, Sijmonsma TP, Verseyden C, Jansen EH, de Koning EJ, Rabelink TJ, Castro Cabezas M. Postprandial recruitment of neutrophils may contribute to endothelial dysfunction. *J Lipid Res* 2003;44:576-583.

[119] Vgontzas AN, Papanicolaou DA, Bixler EO, Lotsikas A, Zachman K, Kales A, Prolo P, Wong ML, Licinio J, Gold PW, Hermida RC, Mastorakos G, Chrousos GP. Circadian interleukin-6 secretion and quantity and depth of sleep. *J Clin Endocrinol Metab* 1999;84:2603-2607.

[120] Rifai N, Ridker PM. High-sensitivity C-reactive protein: a novel and promising marker of coronary heart disease. *Clin Chem* 2001;47:403-411.

[121] Hak AE, Stehouwer CD, Bots ML, Polderman KH, Schalkwijk CG, Westendorp IC, Hofman A, Witteman JC. Associations of C-reactive protein with measures of obesity, insulin resistance, and subclinical atherosclerosis in healthy, middle-aged women. *Arterioscler Thromb Vasc Biol* 1999;19:1986-1991.

[122] Lemieux I, Pascot A, Prud'homme D, Almeras N, Bogaty P, Nadeau A, Bergeron J, Despres JP. Elevated C-reactive protein: another component of the atherothrombotic profile of abdominal obesity. *Arterioscler Thromb Vasc Biol* 2001;21:961-967.

[123] Yudkin JS, Stehouwer CD, Emeis JJ, Coppack SW. C-reactive protein in healthy subjects: associations with obesity, insulin resistance, and endothelial dysfunction: a potential role for cytokines originating from adipose tissue? *Arterioscler Thromb Vasc Biol* 1999;19:972-978.

[124] Munford RS. Statins and the acute-phase response. *N Engl J Med* 2001;344:2016-2018.

[125] Pasceri V, Willerson JT, Yeh ET. Direct proinflammatory effect of C-reactive protein on human endothelial cells. *Circulation* 2000;102:2165-2168.

[126] Wolbink GJ, Brouwer MC, Buysmann S, ten Berge IJ, Hack CE. CRP-mediated activation of complement in vivo: assessment by measuring circulating complement-C-reactive protein complexes. *J Immunol* 1996;157:473-479.

[127] Warne JP. Tumour necrosis factor alpha: a key regulator of adipose tissue mass. *J Endocrinol* 2003;177:351-355.

[128] Moller DE. Potential role of TNF-alpha in the pathogenesis of insulin resistance and type 2 diabetes. *Trends Endocrinol Metab* 2000;11:212-217.

[129] Arner P. The adipocyte in insulin resistance: key molecules and the impact of the thiazolidinediones. *Trends Endocrinol Metab* 2003;14:137-145.

[130] Shimokawa H. Primary endothelial dysfunction: atherosclerosis. *J Mol Cell Cardiol* 1999;31:23-37.

[131] Hsueh WA, Quinones MJ. Role of endothelial dysfunction in insulin resistance. *Am J Cardiol* 2003;92:10J-17J.

[132] Draijer R, Atsma DE, van der LA, van Hinsbergh VW. cGMP and nitric oxide modulate thrombin-induced endothelial permeability. Regulation via different pathways in human aortic and umbilical vein endothelial cells. *Circ Res* 1995;76:199-208.

[133] Langford EJ, Wainwright RJ, Martin JF. Platelet activation in acute myocardial infarction and unstable angina is inhibited by nitric oxide donors. *Arterioscler Thromb Vasc Biol* 1996;16:51-55.

[134] Lefer AM. Nitric oxide: nature's naturally occurring leukocyte inhibitor. *Circulation* 1997;95:553-554.

[135] Sarkar R, Meinberg EG, Stanley JC, Gordon D, Webb RC. Nitric oxide reversibly inhibits the migration of cultured vascular smooth muscle cells. *Circ Res* 1996;78:225-230.

[136] Albertini JP, Valensi P, Lormeau B, Vaysse J, Attali JR, Gattegno L. Soluble L-selectin level is a marker for coronary artery disease in type 2 diabetic patients. *Diabetes Care* 1999;22:2044-2048.

[137] Blann AD, Dobrotova M, Kubisz P, McCollum CN. von Willebrand factor, soluble P-selectin, tissue plasminogen activator and plasminogen activator inhibitor in atherosclerosis. *Thromb Haemost* 1995;74:626-630.

[138] Jang Y, Lincoff AM, Plow EF, Topol EJ. Cell adhesion molecules in coronary artery disease. *J Am Coll Cardiol* 1994;24:1591-1601.

[139] Morise T, Takeuchi Y, Kawano M, Koni I, Takeda R. Increased plasma levels of immunoreactive endothelin and von Willebrand factor in NIDDM patients. *Diabetes Care* 1995;18:87-89.

[140] Gnasso A, Carallo C, Irace C, De Franceschi MS, Mattioli PL, Motti C, Cortese C. Association between wall shear stress and flow-mediated vasodilation in healthy men. *Atherosclerosis* 2001;156:171-176.

[141] Hijmering ML, Stroes ES, Pasterkamp G, Sierevogel M, Banga JD, Rabelink TJ. Variability of flow mediated dilation: consequences for clinical application. *Atherosclerosis* 2001;157:369-373.

[142] Kanters SD, Algra A, van Leeuwen MS, Banga JD. Reproducibility of in vivo carotid intima-media thickness measurements: a review. *Stroke* 1997;28:665-671.

[143] Anderson TJ, Uehata A, Gerhard MD, Meredith IT, Knab S, Delagrange D, Lieberman EH, Ganz P, Creager MA, Yeung AC. Close relation of endothelial function in the human coronary and peripheral circulations. *J Am Coll Cardiol* 1995;26:1235-1241.

[144] Sarabi M, Vessby B, Millgard J, Lind L. Endothelium-dependent vasodilation is related to the fatty acid composition of serum lipids in healthy subjects. *Atherosclerosis* 2001;156:349-355.

[145] Steinberg HO, Tarshoby M, Monestel R, Hook G, Cronin J, Johnson A, Bayazeed B, Baron AD. Elevated circulating free fatty acid levels impair endothelium-dependent vasodilation. *J Clin Invest* 1997;100:1230-1239.

[146] van Etten RW, de Koning EJ, Verhaar MC, Gaillard CA, Rabelink TJ. Impaired NO-dependent vasodilation in patients with Type II (non-insulin-dependent) diabetes mellitus is restored by acute administration of folate. *Diabetologia* 2002;45:1004-1010.

[147] Anderson RA, Evans ML, Ellis GR, Graham J, Morris K, Jackson SK, Lewis MJ, Rees A, Frenneaux MP. The relationships between post-prandial lipaemia, endothelial function and oxidative stress in healthy individuals and patients with type 2 diabetes. *Atherosclerosis* 2001;154:475-483.

[148] Title LM, Cummings PM, Giddens K, Nassar BA. Oral glucose loading acutely attenuates endothelium-dependent vasodilation in healthy adults without diabetes: an effect prevented by vitamins C and E. *J Am Coll Cardiol* 2000;36:2185-2191.

[149] van Oostrom AJ, Castro Cabezas M, Rabelink TJ. Insulin resistance and vessel endothelial function. *J R Soc Med* 2002;95 Suppl 42:54-61.

[150] de Kreutzenberg SV, Crepaldi C, Marchetto S, Calo L, Tiengo A, Del Prato S, Avogaro A. Plasma free fatty acids and endothelium-dependent vasodilation: effect of chain-length and cyclooxygenase inhibition. *J Clin Endocrinol Metab* 2000;85:793-798.

[151] Brown AA, Hu FB. Dietary modulation of endothelial function: implications for cardiovascular disease. *Am J Clin Nutr* 2001;73:673-686.

[152] Jagla A, Schrezenmeir J. Postprandial triglycerides and endothelial function. *Exp Clin Endocrinol Diabetes* 2001;109:S533-S547.

[153] Grundy SM, Denke MA. Dietary influences on serum lipids and lipoproteins. *J Lipid Res* 1990;31:1149-1172.

[154] De Caterina R, Liao JK, Libby P. Fatty acid modulation of endothelial activation. *Am J Clin Nutr* 2000;71:213S-223S.

[155] Huo Y, Ley K. Adhesion molecules and atherogenesis. *Acta Physiol Scand* 2001;173:35-43.

[156] Murohara T, Buerke M, Lefer AM. Polymorphonuclear leukocyte-induced vasocontraction and endothelial dysfunction. Role of selectins. *Arterioscler Thromb* 1994;14:1509-1519.

[157] Worthylake RA, Burridge K. Leukocyte transendothelial migration: orchestrating the underlying molecular machinery. *Curr Opin Cell Biol* 2001;13:569-577.

[158] Wanten G, Emst-De Vries S, Naber T, Willems P. Nutritional lipid emulsions modulate cellular signaling and activation of human neutrophils. *J Lipid Res* 2001;42:428-436.

[159] Wanten GJ, Geijtenbeek TB, Raymakers RA, van Kooyk Y, Roos D, Jansen JB, Naber AH. Medium-chain, triglyceride-containing lipid emulsions increase human neutrophil beta2 integrin expression, adhesion, and degranulation. *JPEN J Parenter Enteral Nutr* 2000;24:228-233.

[160] Moers A, Schrezenmeir J. Palmitic acid but not stearic acid inhibits NO-production in endothelial cells. *Exp Clin Endocrinol Diabetes* 1997;105 Suppl 2:78-80.

[161] Okuda Y, Kawashima K, Sawada T, Tsurumaru K, Asano M, Suzuki S, Soma M, Nakajima T, Yamashita K. Eicosapentaenoic acid enhances nitric oxide production by cultured human endothelial cells. *Biochem Biophys Res Commun* 1997;232:487-491.

[162] Moreno JJ, Mitjavila MT. The degree of unsaturation of dietary fatty acids and the development of atherosclerosis (review). *J Nutr Biochem* 2003;14:182-195.

[163] Agren JJ, Hanninen O, Julkunen A, Fogelholm L, Vidgren H, Schwab U, Pynnonen O, Uusitupa M. Fish diet, fish oil and docosahexaenoic acid rich oil lower fasting and postprandial plasma lipid levels. *Eur J Clin Nutr* 1996;50:765-771.

[164] Williams CM. Postprandial lipid metabolism: effects of dietary fatty acids. *Proc Nutr Soc* 1997;56:679-692.

[165] Kang JX, Leaf A. Antiarrhythmic effects of polyunsaturated fatty acids. Recent studies. *Circulation* 1996;94:1774-1780.

[166] Marchioli R, Barzi F, Bomba E, Chieffo C, Di Gregorio D, Di Mascio R, Franzosi MG, Geraci E, Levantesi G, Maggioni AP, Mantini L, Marfisi RM, Mastrogiuseppe G, Mininni N, Nicolosi GL, Santini M, Schweiger C, Tavazzi L, Tognoni G, Tucci C, Valagussa F. Early protection against sudden death by n-3 polyunsaturated fatty acids after myocardial infarction: time-course analysis of the results of the Gruppo Italiano per lo Studio della Sopravvivenza nell'Infarto Miocardico (GISSI)-Prevenzione. *Circulation* 2002;105:1897-1903.

[167] Morris MC, Sacks F, Rosner B. Does fish oil lower blood pressure? A meta-analysis of controlled trials. *Circulation* 1993;88:523-533.

[168] Newman WP, Middaugh JP, Propst MT, Rogers DR. Atherosclerosis in Alaska Natives and non-natives. *Lancet* 1993;341:1056-1057.

[169] Dyerberg J, Bang HO, Hjorne N. Fatty acid composition of the plasma lipids in Greenland Eskimos. *Am J Clin Nutr* 1975;28:958-966.

[170] Kagawa Y, Nishizawa M, Suzuki M, Miyatake T, Hamamoto T, Goto K, Motonaga E, Izumikawa H, Hirata H, Ebihara A. Eicosapolyenoic acids of serum lipids of Japanese islanders with low incidence of cardiovascular diseases. *J Nutr Sci Vitaminol (Tokyo)* 1982;28:441-453.

[171] Hu FB, Willett WC. Optimal diets for prevention of coronary heart disease. *JAMA* 2002;288:2569-2578.

[172] Daviglus ML, Stamler J, Orencia AJ, Dyer AR, Liu K, Greenland P, Walsh MK, Morris D, Shekelle RB. Fish consumption and the 30-year risk of fatal myocardial infarction. *N Engl J Med* 1997;336:1046-1053.

[173] Marckmann P, Gronbaek M. Fish consumption and coronary heart disease mortality. A systematic review of prospective cohort studies. *Eur J Clin Nutr* 1999;53:585-590.

[174] Burr ML, Fehily AM, Gilbert JF, Rogers S, Holliday RM, Sweetnam PM, Elwood PC, Deadman NM. Effects of changes in fat, fish, and fibre intakes on death and myocardial reinfarction: diet and reinfarction trial (DART). *Lancet* 1989;2:757-761.

[175] Hu FB, Cho E, Rexrode KM, Albert CM, Manson JE. Fish and long-chain omega-3 fatty acid intake and risk of coronary heart disease and total mortality in diabetic women. *Circulation* 2003;107:1852-1857.

[176] Bae JH, Bassenge E, Kim KB, Kim YN, Kim KS, Lee HJ, Moon KC, Lee MS, Park KY, Schwemmer M. Postprandial hypertriglyceridemia impairs endothelial function by enhanced oxidant stress. *Atherosclerosis* 2001;155:517-523.

[177] Betteridge DJ. What is oxidative stress? *Metabolism* 2000;49:3-8.

[178] Dart AM, Chin-Dusting JP. Lipids and the endothelium. *Cardiovasc Res* 1999;43:308-322.

[179] Doi H, Kugiyama K, Oka H, Sugiyama S, Ogata N, Koide SI, Nakamura SI, Yasue H. Remnant lipoproteins induce proatherothrombogenic molecules in endothelial cells through a redox-sensitive mechanism. *Circulation* 2000;102:670-676.

[180] Evans JL, Goldfine ID, Maddux BA, Grodsky GM. Oxidative stress and stress-activated signaling pathways: a unifying hypothesis of type 2 diabetes. *Endocr Rev* 2002;23:599-622.

[181] Hennig B, Meerarani P, Ramadass P, Watkins BA, Toborek M. Fatty acid-mediated activation of vascular endothelial cells. *Metabolism* 2000;49:1006-1013.

[182] Nishikawa T, Edelstein D, Du XL, Yamagishi S, Matsumura T, Kaneda Y, Yorek MA, Beebe D, Oates PJ, Hammes HP, Giardino I, Brownlee M. Normalizing mitochondrial superoxide production blocks three pathways of hyperglycaemic damage. *Nature* 2000;404:787-790.

[183] Collins T, Cybulsky MI. NF-kappaB: pivotal mediator or innocent bystander in atherogenesis? *J Clin Invest* 2001;107:255-264.

[184] Dichtl W, Nilsson L, Goncalves I, Ares MP, Banfi C, Calara F, Hamsten A, Eriksson P, Nilsson J. Very low-density lipoprotein activates nuclear factor-kappaB in endothelial cells. *Circ Res* 1999;84:1085-1094.

[185] Honing ML, Morrison PJ, Banga JD, Stroes ES, Rabelink TJ. Nitric oxide availability in diabetes mellitus. *Diabetes Metab Rev* 1998;14:241-249.

[186] Issemann I, Prince RA, Tugwood JD, Green S. The peroxisome proliferator-activated receptor:retinoid X receptor heterodimer is activated by fatty acids and fibrate hypolipidaemic drugs. *J Mol Endocrinol* 1993;11:37-47.

[187] Jensen MD. Diet effects on fatty acid metabolism in lean and obese humans. *Am J Clin Nutr* 1998;67:531S-534S.

[188] Karason K, Wikstrand J, Sjostrom L, Wendelhag I. Weight loss and progression of early atherosclerosis in the carotid artery: a four-year controlled study of obese subjects. *Int J Obes Relat Metab Disord* 1999;23:948-956.

[189] Busetto L. Visceral obesity and the metabolic syndrome: effects of weight loss. *Nutr Metab Cardiovasc Dis* 2001;11:195-204.

[190] Heilbronn LK, Noakes M, Clifton PM. Energy restriction and weight loss on very-low-fat diets reduce C-reactive protein concentrations in obese, healthy women. *Arterioscler Thromb Vasc Biol* 2001;21:968-970.

[191] Bruun JM, Verdich C, Toubro S, Astrup A, Richelsen B. Association between measures of insulin sensitivity and circulating levels of interleukin-8, interleukin-6 and tumor necrosis factor-alpha. Effect of weight loss in obese men. *Eur J Endocrinol* 2003;148:535-542.

[192] Duncan GE, Perri MG, Theriaque DW, Hutson AD, Eckel RH, Stacpoole PW. Exercise training, without weight loss, increases insulin sensitivity and postheparin plasma lipase activity in previously sedentary adults. *Diabetes Care* 2003;26:557-562.

[193] Ginsberg HN, Kris-Etherton P, Dennis B, Elmer PJ, Ershow A, Lefevre M, Pearson T, Roheim P, Ramakrishnan R, Reed R, Stewart K, Stewart P, Phillips K, Anderson N. Effects of reducing dietary saturated fatty acids on plasma lipids and lipoproteins in healthy subjects: the DELTA Study, protocol 1. *Arterioscler Thromb Vasc Biol* 1998;18:441-449.

[194] Summers LK, Fielding BA, Bradshaw HA, Ilic V, Beysen C, Clark ML, Moore NR, Frayn KN. Substituting dietary saturated fat with polyunsaturated fat changes abdominal fat distribution and improves insulin sensitivity. *Diabetologia* 2002;45:369-377.

[195] Bremer J. The biochemistry of hypo- and hyperlipidemic fatty acid derivatives: metabolism and metabolic effects. *Prog Lipid Res* 2001;40:231-268.

[196] Nolte RT, Wisely GB, Westin S, Cobb JE, Lambert MH, Kurokawa R, Rosenfeld MG, Willson TM, Glass CK, Milburn MV. Ligand binding and co-activator assembly of the peroxisome proliferator-activated receptor-gamma. *Nature* 1998;395:137-143.

[197] Daynes RA, Jones DC. Emerging roles of PPARs in inflammation and immunity. *Nat Rev Immunol* 2002;2:748-759.

[198] Gurnell M, Savage DB, Chatterjee VK, O'Rahilly S. The metabolic syndrome: peroxisome proliferator-activated receptor gamma and its therapeutic modulation. *J Clin Endocrinol Metab* 2003;88:2412-2421.

[199] Lee CH, Olson P, Evans RM. Minireview: lipid metabolism, metabolic diseases, and peroxisome proliferator-activated receptors. *Endocrinology* 2003;144:2201-2207.

[200] Martens FM, Visseren FL, Lemay J, de Koning EJ, Rabelink TJ. Metabolic and additional vascular effects of thiazolidinediones. *Drugs* 2002;62:1463-1480.

[201] Vosper H, Khoudoli GA, Graham TL, Palmer CN. Peroxisome proliferator-activated receptor agonists, hyperlipidaemia, and atherosclerosis. *Pharmacol Ther* 2002;95:47-62.

[202] Chinetti G, Lestavel S, Bocher V, Remaley AT, Neve B, Torra IP, Teissier E, Minnich A, Jaye M, Duverger N, Brewer HB, Fruchart JC, Clavey V, Staels B. PPAR-alpha and PPAR-gamma activators induce cholesterol removal from human macrophage foam cells through stimulation of the ABCA1 pathway. *Nat Med* 2001;7:53-58.

[203] Chinetti G, Lestavel S, Fruchart JC, Clavey V, Staels B. Peroxisome proliferator-activated receptor alpha reduces cholesterol esterification in macrophages. *Circ Res* 2003;92:212-217.

[204] Marx N, Libby P, Plutzky J. Peroxisome proliferator-activated receptors (PPARs) and their role in the vessel wall: possible mediators of cardiovascular risk? *J Cardiovasc Risk* 2001;8:203-210.

[205] Michalik L, Desvergne B, Tan NS, Basu-Modak S, Escher P, Rieusset J, Peters JM, Kaya G, Gonzalez FJ, Zakany J, Metzger D, Chambon P, Duboule D, Wahli W. Impaired skin wound healing in peroxisome proliferator-activated receptor (PPAR)alpha and PPARbeta mutant mice. *J Cell Biol* 2001;154:799-814.

[206] Larsen TM, Toubro S, Astrup A. PPARgamma agonists in the treatment of type II diabetes: is increased fatness commensurate with long-term efficacy? *Int J Obes Relat Metab Disord* 2003;27:147-161.

[207] Wilmsen HM, Ciaraldi TP, Carter L, Reehman N, Mudaliar SR, Henry RR. Thiazolidinediones upregulate impaired fatty acid uptake in skeletal muscle of type 2 diabetic subjects. *Am J Physiol Endocrinol Metab* 2003;285:E354-E362.

[208] Koeffler HP. Peroxisome proliferator-activated receptor gamma and cancers. *Clin Cancer Res* 2003;9:1-9.

[209] Guerin M, Le Goff W, Frisdal E, Schneider S, Milosavljevic D, Bruckert E, Chapman MJ. Action of ciprofibrate in type IIb hyperlipoproteinemia: modulation of the atherogenic lipoprotein phenotype and stimulation of high-density lipoprotein-mediated cellular cholesterol efflux. *J Clin Endocrinol Metab* 2003;88:3738-3746.

[210] Ruotolo G, Ericsson CG, Tettamanti C, Karpe F, Grip L, Svane B, Nilsson J, de Faire U, Hamsten A. Treatment effects on serum lipoprotein lipids, apolipoproteins and low density lipoprotein particle size and relationships of lipoprotein variables to

progression of coronary artery disease in the Bezafibrate Coronary Atherosclerosis Intervention Trial (BECAIT). *J Am Coll Cardiol* 1998;32:1648-1656.

[211] Capell WH, DeSouza CA, Poirier P, Bell ML, Stauffer BL, Weil KM, Hernandez TL, Eckel RH. Short-term triglyceride lowering with fenofibrate improves vasodilator function in subjects with hypertriglyceridemia. *Arterioscler Thromb Vasc Biol* 2003;23:307-313.

[212] Attia N, Durlach V, Roche D, Paul JL, Soni T, Zahouani A, Landron F, Labrousse F, Leutenegger M, Girard-Globa A. Post-prandial metabolism of triglyceride-rich lipoproteins in non-insulin-dependent diabetic patients before and after bezafibrate treatment. *Eur J Clin Invest* 1997;27:55-63.

[213] Vega GL, Cater NB, Hadizadeh DR, III, Meguro S, Grundy SM. Free fatty acid metabolism during fenofibrate treatment of the metabolic syndrome. *Clin Pharmacol Ther* 2003;74:236-244.

[214] Delerive P, De Bosscher K, Besnard S, Vanden Berghe W, Peters JM, Gonzalez FJ, Fruchart JC, Tedgui A, Haegeman G, Staels B. Peroxisome proliferator-activated receptor alpha negatively regulates the vascular inflammatory gene response by negative cross-talk with transcription factors NF-kappaB and AP-1. *J Biol Chem* 1999;274:32048-32054.

[215] van Wijk JP, de Koning EJ, Martens EP, Rabelink TJ. Thiazolidinediones and Blood Lipids in Type 2 Diabetes. A Summary Analysis. *Arterioscler Thromb Vasc Biol* 2003, in press.

[216] Collins AR, Meehan WP, Kintscher U, Jackson S, Wakino S, Noh G, Palinski W, Hsueh WA, Law RE. Troglitazone inhibits formation of early atherosclerotic lesions in diabetic and nondiabetic low density lipoprotein receptor-deficient mice. *Arterioscler Thromb Vasc Biol* 2001;21:365-371.

[217] Li AC, Brown KK, Silvestre MJ, Willson TM, Palinski W, Glass CK. Peroxisome proliferator-activated receptor gamma ligands inhibit development of atherosclerosis in LDL receptor-deficient mice. *J Clin Invest* 2000;106:523-531.

[218] Takagi T, Akasaka T, Yamamuro A, Honda Y, Hozumi T, Morioka S, Yoshida K. Troglitazone reduces neointimal tissue proliferation after coronary stent implantation in patients with non-insulin dependent diabetes mellitus: a serial intravascular ultrasound study. *J Am Coll Cardiol* 2000;36:1529-1535.

[219] Tack CJ, Ong MK, Lutterman JA, Smits P. Insulin-induced vasodilatation and endothelial function in obesity/insulin resistance. Effects of troglitazone. *Diabetologia* 1998;41:569-576.

[220] Walker AB, Chattington PD, Buckingham RE, Williams G. The thiazolidinedione rosiglitazone (BRL-49653) lowers blood pressure and protects against impairment of endothelial function in Zucker fatty rats. *Diabetes* 1999;48:1448-1453.

[221] Berger JP, Petro AE, Macnaul KL, Kelly LJ, Zhang BB, Richards K, Elbrecht A, Johnson BA, Zhou G, Doebber TW, Biswas C, Parikh M, Sharma N, Tanen MR, Thompson GM, Ventre J, Adams AD, Mosley R, Surwit RS, Moller DE. Distinct properties and advantages of a novel peroxisome proliferator-activated protein [gamma] selective modulator. *Mol Endocrinol* 2003;17:662-676.

[222] Rocchi S, Picard F, Vamecq J, Gelman L, Potier N, Zeyer D, Dubuquoy L, Bac P, Champy MF, Plunket KD, Leesnitzer LM, Blanchard SG, Desreumaux P, Moras D, Renaud JP, Auwerx J. A unique PPARgamma ligand with potent insulin-sensitizing yet weak adipogenic activity. *Mol Cell* 2001;8:737-747.

[223] Brand CL, Sturis J, Gotfredsen CF, Fleckner J, Fledelius C, Hansen BF, Andersen B, Ye JM, Sauerberg P, Wassermann K. Dual PPARalpha /gamma activation provides enhanced improvement of insulin sensitivity and glycemic control in ZDF rats. *Am J Physiol Endocrinol Metab* 2003;284:E841-E854.

[224] Sauerberg P, Pettersson I, Jeppesen L, Bury PS, Mogensen JP, Wassermann K, Brand CL, Sturis J, Woldike HF, Fleckner J, Andersen AS, Mortensen SB, Svensson LA, Rasmussen HB, Lehmann SV, Polivka Z, Sindelar K, Panajotova V, Ynddal L, Wulff EM. Novel tricyclic-alpha-alkyloxyphenylpropionic acids: dual PPARalpha/gamma agonists with hypolipidemic and antidiabetic activity. *J Med Chem* 2002;45:789-804.

[225] Brown MS, Goldstein JL. A receptor-mediated pathway for cholesterol homeostasis. *Science* 1986;232:34-47.

[226] Grundy SM. Consensus statement: Role of therapy with "statins" in patients with hypertriglyceridemia. *Am J Cardiol* 1998;81:1B-6B.

[227] Castro Cabezas M, Erkelens DW, Kock LA, de Bruin TW. Postprandial apolipoprotein B100 and B48 metabolism in familial combined hyperlipidaemia before and after reduction of fasting plasma triglycerides. *Eur J Clin Invest* 1994;24:669-678.

[228] Castro Cabezas M, de Bruin TW, Kock LA, Kortlandt W, Linde-Sibenius TM, Jansen H, Erkelens DW. Simvastatin improves chylomicron remnant removal in familial combined hyperlipidemia without changing chylomicron conversion. *Metabolism* 1993;42:497-503.

[229] Grip O, Janciauskiene S, Lindgren S. Atorvastatin activates PPAR-gamma and attenuates the inflammatory response in human monocytes. *Inflamm Res* 2002;51:58-62.

[230] Kandoussi A, Martin F, Hazzan M, Noel C, Fruchart JC, Staels B, Duriez P. HMG-CoA reductase inhibition and PPAR-alpha activation both inhibit cyclosporin A induced endothelin-I secretion in cultured endothelial cells. *Clin Sci (Lond)* 2002;103 Suppl 48:81S-83S.

[231] Palinski W, Napoli C. Unraveling pleiotropic effects of statins on plaque rupture. *Arterioscler Thromb Vasc Biol* 2002;22:1745-1750.

[232] Sowers JR. Effects of statins on the vasculature: Implications for aggressive lipid management in the cardiovascular metabolic syndrome. *Am J Cardiol* 2003;91:14B-22B.

[233] Maron DJ, Fazio S, Linton MF. Current perspectives on statins. *Circulation* 2000;101:207-213.

[234] Luan Z, Chase AJ, Newby AC. Statins Inhibit Secretion of Metalloproteinases-1, -2, -3, and -9 From Vascular Smooth Muscle Cells and Macrophages. *Arterioscler Thromb Vasc Biol* 2003;23:769-775.

[235] Walter DH, Rittig K, Bahlmann FH, Kirchmair R, Silver M, Murayama T, Nishimura H, Losordo DW, Asahara T, Isner JM. Statin therapy accelerates reendothelialization:

a novel effect involving mobilization and incorporation of bone marrow-derived endothelial progenitor cells. *Circulation* 2002;105:3017-3024.

[236] Waehre T, Damas JK, Gullestad L, Holm AM, Pedersen TR, Arnesen KE, Torsvik H, Froland SS, Semb AG, Aukrust P. Hydroxymethylglutaryl coenzyme a reductase inhibitors down-regulate chemokines and chemokine receptors in patients with coronary artery disease. *J Am Coll Cardiol* 2003;41:1460-1467.

[237] Waldman A, Kritharides L. The pleiotropic effects of HMG-CoA reductase inhibitors: their role in osteoporosis and dementia. *Drugs* 2003;63:139-152.

[238] Chan KK, Oza AM, Siu LL. The statins as anticancer agents. *Clin Cancer Res* 2003;9:10-19.

[239] Polman CH, Uitdehaag BM. New and emerging treatment options for multiple sclerosis. *Lancet Neurol* 2003;2:563-566.

[240] Randomised trial of cholesterol lowering in 4444 patients with coronary heart disease: the Scandinavian Simvastatin Survival Study (4S). *Lancet* 1994;344:1383-1389.

[241] Sacks FM, Pfeffer MA, Moye LA, Rouleau JL, Rutherford JD, Cole TG, Brown L, Warnica JW, Arnold JM, Wun CC, Davis BR, Braunwald E. The effect of pravastatin on coronary events after myocardial infarction in patients with average cholesterol levels. Cholesterol and Recurrent Events Trial investigators. *N Engl J Med* 1996;335:1001-1009.

[242] Shepherd J, Cobbe SM, Ford I, Isles CG, Lorimer AR, Macfarlane PW, McKillop JH, Packard CJ. Prevention of coronary heart disease with pravastatin in men with hypercholesterolemia. West of Scotland Coronary Prevention Study Group. *N Engl J Med* 1995;333:1301-1307.

[243] Downs JR, Clearfield M, Weis S, Whitney E, Shapiro DR, Beere PA, Langendorfer A, Stein EA, Kruyer W, Gotto AM, Jr. Primary prevention of acute coronary events with lovastatin in men and women with average cholesterol levels: results of AFCAPS/TexCAPS. Air Force/Texas Coronary Atherosclerosis Prevention Study. *JAMA* 1998;279:1615-1622.

[244] Weigle DS. Pharmacological therapy of obesity: past, present, and future. *J Clin Endocrinol Metab* 2003;88:2462-2469.

[245] Reitsma JB, Castro Cabezas M, de Bruin TW, Erkelens DW. Relationship between improved postprandial lipemia and low-density lipoprotein metabolism during treatment with tetrahydrolipstatin, a pancreatic lipase inhibitor. *Metabolism* 1994;43:293-298.

[246] Miller M. Niacin as a component of combination therapy for dyslipidemia. *Mayo Clin Proc* 2003;78:735-742.

In: Focus on Atherosclerosis Research
Editor: Leon V.Clark, pp. 165-183
ISBN 1-59454-044-6

Chapter VIII

Involvement of Endothelial Dysfunction in Atherogenesis

Pavel Poredos
Department for Vascular Disease, University Medical Centre Ljubljana, Slovenia

Abstract

Healthy endothelium plays a central role in cardiovascular control. Therefore, endothelial dysfunction (ED) may have a particularly significant role in the pathogenesis of atherosclerosis. ED is a consequence of the harmful effects of risk factors of atherosclerosis on the vessel wall and is closely related to the number of risk factors, to their intensity and their duration. ED has been demonstrated in subjects with hypercholesterolemia, diabetes, hypertension, smoking and in patients with atherosclerotic disease (coronary, peripheral arterial). It was also shown that ED is an early event in type I and II diabetes and that it is related to the development and progression of diabetic vascular complications. It was shown that ED (demonstrated by flow mediated endothelium dependent dilation) is inversely related to the extent of microalbuminuria.

The involvement of risk factors in ED is also supported by results of intervention studies that showed regression of ED with treatment of risk factors. Improvement of ED may be achieved by elimination of risk factors, by substitution of natural protective endothelial substances (e. g, L-arginine), inhibitors of endothelium-derived contracting factors (e.g., ACE inhibitors, angiotensin II receptor antagonists), cytoprotective agents (e.g., free-radical scavengers such as superoxide dismutase), lipid-lowering drugs or diets and by physical exercise. All these data show that ED is reversible and by treatment of risk factors it is possible to restore vascular function.

ED promotes progression of atherosclerosis and probably plays an important role in the development of thrombotic complications in the late stages of the disease. As ED is a key underlying factor in the atherosclerotic process, markers of endothelial abnormalities have been sought, particularly those involving disturbed endothelium-dependent vasomotion or related cellular products – circulating markers. Using these tests it is possible to follow the dose - response of harmful effects of risk factors, and the effects of preventive procedures on vessel wall function. Determination of ED also has important clinical implications. It was shown that ED is significantly and directly correlated with the occurrence of cardiac events and that cardiac events increased as ED worsens.

Normal Endothelial Function

The vascular endothelium is the one cell-layer, leaning of the vessel wall that is an important regulatory organ in maintaining cardiovascular homeostasis. The endothelium, which was once thought to be a simple barrier, separating the blood and vessel wall, is now known to sense changes in haemodynamic forces, and blood-borne signals. The endothelium responds to physical and chemical stimuli by synthesis and release of variety of vasoactive, thromboregulatory as well as growth factors. Substances released by the endothelium contribute to the local vascular regulation. Endothelial cells control vascular tone by releasing of nitric oxide (NO), prostacyclin and contracting factors, such as endothelin. Further, endothelium influences vascular tone by converting angiotensin-I into angiotensin-II, by inactivating bradykinin, norepinephrine and serotonin [1].

The endothelium controls underlying smooth muscle tone in response to certain pharmacological and physiological stimuli. The existence of an endothelium-derived relaxing factor was first suggested by Furchgott and Zawadzki, when they observed that rabbit aortic rings relaxed to acetylcholine only in the presence of an intact endothelium. This biological effect was prescribed to the endothelium-derived relaxing factor (EDRF), which structure was later identified as nitrogen oxide. NO is synthesized from L-arginine by an enzyme, NO synthase. NO modulates vascular tone at rest in both, the systemic and pulmonary circulations. In addition, NO-release is stimulated by increased flow (leading to increased shear stress on the endothelium) and by bradykinin, thrombin, acetylcholine, serotonin, and a variety of other circulating agents, which increase NO-release via activation of specific endothelial cell-receptors. In smooth muscle cells, this causes a reduction of intracellular calcium concentration, leading to vasorelaxation. In addition, NO inhibits the interaction between blood-cells and the vessel wall. It is probably also an endogenous inhibitor of tumor necrosis factor and the expression of proinflammatory molecules, such as the vascular cell-adhesion molecule (VCAM-1) and chemoatractant protein-1 (MCP-1) [2, 3].

Healthy endothelium has anticoagulant, antiplatelet, and fibrinolytic properties. On endothelial cells different anticoagulant reaction involving thrombin formation take place. Platelet adhesion to endothelial cells is inhibited by the endothelium-derived archidonic acid metabolite prostacyclin, different stimuli release also tissue-type plazminogen activator (TPA) from the endothelium, which allows the normal endothelium to limit the extent of thrombus formation.

Normal endothelium also plays an important role in vascular growth, immunological regulation, lipoprotein metabolism, and integration and transduction of blood-borne signals [4].

Endothelial Dysfunction

Endothelial damage can result in endothelial dysfunction that is characterized by an imbalance between relaxing and contracting factors, between anti- and procoagulant mediators or growth-inhibiting and-promoting factors. Endothelial dysfunction can result from mechanical or biochemical injury to the endothelium. Physical damage of endothelium is mostly caused by hypertension, experimentally induced acute hypertension has been shown to disturb endothelial integrity. Several other risk-factors like hypercholesterolemia, diabetes, and smoking probably cause injury to the endothelium through biochemical mechanisms. A common denominator for all these harmful conditions is probably increased oxidative stress, which has been suggested as a direct cause of endothelial dysfunction. Therefore, vessel wall damage results in endothelial dysfunction, which is the first measurable step in atherogenesis [5, 6].

Indicators of Endothelial Dysfunction

After the recognition that atherosclerosis develops over decades and therefore has a long preclinical (silent) phase prior to the onset of clinical symptoms, and that in this early phase changes of the arterial wall are mostly reversible and preventable, there has been considerable interest in developing diagnostic tools for detecting and monitoring early vascular changes in asymptomatic subjects. As endothelial dysfunction is a key underlying factor in the atherosclerotic process, markers of endothelial abnormalities have been sought, particularly those involving disturbed endothelium-dependent vasomotion or related cellular products [7]. Ideally, testing for arterial endothelial dysfunction should involve methods that are safe, non-invasive, economical, reliable, and reproducible, that correlate with the extent of subclinical atherosclerosis, predict the subsequent risk and respond to therapy. Because endothelium has different functions, different tests would be required to measure several different aspects of endothelial dysfunction. Areas of potential interest for detection of endothelial dysfunction include tests of endothelium-dependent vasomotion, as well as circulating markers of endothelial function.

Functional Methods for Studying Endothelial Function - NO Dependent Vasodilation

Most of the functional methods for in vivo endothelial testing examine the ability of the endothelium to cause vasodilation in response to pharmacological and physiological stimuli that increase the endothelial release of NO.

In vivo assessment of coronary endothelial function in humans was first reported by Ludmer and co-workers in 1986 [8]. The coronary artery diameter was measured by quantitative angiography before and after intracoronary infusion of acetylcholine. Later on, these techniques provided valuable insights into the risk factors of endothelial dysfunction in coronary arteries and into endothelial dysfunction preceding overt atherosclerosis. The potential for reversibility of endothelial dysfunction in the coronary arteries by therapeutic intervention has also been assessed by this methodology [9]. The major disadvantage of intra-arterial tests is their invasive nature, which makes them generally unsuitable for routine use in subjects who are at risk of atherosclerosis but who have no clinical symptoms or signs of disease. Therefore, noninvasive or minimally invasive clinical tests of endothelial function have been developed.

Positron emission tomography (PET) is a unique method that enables noninvasive quantization of myocardial blood flow, which is dependent on the endothelial function of macro- and microvessels. By measuring blood flow at rest and after pharmacological stimulation with dipyridamole or adenosine it is possible to calculate the coronary blood flow reserve. In cases with impairment in coronary reactivity the blood flow reserve is decreased and may represent an early marker of endothelial dysfunction and subclinical coronary atherosclerosis [10].

Endothelial function may also be tested noninvasively in the peripheral conduit arteries using high-resolution external vascular ultrasound. Measurement of brachial artery diameter has been used to study the vasodilator response induced by reactive hyperemia. A 5-minute occlusion of flow to the upper extremity produced by inflation of a blood pressure cuff, followed by a release of the occlusion, results in an immediate 5-10 fold increase in blood flow as a result of resistance vessel dilation (reactive hyperemia). This marked increase in both flow and shear stress serves as a stimulus for the release of NO with subsequent conduit vessel dilation. Dilation of conduit vessels in response to an increase in flow is mediated by endothelium-derived NO. In individuals without atherosclerosis and risk factors of atherosclerosis an increase in flow-induced by exercise, cold, pressure testing all resistance vessel vasodilators results in a 15-20% dilation in the conduit vessel. This response is attenuated in subjects with risk factors and atherosclerotic disease. Taken together, this data convincingly demonstrate that flow-mediated vasodilation of the brachial artery is dependent on the endothelial release of NO. With application of nitroglycerine that is a potent smooth muscle cell dilator it is possible to measure endothelium-independent vasodilation.

There is growing evidence that endothelial dysfunction is a systemic disease, therefore vasodilator response of a peripheral conduit artery parallels the endothelium-dependent coronary artery vasomotion. Thus, the non-invasive evaluation of peripheral arteries may be useful in screening patients for deteriorative effects of risk factors on endothelial function of coronary arteries and might potentially serve as a »window« to coronary and systemic functional disturbances related to atherosclerosis.

Celermajer and colleagues have demonstrated the potential utility of this technique and their results were confirmed in other studies. The coefficient of variation in the measurement of the baseline diameter accounted between 1,4 and 2,6%. Using a technique with a 7,5-10,0 MHz probe and off-line analysis it is possible to demonstrate a coefficient of variation of less than 2% for intraobserver variability and replicability [11, 12, 13].

Being noninvasive, the ultrasound method has been applied widely to asymptomatic subject groups. These studies have provided information on the effects of various risk factors of early atherogenesis. As these techniques are accurate and relatively reproducible, serial studies may be performed, allowing trials of reversibility of endothelial dysfunction in asymptomatic subjects at high risk of arterial atherosclerotic disease.

Circulating Markers of Endothelial Function

Endothelial injury may result in the release of various factors that can be detected in the circulation and can be potentially used as markers of endothelial dysfunction.

Endothelin-1

Endothelin-1 is an endothelium-derived peptide that has powerful vasoconstrictor properties. Increased levels ofendothelin-1 have been demonstrated in conditions associated with endothelial dysfunction, such as atherosclerosis [14], hypercholesterolemia [15], and cigarette smoking [16]. Lerman and co-workers demonstrated increased endothelin-1 in subjects with endothelial dysfunction of the coronary arteries [17]. Furthermore, higher basal levels and exercise-induced increases of endothelin-1 were demonstrated in patients with peripheral arterial occlusive disease (PAOD) [18]. Acute ischemia during claudication induces injury to vascular endothelial cells and probably triggers the release of endothelin-1. Therefore, it has been suggested that the levels of circulating endothelin in plasma may represent a marker of endothelial dysfunction. As endothelin-1 has mitogenic properties as well, it may also play a role in the development of premature atherosclerosis. Involvement of endothelin-1 in atherogenesis is also supported by the findings of Mangiafico and co-workers who demonstrated an interrelationship between increased endothelin-1 release and augmented insulin release in PAOD patients [19].

Von Willebrand Factor

The von Willebrand factor (vWF) is a glycoprotein synthesized mainly by vascular endothelial cells. It has important functions in haemostasis, participating in the coagulation and in the formation of platelet plugs at sites of endothelial damage. vWF is elevated in patients with hypercholesterolemia, and reduction of total cholesterol has been shown to be associated with a reduction in vWF levels [20]. Its levels are elevated in situations characterized by vascular damage with denudation of the endothelium, and further is has been suggested that injured endothelial cells leak vWF, leading to increased plasma levels. Recent studies have shown that elevated levels of vWF may be predictive of recurrent cardiovascular events in patients with known cardiovascular disease [21].

TPA and PAI-1

Tissue plasminogen activator (TPA) is a protein released by endothelial cells. It activates the reaction in which plasminogen is converted to plazmin. TPA regulates the fibrinolytic activity of blood in balance with plasminogen activator inhibitor-1 (PAI-1), another product of endothelial cells, serving as the primary inhibitor of TPA. These circulating vascular endothelial markers are elevated in hypercholesterolemia [22]. Furthermore, prospective studies have shown that high concentrations of TPA antigen in healthy men may be associated with a subsequent risk of myocardial infarction and stroke [23]. Data from the Atherosclerosis Risk in Communities (ARIC) study have shown that increased levels of TPA and PAI-1 are associated with subclinical carotid atherosclerosis, measured as increased intima-media thickness [24].

Therefore, the different proteins involved in haemostasis and synthesized by endothelial cells have been proposed as markers of endothelial dysfunction. The main problem is the relatively poor specificity of these markers, which may hamper their use in detection or monitoring of endothelial dysfunction.

Adhesion Molecules

Binding of circulating leukocytes to the vascular endothelium and further leukocyte migration into the subendothelial space are major processes in the development of atherosclerosis. These events are mediated through diverse cellular adhesion molecules that are expressed on the surface of vascular endothelial cells: vascular cell adhesion molecule-1 (VCAM-1), endothelial-leukocyte adhesion molecule-1 (E-selectin), intercellular adhesion molecule-1 (ICAM-1) and P-selectin. Circulating soluble forms of these adhesion molecules have been detected in plasma. The origin and function of these circulating soluble adhesion molecules are unclear, but they most likely arise from endothelial cells.

Several studies have suggested that circulating adhesion molecules may serve as markers of endothelial damage and/or atherosclerosis. In the ARIC study it was shown that E-selectin and ICAM-1 are associated with carotid artery intima-media thickness and are independent predictors of incident coronary artery disease [25]. This study also showed higher levels of ICAM-1 in smokers. Squadrito and co-workers observed higher levels of circulating-ICAM-1 and E-selectin in patients with acute myocardial infarction [26]. Increased levels of plasma P-selectin have been observed in unstable angina [27], myocardial infarction [28] and hypercholesterolemia [29]. Elevated levels of ICAM-1 were also found in patients with peripheral vascular disease [30]. Furthermore, ICAM-1 levels may predict cardiovascular event rates in apparently healthy men (31].

Of the risk factors, serum lipoproteins seem to be especially related to the levels of circulation adhesion molecules. In the study by Hackman and co-workers, dyslipidemia was associated with elevated levels of ICAM-1, VCAM-1, and E-selectin [32]. Successful lipid-lowering treatment with statins significantly reduced the levels of circulating E-selectin. Therefore, these studies suggest that cell adhesion molecules may serve as possible markers of the efficacy of lipid-lowering therapy.

Risk Factors of Atherosclerosis and Endothelial Dysfunction

Endothelial dysfunction has been demonstrated in subjects with different risk factors for atherosclerosis, such as hypercholesterolemia, diabetes, hypertension, smoking, and in patients with atherosclerotic disease (coronary, peripheral arterial). Further, numerous cardiovascular disorders have been shown to be associated with altered endothelium-dependent relaxation such as atherosclerosis, heart failure and post-menopause state.

Hypercholesterolemia

Hypercholesterolemia is characterized by long-lasting and moderate increase in serum cholesterol. Clinical studies investigating the effects of hypercholesterolemia on endothelial function in the human circulation have, in general, confirmed experimental observations that it is associated with impaired endothelium-dependent vasodilation of coronary and peripheral conduit and resistance vessels. In hypercholesterolemic patients with angiographically smooth coronary arteries a striking vasoconstriction in response to intracoronary infusion of acetylcholine can usually be observed. Since a recent data suggested that vasomotor response of acetylcholine is at least in part mediated by the release of NO, the abnormal response to acetylcholine probably reflects the diminished availability of endothelial mediators including NO (4).

In general, the degree of endothelial dysfunction of coronary microcirculation correlates with total serum cholesterol level [33]. Indeed, the ratio of LDL/HDL rather than absolute values of LDL determines the degree of endothelial dysfunction in the coronary microvasculature, indicating that HDL exerts a protective effect on endothelial function in patients with hypercholesterolemia [34]. Moreover, plasma autoantibody levels against oxidized LDL are closely related to endothelium – dependent relaxation suggesting that oxidized LDL may be more toxic to the endothelium than native LDL [35].

Several possible mechanisms have been proposed by which hypercholesterolemia impairs endothelial function:

a) reduced synthesis of endothelium – derived relaxing factor (NO),
b) altered membrane receptor coupling mechanisms affecting the release of NO and,
c) impaired diffusion or augmented use of NO in the vessel wall.

The rate limiting factor for the production of nitric oxide in hypercholesterolemia could be intracellular availability of L – arginine, in several studies namely treatment of hypercholesterolemia with L – arginine showed beneficial effect on endothelial function [36, 37].

Hypertension

Hypertension – a large body of evidence indicates that endothelial dysfunction is a characteristic of patients with essential hypertension (38, 39, 40, 41]. Endothelial dysfunction associated with essential hypertension is characterized by impaired NO availability, determined by oxidative stress production, which causes NO breakdown [42]. Reactive oxygen species, mainly superoxide anions, combine and destroy NO, producing peroxide nitrates, which have several negative effects on vascular function and structure. Further, an interaction between NO-system and endotelin-1 can participate in the pathogenesis of endothelial dysfunction, although plasma levels of endothelin-1 do not seem to be increased in essential hypertension, it is worth noting that the vasoconstrictor activity has been found to be increased in parallel with diminished NO availability. In patients with essential hypertension endothelium-dependent vasodilation is impaired to acetylcholine and bradykinin application. These findings indicate that the endothelial dysfunction in this condition is not related to a specific defect of a single intracellular signal-transduction pathway and suggests a more generalized abnormality of endothelial vasodilator function. These results contrast with those is hypercholesterolemic patients in whom response only to acetylcholine is impaired [43].

It is presently unclear whether endothelial dysfunction is a primary abnormality or a consequence of the elevated blood pressure values [44]. Data in animal models of genetic hypertension are in favour of direct damage of the endothelial cells secondary to a prolonged increase of the haemodynamic load and shear stress. In patients with phaeochromocytoma, primary aldosteronism or renovascular hypertension, an impairment of the vasodilator response to acetylcholine was found despite there being no genetic predisposition to hypertension. Further, the endothelial dysfunction observed in hypertension appears to be a consequence of high blood pressure since a variety of antihypertensive treatment normalizes this responses. However, in normotensive offsprings of hypertensive parents, an early alteration of endothelium-dependent vasodilation in the forearm was observed [45]. We also found impaired endothelial function in offsprings of parents with essential hypertension although their blood pressure was in normal range [46]. These data suggest a possibly relevant role of genetic mechanisms in the genesis of endothelial dysfunction in hypertension. Recent observations also favour the hypothesis that a dysfunctional endothelial NO-synthase gene polymorphism may be associated with essential hypertension, although this was not consistently observed in all populations [47].

Diabetes Mellitus

Evidence that endothelial dysfunction occurs in diabetic patients is found by an elevated plasma concentration of the von-Willebrand factor, impaired prostacyclin and plasminogen activator release, and decreased lipoprotein lipase activity. It has also been reported that in patients with insulin-dependent and insulin-independent diabetes mellitus some aspects of endothelial dysfunction related to vasodilator response are present [48, 49]. It was shown that flow mediated endothelial vasodilation capability of peripheral arteries is closely related to microalbuminuria [50, 51, 52, 53]. It is possible that endothelial dysfunction contributes to the overt vascular complication in diabetes. A considerable body of evidence indicates that endothelial dysfunction is closely associated with a development of diabetic retinopathy, nephropathy and is involved in pathogenesis of accelerated forms of atherosclerosis in diabetic patients [54]. The appearance of endothelial dysfunction is not related to the duration of the diabetes. In some patients it is preserved regardless of disease duration, provided urinary albumin excretion is normal. There is a substantial evidence that endothelium-dependent vasodilation is abnormal in both conduit arteries and resistance vessels of diabetic animals. In patients with insulin-dependent diabetes mellitus the abnormal endothelium dependent vasodilation has been shown to be caused by decreased release or activity of NO [39].

Smoking

Experimental evidence has suggested that smoking induces endothelial dysfunction. Different studies showed that endothelium dependent (flow-mediated) dilation of brachial artery is significantly impaired in smokers as a whole compared to control subjects. Endothelial dysfunction progresses with the duration of smoking and with the number of smoked cigarettes [55]. This dose-dependence of smoking-related endothelial dysfunction and its reversibility after cessation of smoking support the close interrelationship between smoking and endothelial function [56]. Decreased flow-mediated dilation in smokers could be the result of inactivation of NO by oxygen-derived free radicals or by enhanced release of endothelin-1, recognized as a potent vasoconstrictor. The exact components of the cigarette smoke responsible for endothelial dysfunction and vasculopathy are not known, although nicotine and carbon monoxide have been implicated. Changes in lipid metabolism including HDL decrease and LDL increase may also contribute to deteriorated endothelial function in smokers. Further, it was shown that the effect of smoking on endothelial function is not an acute phenomenon but it is most probably a consequence of chronic influence of nicotine [57].

Different studies have also reported endothelial dysfunction in subjects with other non-classical risk factors of atherosclerosis, like hyperhomocysteinemia. Flow-mediated dilation capability of brachial artery was reduced in subjects with insulin resistance and polimetabolic syndrome, in overweight subjects and is closely related to systemic inflammatory markers (CRP) and adhesion molecules [54, 58].

We demonstrated that endothelial dysfunction progresses with the duration of hypertension, diabetes or smoking and that a close relationship exists between the intensity of an individual risk factor or the number of presented risk factors and endothelial function [46, 50, 55, 59]. Therefore, a dose-response relation exists between risk factors of atherosclerosis and endothelial dysfunction.

Mechanism of the Involvement of Risk Factors in Endothelial Dysfunction and Atherogenesis

The mechanisms whereby risk factors cause endothelial dysfunction are largely unknown; a common denominator for all these conditions is probably increased oxidative stress, which has therefore been suggested as an important cause of endothelial dysfunction. Most known risk factors cause excessive production of superoxide anions, with consequent degradation of NO before it can reach target tissues [60, 61]. Decreased bioavailability of NO in the presence of risk factors is most probably also caused by decreased expression of nitrogen oxide synthase activity, although some findings suggested that appointed risk factors, like hypercholesterolemia, might paradoxically increase the activity and/or expression of NO synthase [62]. These findings suggest the possibility that the production of No is increased or unaltered, but it is destroyed by the superoxide radicals before having an opportunity to leave the endothelium. As NO acts as a vasodilator and inhibits platelet adherence and aggregation, smooth muscle proliferation, and endothelial cell-leukocyte interaction, decreased NO activity may contribute importantly to the initiation and progression of atherosclerotic lesions. The consequences of mechanical or chemical damage of the endothelium by different risk factors are also several cellular processes, such as inflammation, that maintain endothelial dysfunction and promote atherosclerosis.

Furthermore, is has been shown that endothelial dysfunction that precedes the early morphologic atherosclerotic changes in the arterial wall plays an important role in the development and growth of atherosclerotic lesions, and in the development of ischemia and thrombosis in the late stages of the disease.

In recent years it has become evident that an inflammatory process is involved in atherosclerosis and there is mounting evidence that inflammation plays a role in the development of coronary and other atherosclerotic. Elevated concentrations of acute phase reactants, such as C-reactive protein (CRP) and proinflammatory cytokine interleukin-6 (IL-6) were found in patients with acute coronary syndromes [63, 64] and peripheral arterial disease [65]. Further, it was shown that increased levels of CRP are associated with an increased risk of ischemic stroke [66] and with a worse outcome in these patients [67]. CRP may play a direct role in promoting the inflammatory component of atherosclerosis. Unfortunately, it was shown that CRP, at concentrations frequently observed in high-risk subjects, induces significant expression of adhesion molecules by human endothelial cells [68]. Studies also showed that an elevated CRP level is not only related to the increased levels of circulating markers of endothelial dysfunction but also to abnormal systemic endothelial vascular reactivity [69]. Thus, the identification of elevated CRP levels as an independent predictor of endothelial dysfunction might provide an important clue to link systemic markers of inflammatory processes to endothelial dysfunction and atherosclerotic disease progression.

Study of endothelial dysfunction raised the question if it is a cause or consequence of atherosclerosis. Endothelial dysfunction is most probably a consequence of the harmful effects of risk factors of atherosclerosis on the vessel wall. However, recent observations favour the hypothesis that endothelial dysfunction could also be a primary defect and as such directly inherited. However, some observations showed that dysfunctional endothelial NO synthase gene polymorphism is associated with some risk factors such as hypertension and is therefore not a consequence, but rather a primary abnormality.

Irrespective of the sequence of events, endothelial dysfunction probably promotes atherogenesis through different mechanisms: increased adherence of monocytes, enhanced permeability of the endothelial layer to monocytes/macrophages and lipoproteins which then accumulate in the vessel wall [70]. Endothelial dysfunction is also related to increased platelet adherence, and smooth muscle cell migration and proliferation - both involved in atherogenesis [71]. Because one of the earliest events in atherogenesis is the adherence of circulating monocytes to intact endothelial cells, atherosclerosis behaves as a chronic inflammatory process. However, this presumption does not preclude involvement of endothelial dysfunction, which most probably initiates different pathological processes, including inflammation.

Endothelial Dysfunction and Cardiovascular Disease

Coronary Heart Disease

Because of endothelial dysfunction in atherosclerotic human coronary arteries endothelium-mediated relaxation is impaired and shifting the balance in favour of vasoconstriction in response to a variety of stimuli as exercise, mental stress, and cold exposure. Subsequent clinical studies demonstrated that abnormal constrictor responses to acetylcholine were not only observed in angiographically diseased epicardial arteries, but also in subjects with entirely smooth epicardial arteries and the presence of risk factors for coronary artery disease [72, ,73].

In addition, there is evidence that endothelial dysfunction of the epicardial arteries progresses in a hierarchical fashion, with receptor – dependent responses (acetylcholine, serotonin) being impaired before receptor - independent (FMD, cold pressure induced) vasodilation. Studies demonstrated that the acetylcholine response was converted into an abnormal constrictor response even in the presence of minor atherosclerotic arterial wall thickening, whereas flow-mediated dilation was well preserved. Thus, an abnormal vasomotor response to acetylcholine does indeed provide the earliest evidence of endothelial vasoactive dysfunction in the human coronary circulation [74, 75].

Atherosclerotically changed epicardial arteries are exhibiting abnormal constrictor responses also to a variety of common daily life stimuli like exercise, mental stress, or exposure to cold. Thus, endothelial vasodilator dysfunction appears to render the atherosclerotic vessels more sensitive to the constrictive effects of cateholamines and results in the pathogenesis of inappropriate vasoconstriction, which represents the fundamental functional disturbance in the vascular biology of atherosclerosis [76, 77].

In addition to impairment of flow mediated dilation of the epicardial conductance vessels, in atherosclerosis the vasodilator function of the coronary resistance vessel is adversely affected. A defective endothelium-mediated vasodilator function may therefore potentate known trigger mechanisms of myocardial ischemia and thereby induce a mismatch between myocardial oxygen supply and demand. Even through the changes in luminal diameter produced by inappropriate vasoconstriction of the epicardial vessel in response to sympathetic activation are usually less than 30 %, such an increase in the arterial tone might be enough to convert a subcritical stenosis into a critical one with an ensuing decrease in blood flow and impaired FMD of the resistance vessels will further reduce coronary flow reserve [72].

Endothelial dysfunction might also be related to circadian variation in transient ischemic episodes, being most frequent in the morning hours. Indeed, vascular resistance has been shown to be elevated in the morning hours.

In unstable angina pectoris, which is mostly caused by plaque rupture, a number of vasoactive substances are released into the coronary circulation, most notably serotonin and thrombin. Both substances have been shown to exert potent vasoconstrictor effects in the presence of dysfunctional endothelium. Thus, endothelial dysfunction may cause a constrictor response and importantly magnify the ischemic response in the distal vascular bed and whatever the mechanism of acute myocardial infarction, inappropriate dilation of resistance vessels distal to the site of coronary thrombosis could influence the size of myocardial necrosis. This is one of the reasons that acute events and consequences are unrelated to stenosis severity. Further in the vascular bed of the non-infarct related arteries in the presence of endothelial dysfunction, there is also and enhanced response of resistance vessels to systemic and local neurohormonal constrictor stimuli, which could increase the extent of the ischemia at the periphery of the infracted area and reduce collateral flow to the infarct-related arterial bed, thus contributing to the acute impairment of ventricular function and to the extension of necrosis [78].

Endothelial dysfunction could also be responsible for the no-reflow phenomenon. Endothelial cells are at least as susceptible as myocytes to acute ischemic injury and reperfusion damage. This is evident from the profound loss of capacity of myocardium to be reperfused following severe ischemia which is known as the no-reflow phenomenon. However, this inability to perfuse dead tissue compromises also perfusion of the immediately adjacent viable tissue, which is a region of »low flow« in which, despite their dilatation, many capillaries are incompetent.

There also exists interrelationship between endothelial dysfunction and microvascular angina (syndrome X). This clinical setting is characterized by angina-like pain and abnormal exercise ECG changes in the presence of a normal coronary angiogram. The role of endothelial dysfunction in this setting is controversial. It has been shown that patients with microvascular angina have endothelial dysfunction in the resistance vessels, possibly as a result of diminished formation of NO. However, other studies on coronary epicardial vasodilatation in patients with syndrome X were unable to find signs of endothelial dysfunction [73].

Therefore, endothelial dysfunction has different clinical implications. Its importance was also confirmed by study of Schächinger who found close relationship between endothelial dysfunction and the occurrence of coronary events.

Congestive Heart Failure

The pathogenesis of heart failure is determined by the ventricular and vascular responses to myocellular injury. Abnormalities of vascular endothelial function may be of particular relevance to the pathophysiology of congestive heart failure, which is characterized by increased basal vasomotor tone and reduced vasodilatory reserve. Experimental and clinical studies suggest that the vascular endothelium may play an important role in modulating progression of ventricular and vascular remodeling in heart failure. Endothelial cell dysfunction has been described in the coronary and skeletal muscle circulations of patients with heart failure and appears to be characterized by decreased endothelial synthesis of nitric oxide and increased production of endothelin-1. The pathogenesis of endothelial dysfunction in heart failure is unknown, but may be related to increased oxidative stress, abnormal regional flow conditions, and cytokine and neurohormonal activation. The specific role of endothelial dysfunction in the pathogenesis of heart failure remains to be determined. If endothelial dysfunction does contribute to progression of disease in early heart failure, specific therapies to enhance endothelial dysfunction may improve long-term morbidity and mortality [79].

Reversibility of Endothelial Dysfunction and Therapeutic Aspects

Endothelial dysfunction may be improved by substitution of natural protective endothelial substances (e.g., stable prostacyclin analogues, nitrovasodilators, L-arginine), inhibitors of antagonists of endothelium-derived contracting factors (e.g., ACE inhibitors, angiotensin II receptor antagonists, thromboxane receptors antagonists, endothelin antagonists), cytoprotective agents (e.g., free-radical scavengers, such as superoxide dismutase) and lipid-lowering drugs or diets and physical exercise.

L-arginine, the precursor of NO, has been reported to improve vascular responses to endothelium-dependent vasodilatators, both in hypercholesterolemic animals and in humans. L-arginine has a clear antiatherogenic effect, but this effect is associated with rather slight changes in vascular reactivity that are suggestive of a slight increase in NO elaboration by the endothelium. These suggest that the antiatherogenic effects of L-arginine may not be mediated entirely by its effect on the endothelium [80].

Different studies showed that treating risk factors results in improvement of endothelial dysfunction. It has been shown that treatment of hypercholesterolemia with different statins improves endothelial function [81, 82]. Further, regression of endothelial dysfunction was observed during treatment of arterial hypertension with various drugs [83]. Several antihypertensive agents can prevent and reverse impaired endothelium-dependent relaxation, in large conduit as well as in resistance arteries of hypertensive rats [84]. In contrast, data from studies in hypertensive patients are still controversial. Calcium channel antagonists seem to mediate part of their pharmacological effect by increasing endothelial NO – production with consequent improvement of endothelial function [85]. Angiotensin-converting enzyme (ACE) inhibitors, beyond inhibiting the renin-angiotensin system, diminish the inactivation of bradykinin, thus leading to an augmentation of NO release. In addition, these compounds stabilize the β2 receptors, and reduce oxidative stress [86]. ACE inhibitors seem to improve endothelial function in subcutaneous arteries [87] and the renal circulation [88]. In the forearm circulation, on the other hand, treatment with captopril and enalapril [89] or cilazil [90] failed to improve vasodilatation to muscarinic agonists, while lisinopril selectively improved the vasodilating response to bradykinin without restoring NO bioavailability (91, 92). Further, the available data underlie the role of endothelial dysfunction and the potential benefit of ACE inhibition in the field of ischemic heart disease. ACE inhibition of kininase inhibits the breakdown of bradykinin resulting in increased release of NO, consequently pathological vasoconstriction is prevented and vasodilation of the coronary arteries occurs [93, 94, 95].

We also observed improvement of endothelial dysfunction during physical training of patients with cardiac insufficiency and polymetabolic syndrome [96]. Further improvement of endothelial dysfunction was registered during growth hormone replacement in growth hormone deficient patients [97]. All these data show that endothelial dysfunction is reversible and by treating risk factors it is possible to improve endothelial dysfunction and to restore vascular function.

References

[1] Drexler H. Factors involved in the maintenance of endothelial function. *Am J Cardiol* 1998; 82: 3S-4S.

[2] Fleming I, Busse R. NO: the primary EDRF. *J Moll Cell Cardiol* 1999; 31: 5-14.

[3] Vanhoutte PM. Endothelium-dependent contractions: from superoxide anions to TP-receptor agonists. *Dialogues Cardiovasc Med* 2002; 7: 211-2.

[4] Drexler H, Hornig B. Endothelial dysfunction in human disease. *J Mol Cell Cardiol* 1999; 31: 51-60.

[5] Rubanyi GM. The role of endothelium in cardiovascular homeostasis and diseases. *J Cardiovasc Pharmacol* 1993; 22: S1-14.

[6] Shimokawa H. Primary endothelial dysfunction: atherosclerosis. *J Mol Cell Cardiol* 1999; 41: 23-37.

[7] Raitakari OT, Celermajer DS. Testing for endothelial dysfunction. *Ann Med* 2000; 32: 293-304.

[8] Ludmer PI, Selwyn AP, Shook TL, Wayne RR, Mudge GH, Alexander RW et al. Paradoxical vasoconstriction induced by acetylcholine in atherosclerotic coronary arteries. *N Engl J Med* 1986; 315: 1046-51.

[9] Anderson TJ, Uehata A, Gerhard MD, Meredith IT, Knab S, Delagrange D et al. Close relationship of endothelial function in the human coronary and peripheral circulation. *J Am Coll Cardiol* 1995; 26: 1235-41.

[10] Pitkanen OP, Raitakari OT, Niinikoski H, Nuutila P, Lida H, Voipio-Pukkis LM et al. Coronary flow reserve is impaired in young men with hypercholesterolemia. *J Am Coll Cardiol* 1996; 28: 1705-11.

[11] Anderson TJ, Gerhard MD, Meredith IT, Charbonneau F, Delagrange D, Creager MA, Selwyn AP, Ganz P. Systemic nature of endothelial dysfunction in atherosclerosis. *Am J Cardiol* 1995; 75: 718-48.

[12] Uehata A, Gerhard MD, Meredith IT, Lieberman EH, Selwyn AP, Creager M, Polak J, Ganz P, Yeung AC, Anderson TJ. Close relationship of endothelial dysfunction in coronary and brachial artery. (Abstr.) *Circulation* 1993; 88: 1-618.

[13] Celermajer DS, Sorensen KE, Gooch VM, Spiegelhalter DJ, Miller OI, Sullivan ID et al. Non-invasive detection of endothelial dysfunction in children and adults at risk of atherosclerosis. *Lancet* 1992; 340: 1111-5.

[14] Lerman A, Edwards BS, Hallett JW, Heublein DM, Sandberg SM, Burnett JCJ. Circulation and tissue endothelin immunoreactivity in advanced atherosclerosis. *N Engl J Med* 1991; 325: 997-1001.

[15] Haak T, Marz W, Jungmann E, Hauser S, Siekeier R, Gross W et al. Elevated endothelin levels in patients with hyperlipoproteinemia. *Clin Investig* 1994; 72: 580-4.

[16] Haak T, Jungmann E, Raab C, Usadel KH. Elevated endothelin-1 levels after cigarette smoking. *Metabolism* 1994; 43: 267-9.

[17] Lerman A, Holmes DRJ, Bell MR, Garratt KN, Nishimura RA, Burnett JCJ. Endothelin in coronary endothelial dysfunction and early atherosclerosis in humans. *Circulation* 1995; 92: 2426-31.

[18] Mangiafico RA, Malatino LS, Spada RS, Santonocito M, Messina R, Dellarte S, et all. Treadmill exercise-induced release of endothelin-1 in patients with peripheral arterial occlusive disease at Fontaine stage IIb. *Int Angiol* 2000; 19: 14-7.
[19] Mangiafico RA, Malatino LS, Santonocito M, Spada RS. Plasma endothelin-1 concentrations in non-insulin-dependent diabetes mellitus and nondiabetic patients with chronic arterial obstructive disease of the lower limbs. *Int Angiol* 1999; 18: 77-9.
[20] Blann AD, Davies A, Miller JP, McCollum CN. Reduction in cholesterol is associated with reduction in von Willebrand factor. *Atherosclerosis* 1993; 103: 299-304.
[21] Cortellaro M, Boschetti C, Cofrancesco E, Zanussi C, Catalano M, de Gaetano G et al. The PLAT Study: haemostatis function in relation to atherothrombotic ischemic events in vascular disease patients: principal results. *Arterioscler Thromb* 1992; 12: 1063-70.
[22] Gepper A, Graf S, Beckmann R, Hornykewycz S, Schuster E, Binder BR et al. Concentration of endogenous tPA antigen in coronary artery disease. Relation to thrombotic events, aspirin treatment, hyperlipidemia, and multivessel disease. *Arterioscler Thromb Vasc Biol* 1998; 18: 1634-42.
[23] Ridker PM, Hennekens CH, Stampfer MJ, Manson AE, Vaughan DE. Prospective study of endogenous tissue plasminogen activator and risk of stroke. *Lancet* 1994; 343: 940-3.
[24] Salomaa V, Stinson V, Kark JD, Folsom AR, Davis CE, Wu KK. Association of fibrinolytic parameters with early atherosclerosis. The ARIC study. *Circulation* 1995; 91: 283-90.
[25] Hwang SJ, Ballantyne CM, Sharrett AR, Smith LC, Davis CE, Gotto AMJ et al. Circulating adhesion molecules VCAM-1, ICAM-1, and E-selectin in carotid atherosclerosis and incident coronary heart disease cases: the Atherosclerosis Risk in Communities (ARIC) Study. *Circulation* 1997; 96: 4219-25.
[26] Squadrito F, Saitta A, Altavilla D, Loculano M, Canale P, Campo GM et al. Thrombolytic therapy with urokinase reduces increased circulating endothelial adhesion molecules in acute myocardial infarction. *Inlamm Res* 1996; 43: 14-9.
[27] Ikeda H, Takajo Y, Ichiki K, Ueno T, Maki S, Noda T et al. Increased soluble form P-selectin in patients with unstable angina. *Circulation* 1995; 92: 1693-6.
[28] Ikeda H, Nakayama H, Oda T, Kuwano K, Muraishi A, Sugi K et al. Soluble form of P-selectin in patients with acute myocardial infarction. *Coron Artery Dis* 1994; 5: 515-8.
[29] Davi G, Romano M, Mezzetti A, Procopio A, Iacobelli S, Antidormi T et al. Increased levels of soluble P-selectin in hypercholesterolemic patients. *Circulation* 1998; 97: 953-7.
[30] Blann AD, McCollum CN. Circulating endothelial cell/leukocyte adhesion molecules in atherosclerosis. *Thromb Haemost* 1994; 72: 151-4.
[31] Ridker PM, Hennekens CH, Roitman-Johnson B, Stampfer MJ, Alien J. Plasma concentration of soluble intercellular adhesion molecule-1 and risk of future myocardial infarction in apparently healthy men. *Lancet* 1998; 351: 88-92.
[32] Hackman A, Abe Y, Insulf WJ, Pownall H, Smith I, Dunn K et al. Levels of soluble cell adhesion molecules in patients with dyslipidemia. *Circulation* 1996; 93: 1334-8.

[33] Zeiher AM, Drexler H, Saurbier B, Just H. Endothelium-mediated coronary blood flow modulation in humans: effects of age, atherosclerosis, hypercholesterolemia, and hypertension. J *Clin Invest* 1993; 92; 652-62.

[34] Ouyyumi A. Effect of atherosclerosis on endothelium-dependent inhibition of platelet activation in humans. *Circulation* 1998; 98: 17-24.

[35] Heitzer T, Ylä-Herttuala S, Luoma J, Kurz S, Münzel T, Just J, Olschewski M, Drexler H. Cigarette smoking potentiates endothelial dysfunction of forearm resistance vessels in patients with hypercholesterolemia. *Circulation* 1996; 93: 1346-64.

[36] Drexler H, Zeiher AM, Meinzer K, Just H. Correction of endothelial dysfunction in coronary microcirculation of hypercholesterolaemic patients by L-arginine. *Lancet* 1991; 338: 1546-50.

[37] Creager MA, Gallagher SJ, Girerd HY, Coleman SM, Dzau VJ, Cooke JP. L-arginine improves endothelium-dependent vasodilation in hypercholesterolemic humans. *J Clin Invest* 1992; 90: 1248-53.

[38] Panza JA, Ouyyumi AA, Brush JE Jr, Epstein SE. Abnormal endothelium dependent vascular relaxation in patients with essential hypertension. *N Engl J Med* 1990; 323: 22-7.

[39] Ouvina SM, La Greca RD, Zanaro NL, Palmer L, Sassetti B. Endothelial dysfunction, nitric oxide and platelet activation in hypertensive and diabetic type II patients. *Thromb Res* 2001; 102: 107-11.

[40] Shimokawa H. Endothelial dysfunction in hypertesnion. *J Atheroscler Thromb* 1998; 4: 118-27.

[41] Mombouli JV, Vanhoutte PM. Endothelial dysfunction: from physiology to therapy. *J Mol Cell Cardiol* 1999; 31: 61-74.

[42] Taddei S, Virdis A, Ghiadoni L, Magagna A, Salvetti A. Vitamin C improves endothelium-dependent vasodilation by restoring nitric oxide activity in essential hypertension. *Circulation* 1998; 97: 2222-9.

[43] Taddei S, Salvetti A. Endothelial dysfunction in essential hypertension: clinical implications. *J Hypertens* 2002; 20: 1671-4.

[44] Rizzoni D. Endothelial dysfunction in hypertension: fact or fantasy? *J Hypertens* 2002; 20: 1479-81.

[45] Tadei S, Virdis A, Mattei P, Ghiadoni L, Sudano I, Salvetti A. Defective L-arginine nitric oxide pathway in offspring of essential hypertensive patients. *Circulation* 1996; 94: 1298-303.

[46] Zizek B, Poredos P, Videcnik V. Endothelial dysfunction in hypertensive patients and in normotensive offspring of subjects with essential hypertenison. Scientific letter. *Heart* 2001; 85: 215-6.

[47] Bonnardeaux A, Napaud S, Charrue A, Jeunemaitre X, Corcol P, Soubrier E. Lack f evidence for linkage of the endothelial cell nitric oxide synthase gene to essential hypertension. *Circulation* 1995; 91: 96-102.

[48] Clarkson P, Celermajer DS, Donald AE, Sampson M, Sorense KE, Adams M, et al. Impaired vascular reactivity in insulin-dependent diabetes mellitus is related to disease duration and low density lipoprotein cholesterol levels. *J Am Coll Cardiol* 1996; 28: 573-9.

[49] Schiekofer S, Balletshofer B, Andrassy M, Bierhaus A, Nawroth PP. Endothelial dysfunction in diabetes mellitus. *Sem Thromb Hemostas* 2000; 26: 503-11

[50] Poredos P, Kek A. Relation of blunted dilation of the brachial artery in insulin-dependent diabetes mellitus to microalbuminuria. *Am J Cardiol* 2000; 86: 364-7.

[51] Fioretto P, Stehouwer CDA, Mauer M, Chiesure-Corona M, Brocco E, Carraro A et al. Heterogeneous nature of microalbuminuria in NIDDM: studies of endothelial function and renal structure. *Diabetologia* 1998; 41: 233-6.

[52] Stehouwer CDA, Nauta JJP, Zeldenrust GC, Hackeng WHL, Donker AJM, Den Ottolan-der GJH. Urinary albumin excretion, cariodvascular disease, and endothelial dysfunction in non-insulin dependent diabetes mellitus. *Lancet* 1992; 340: 319-23.

[53] Yudkin JS, Forrest RD, Jackson CA. Microalbuminuria as predictor of vascular disease in non-diabetic subjects. *Lancet* 1998; 2: 530-3.

[54] Lim Su C, Caballero AE, Smakowski P, LoGerfo FW, Horton ES, Veves A. Soluble intercellular adhesion molecule, vascular cell adhesion molecule, and impaired microvascular reactivity are early markes of vasculopathy in type 2 diabetic individuals without microalbuminuria. *Diabetes Care* 1999; 22: 1863-70.

[55] Poredos P, Orehek M, Tratnik E. Smoking is associated with dose-related increase of intima-media thickness and endothelial dysfunction. *Angiology* 1997; 50: 201-7.

[56] Celermajer DS, Sorensen KE, Georgakopoulos D, et al. Cigarette smoking is associated with dose-related and potentially reversible impairment of endothlium-dependent dilation in healthy young adults. *Circulation* 1983; 88: 2149-55.

[57] Eliasson M, Asplund K, Evrin PE, et al. Relationship of cigarette smoking and snuff dipping to plasma fibrinogen, fibrinolytic variables and serum insulin. The northern Sweden MONICA study. *Atherosclerosis* 1995; 113: 41-53.

[58] De Meyer GRY, Herman AG. Vascular endothelial dysfunction. *Progrr Cardiovasc Disease 1997*; 39: 325-42.

[59] Poredos P, Kek A, Verhovec R. Morphological and functional changes of the arterial wall in subjects at risk of atherosclerosis and in patients with peripheral arterial occlusive disease. *VASA* 1997; 26: 271-6.

[60] Ohara Y, Peterson TE, Harrison DG. Hypercholesterolemia increases endothelial superoxide anion production. *J Clin Invest* 1993; 91: 2546-51.

[61] Rajagopalan S, Kurz S, Mű̈nzel T, Tarpey M, Freeeman BA, Griendling KK, et al. Angiotensin II medited hypertension in the rat increases vascular superoxide production via membrane NADH/NADPH oxidase activation: Contribution to alterations of vasomotor tone. *J. Clin Invest* 1996; 97: 1916-23.

[62] Minor RL, Jr., Myers PR, Guerra R, Jr., Bates JN, Harrison DG. Diet-induced atherosclerosis increases the release of nitrogen oxides from rabbit aorta. *J Clin Invest* 1990; 86: 2109-16.

[63] Liuzzo G, Biasucci LM, Gallimore JR, Grillo RL, Rebuzzi AG, Pepys MB, et al. The prognostic value of C-reactive protein and serum amyloid A protein in severe unstable angina. *N Engl J Med* 1994; 331: 417-24.

[64] Yudkin JS, Kumari M, Humphries SE, Mohamed-Ali V. Inflammation, obesity, stress and coronary heart disease: is interleukin-6 the link? *Atherosclerosis* 2000; 148: 209-14.

[65] Ridker PM, Cushman M, Stampfer MJ, Tracy RP, Hennekens CH. Plasma concentration of C-reactive protein and risk of developing peripheral vascular disease. *Circulation* 1998; 97: 425-8.

[66] Napoli MD, Papa F, Bocola V. C-reactive protein in ischemic stroke. *Stroke* 2001; 32: 917-24.

[67] Napoli MD, Papa F, Bocola V. Prognostic influence of increased C-reactive protein and fibrinogen levels in ischemic stroke. *Stroke* 2001; 32: 133-8.

[68] Pasceri V, Willerson J, Yeh ETH. Direct proinflammatory effect of C-reactive protein on human endothelial cells. *Circulation* 2000; 102: 2165-8.

[69] Fichtlscherer S, Rosenberger G, Walter DH, Breuer S, Dimmeler S, Zeiher A. Elevated C-reactive protein levels and impaired endothelial vasoreactivity in patients with coronary artery disease. *Circulation* 2000; 102: 1000-6.

[70] Steinberg DS. Lipoproteins and the pathogenesis of atherosclerosis. *Circulation* 1987; 76: 508-14.

[71] Henderson AH. Endothelium in control. *Br Heart J* 1991; 65: 116-25.

[72] Zeiher AM, Schächinger V, Saurbier B, Just H. Assessment of endothelial modulation of coronary vasomotr tone: Insights into a fundamental functional disturbance in vascular biology of atherosclerosis. *Basic Res Cardiol* 1994; 89: 115-28.

[73] Angus JA. Role of the endothelium in the genesis of cardiovascular disease. *Clin Exp Pharmacol Physiol* 1996; 23: S16-S22.

[74] Kuo L, Davis MJ, Cannon MS, Chilian WM. Pathophysiological consequences of atherosclerosis extend into the coronary microcirculation. Restoration of endothelium-dependent responses by L-arginine. *Circulation Res* 1992; 70: 465-6.

[75] Sellke FW, Armstrong ML, Harrison DG. Endothelium-dependent vascular relaxation is abnormal in the coronary microcirculation of atherosclerotic primates. *Circulation* 1990; 81: 1586-93.

[76] Vita JA, Tressure CB, Yeung AC, et al. Patients with evidence of coronary endothelial dysfunction as assessed by acetylcholine infusion demonstrate marked increase in sensitivity to constrictor effect of catecholamines. *Circulation* 1992; 85: 1390-7.

[77] Zeiher AM, Drexler H, Wollschläger H, Just H. Endothelial dysfunction of the coronary microvasculature is associated with impaired coronary blood flow regulation in patients with early atherosclerosis. *Circulation* 1991; 84: 1984-92.

[78] Schächinger V, Britten MB, Zeiher AM. Prognostic impact of coronary vasodilator dysfunction on adverse long-term outcome of coronary heart disease. *Circulation* 2000; 101: 1899-906.

[79] Katz SD. Mechanisms and implications of endothelial dysfunction in congestive heart failure. *Curr Opinion Cardiol* 1997; 12: 259-64.

[80] Andreson TJ, Meredith IT, Yeung AC, Frei B, Selwyn AP, Ganz P. The effect of cholesterol-lówering and antioxidant therapy on endothelium-dependent coronary vasomotion. *N Engl J Med* 1995; 332: 488-93.

[81] Treasure CB, Klein L, Weintraub WS, Talley JD, Stillabower ME, Kosinski AS, et al. Beneficial effects of cholesterol-lowering therapy on the coronary endothelium in patients with coronary artery disease. *N Engl J Med* 1995; 332: 481-7.

[82] Higashi Y, Sasaki S, Nakagawa K, Ueda T, Yoshimizu A, Kurisu S, et al. A comparison of angiotensin-converting enzyme inhibitors, calcium antagonists, beta-blockers and diuretic agents on reactive hyperemia in patients with essential hypertension: a multicenter study. *J Am Coll Cardiol* 2000; 35: 284-91.

[83] Lűscher TF, Vanhoute PM, Raij L. Antihypertensive treatment normalizes decreased endothelium-dependent relaxation in rats with salt-induced hypertension. *Hypertension* 1987; 9: 193-7.

[84] Berkels R, Bertsch A, Breitenbach T, Klaus W, Roesen R. The calcium antagonist nifedipine stimulates endothelial NO relase in therapeutical concentrations. *Pharm Pharmacol Lett 1996*; 6: 75-8.

[85] Gainer JV, Morrow JD, Loveland A, King DJ, Brown NK. Effects of bradykinin-receptor blockade on the response to angiotensin-converting-enzyme inhibitor in normotensive and hypertensive subjects. *N Engl J Med* 1998; 339: 1285-92.

[86] Schiffrin EL, Sventek P, Li JS, Turgeon A, Reudelhuber T. Antihypertensive effect of an endothelin receptor antagonist in DOCA-salt spontaneously hypertensive rats. *Br J Pharmacol* 1995; 115: 1377-81.

[87] Mimran A, Ribstein J, DuCailar G. Contrasting effect of antihypertensive treatment on the renal response to L-arginine. *Hypertension* 1995; 26: 937-41.

[88] Creager MA, Roddy MA. Effect of captopril and enalapril on endothelial function in hypertensive patients. *Hypertension* 1994; 24: 499-505.

[89] Kiowski W, Linder L, Nuersch R, Martin B. Effects of cilazapril on vascular structure and function in essential hypertension. *Hypertension* 1996; 27: 371-6.

[90] Taddei S, Virdis A, Ghiadoni L, Mattei P, Salvetti A. Effects of angiotension converting enzyme inhibition on endothelium-dependent vasodilatation in essential hypertensive patients. *J Hypertens* 1998; 16: 447-56.

[91] Ruschitzka F, Noll G, Lűscher TF. Angiotensin converting enzyme inhibitors and vascular protection in hypertension. *J Cardiovasc Pharmacol* 1999; 34: S8-S12.

[92] Mancini GBJ. Role of angiotensin-converting enzyme inhibition in reversal of endothelial dysfunction in coronary artery disease. *Am J Med* 1998; 105: 40S-7S.

[93] Ambrosioni E, Bacchelli S, Esposti DD, Borghi C: Anti-ischemic effect of angiotensin-converting enzyme inhibitors: a future therapeutic perspective. *J Cardiovasc Pharmacol* 2001; 37: S3-S9.

[94] Lazar HL, Bao Y, Rivers S, Colton T, Bernard SA. High tissue affinity angiotensin-converting enzyme inhibitors improve endothelial function and reduce infarct size. *Ann Thorac Surg 2001*; 72: 548-54.

[95] Lavrencic A, Guzic-Salobir B, Keber I. Physical training improves flow-mediated dilatation in patients with the polymetabolic syndrome. *Arteroscler Thromb Vasc Biol* 2000; 20: 138-41.

[96] Pfeifer M, Vehovec R, Zizek B, Prezelj J, Poredos P, Clyton RN. Growth hormone (GH) treatment reverses early atherosclerotic changes in GH-dificient adults. *J Clin Endocrinol Metabol* 1999; 84: 453-7.

In: Focus on Atherosclerosis Research
Editor: Leon V.Clark, pp. 185-213
ISBN 1-59454-044-6

Chapter IX

Oxidative Stress Hypothesis of Atherosclerosis

Fausta Micheletta, Silvia Natoli and Luigi Iuliano*
Department of Internal Medicine, University La Sapienza, Rome, Italy

Abstract

Atherosclerosis is a complex, irreversible disease of the artery tree causing an excess of morbility and most of the deaths due to myocardial infarction and stroke. The disease process initiates as early in childhood and progresses for decades showing a spectrum of lesions in the arteries ranging from the fatty streak to the advanced plaque. Two elements are present in all the stages of the disease: macrophages and lipids. Striking evidences suggest that the very early event of the atherosclerotic process is the accumulation of circulating LDL into macrophages in the intima that leads to foam cell formation. LDL entry in macrophages is mediated by the scavenger receptor whose ligand is not a native LDL but a modified form of LDL believed to occur in vivo by free radical-mediated oxidation. Furthermore, it has been demonstrated that the oxidation of LDL produces an array of oxidized bioactive lipids with a potential contribution in cellular events of atherogenesis, including endothelial dysfunction, platelet aggregation, and apoptosis. The role of oxidative stress in atherosclerosis has also been supported by the inihibitory effect of antioxidants in experimental atherosclerosis.

Introduction

From the initial studies published almost 20 years ago showing the relevance of LDL oxidation in foam cell formation, a large body of data has been accumulated providing a solid background to the theory that the oxidative stress, collectively the uncontrolled free radical

* Correspondence: Luigi Iuliano, M.D., Department of Internal Medicine, University La Sapienza. Via del Policlinico 155. 00161 Rome.Italy, Tel/Fax. 39 06 4938 3333, e-mail: luigi.iuliano@uniroma1.it

reactions escaping the antioxidant defense system, plays a central role in the atherosclerotic process.

Atherosclerosis is a complex, irreversible disease of the artery tree causing an excess of morbility and most of the deaths due to myocardial infarction and stroke. Two elements associated to all the stages of the disease are macrophages and lipids, which have a crucial role in the initiation and progression of the atherosclerotic process. Striking evidences suggest that the atherosclerotic process initiates in the intima with the accumulation into macrophages of lipids [1] derived from circulating LDL, which had penetrated the endothelial barrier. In the intima, LDL are retained and taken up by macrophages which progressively transform into foam cells, and eventually die and liberate their lipid content in the intima. Extracellular lipids, present in the initial lesions, are associated to cell remnants suggesting that extracellular lipids originate from liberation of intracellular lipids after the cell death. While the mechanism leading to LDL entry in the intima is still unknown, a huge number of studies has unraveled the mechanism of LDL entry in macrophages. In contrast to the majority of cells, which display the LDL receptor (LDL-R), macrophages accumulate lipids in their citoplasm by using a different receptor, the scavenger receptor (SR). The LDL-R pathway is down-regulated to keep constant the intracellular cholesterol concentration, contrarily to the SR pathway that is not down-regulated and provokes the lipid engulfment and the appearance of foam cells. An additional functional characteristic of the SR is its low affinity for LDL but high affinity for LDL that had been subjected to structural modification. Modified LDL occur in vivo and for convention it is now widely accepted to distinguish native LDL (n-LDL) from modified LDL (m-LDL), which use the LDL-R pathway and the SR pathway respectively. Studies on mechanisms of LDL modification have provided substantial evidence that free radicals are responsible for generating m-LDL usable for the SR-pathway. Free radicals, both generated chemically *in vitro* or by activated cells, induce peroxidation of LDL lipids with concomitant release of reactive by-products which alter the molecular structure of apo B and induce a shift for receptor recognition: from the LDL-R to the SR receptor. Since the uptake of oxidized LDL (ox-LDL) is not restricted by SR down-regulation, macrophage are engulfed by lipids and transforms into foam cells [2]. Additionally to the role in foam cell formation, ox-LDL possess several biological activities relevant for atherogenesis. Oxidized LDL induces expression of adhesion molecules on endothelial cells that facilitates the recruitment and the uptake of circulating inflammatory cells, induces endothelial dysfunction through impairment of the biological activity of nitric oxide, provokes monocyte chemotaxis and adhesion, cytotoxicity, platelet aggregation and thrombus formation [3-7]. Further support to the role of ox-LDL and free radicals, collectively oxidative stress, in the atherosclerotic process was also furnished by epidemiological studies and by the use of specific addressed therapy based on antioxidants. Studies concerning experimenatal atherosclerosis showed the effectiveness of antioxidants in inhibiting the lesion formation *in vivo* in animals. The oxidative stress hypothesis of atherogenesis has been finally transferred at clinical level where a number of trials tested antioxidants in several disease settings. In contrast to experimental atherosclerosis, results from clinical trials with antioxidants are inconclusive as some trials demonstrated the effectiveness of antioxidants in reducing the end-points while others showed neutral results. In this chapter, the relevance of oxidative stress in atherogenesis is discussed by focusing on the history of studies on LDL oxidation, and on studies with antioxidants in experimenatal atherosclerosis and in human.

Atherogenesis, Early and Late Events

Atherosclerosis is a complex process involving entering, modification and accumulation of plasma lipoproteins, and recruitment and proliferation of cells in the arterial wall [1, 8, 9]. The process advances through a series of stages beginning with the focal accumulation of lipids in arterial intima and the appearance of the fatty streak lesion, composed largely of foam cells which are lipid-engulfed macrophages. The fatty streak evolves towards the complex atherosclerotic plaque consisting of an extracellular lipid core covered by a fibrous cap and areas rich in inflammatory cells. The initiating event of the disease is the appearance of intracellular lipid deposits in the arterial wall far from the macroscopic appearance of fatty streaks [10]. Thus, lipid accumulation in macrophages and formation of foam cells is a key event.

Accumulation of lipids in isolated groups of macrophages in arterial intima was shown in human fetal aortas [11] and infants [10] and are considered as "preatheroma" lesions. The fatty streak is the macroscopic appearance of the early lesion, which is present in most children around puberty. Fatty streaks consist of numerous macrophages foam cells, stratified in layers, small lipid droplets and cell remnants, mostly derived from macrophage foam cells that died. Fatty streaks eventually enlarges to generate advanced lesions that may transform into vulnerable plaque responsible for vascular accidents. Advanced plaques uphold the avidity for circulating LDL as can be shown *in vivo* by LDL scintigraphy [12]. In human, LDL trafficking from blood to plaque-resident macrophages has been demonstrated *in vivo* by autoradiography and immunoistochemistry studies of plaque taken at surgery from patients undergoing carotid endoarterectomy previously injected with radiolabeled LDL [13].

Thus, the simplest evolution scheme of atherosclerotic lesion is the progressive entrapment of lipoproteins in macrophages within the arterial intima, formation of foam cells, and death of foam cells. Lipids and cell debris resulting from cell death generate the lipid core of atheromas. While multiple factors playing a role in this process have been identified, such as release of metalloproteinases, cell recruiting factors, growth factor, cell migration inhibitory factors, apoptotic and necrotic signals, the hallmark of atherosclerosis is the accumulation in macrophages of lipids vehicled in the intima through low density lipoproteins.

Structure of LDL Particles and Metabolism

LDL, the major cholesterol carrier in plasma, is a large, heterogeneous and complex molecule composed by the apoprotein B100 (apoB), various lipid classes including phospholipids and cholesterol, and antioxidants (Figure 1) [14]. For convention LDL particles are defined by the density range 1.019-1.063 g/ml, which is also the target density of most widespread method of ultracentrifugation to obtain LDL from plasma. LDL particles are globular in shape with an average diameter of 22 nm, and are composed of an hydrophobic core and an external coat of polar lipids anchoring apoB. LDL carries about 60% of total serum cholesterol [15]; 2/3 of unesterified cholesterol is located in the surface layer intercalated to phospholipids, while the remaining 1/3 is located in the core [14].

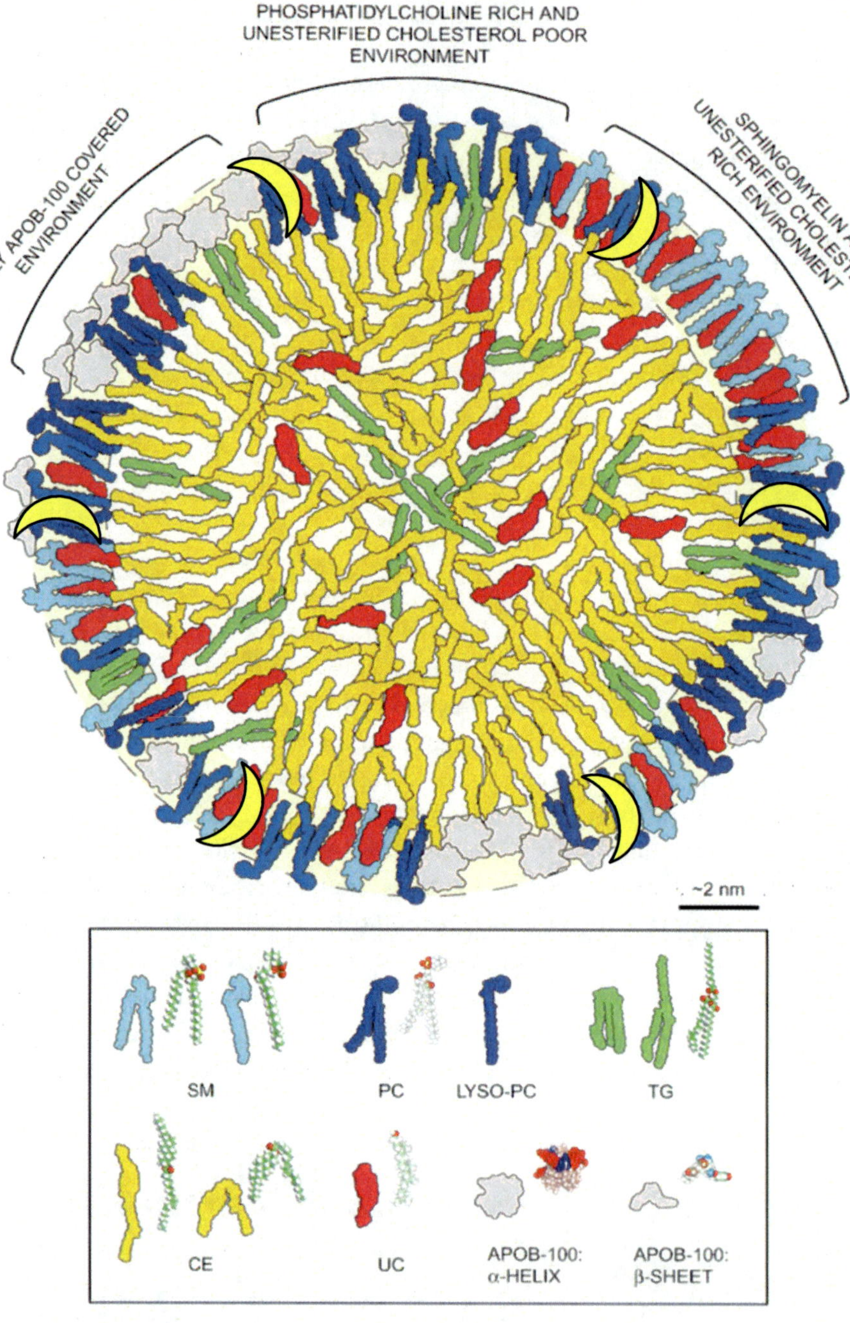

Figure 1 Schematic molecular model of LDL. The molecular components of the particle are drawn in both the correct percentage and size ratios, except for the six α-tocopherol molecules (yellow arcs).Kindly provided by Kovanen PT; modified and reprinted from P[14], with permission from Elsevier".

The lipophilic core contains about 1600 molecules of cholesteryl ester and 170 molecules of triglycerides. The core is coated by a monolayer composed by about 700 phospholipid molecules (about 65% is phosphatidylcholine and 26% is sphingomyelin) whose polar heads help to solubilize LDL in aqueous phase. LDL particles carries also lipophilic antioxidants which highly protect oxidizable lipids, such as polyunsaturated fatty acids (PUFA), against free radical attack and oxidation. These antioxidants include alpha- and gamma-tocopherol, lycopene, alpha- and beta-carotene, crytpoxanthin, cantaxanthin, zeaxanthin, lutein and ubiquinol. On a molar basis, alpha-tocopherol is the major antioxidant corresponding to 82% of LDL antioxidants. Six molecules of alpha-tocopherol are present in each LDL molecule [15]. A single copy of apoB, one of the largest monomeric protein known (500 KD) with large amphipatic regions for lipid association, is present to each LDL particle. A site located in the residues 3359-3369 [16] serves for binding the LDL receptor that mediates cholesterol influx in cells.

LDL are secreted by the liver as the larger precursor VLDL (diameter 80-200 nm) that contains higher amounts of triglycerides and additional apoproteins. In the peripheral circulation (muscle and adipose tissue capillaries) triglycerides are removed and proteins are exchanged with other lipoproteins. Thus, the LDL molecule is generated from VLDL after size reduction, loss of triglycerides and preservation of a single apoB protein copy.

LDL Receptor

Cells possess a metabolic pahway for de-novo cholesterol synthesis, and circulating LDL is an additional source for cell requirements of cholesterol as essential constituent of membranes. The receptor-mediated mechanism tightly controls LDL cholesterol uptake. This has been known since 1974 when Brown and Goldstein began to publish a large series of data describing how the LDL receptor works and influences the circulating levels of LDL [17, 18]. LDL receptors are located on coated pits, specialized indentations of cell membrane, where the LDL-receptor complex is closed in microvescicles, internalized and transported to lysosomes, here the LDL particle is disassembled and cholesterol is delivered to intracellular pool for metabolic and membrane structure purposes. Specialized pathways of cholesterol metabolism are present in liver cells (lipoprotein and bile acid metabolism) and in endocrine cells (hormone metabolism in adrenal cortex, testis and ovary) which possess very high LDL receptor activity. This can be visualized *in vivo* in humans by total body scintigraphy imaging after infusion of radiolabelled LDL which show fast and persistent uptake of radiotracer in the liver and adrenal glands [12]. All the other tissues express LDL receptors for taking up cholesterol necessary for membrane structure. Intracellular cholesterol homeastasis is tightly controlled by feed-back signals acting at two levels, the LDL receptor and the cholesterol biosynthesis pathway. Thus, LDL influx in cells induces LDL-receptor down regulation and inhibition of cellular cholesterol biosynthesis. Blockade of the cholesterol influx in cells as occurs in genetic defects that impair LDL binding to the receptor, induces increase in circulating LDL levels. Among four aberrant genes causing high blood colesterol levels, one causes a deficit in the LDL receptor and one causes a mutation in apoB leading to impairment of protein-receptor interaction and consequent reduction in plasma clearance of LDL [19].

The discovery of the LDL receptor system has had a deep impact in medicine introducing the development of statins, drugs that inhibit the 3-hydroxy-3-methylglutaryl-CoA reductase enzyme. The inhibition of this rate-controlling enzyme in cholesterol biosynthesis induces LDL receptor expression, increased cell LDL uptake and consequently the plasma cholesterol concentration is reduced. Being the major cholesterol carrying protein in plasma LDL is implicated in cardiovascular disease as demonstrated by epidemiological, genetic and animal studies, and trials with statins [20, 21].

LDL Receptor Paradox

In the years in which Brown and Goldestein unraveled the LDL receptor pathway, it became evident the paradox of lipids laden macrophages in atherosclerotic lesions. How does macrophage accumulate lipids and transform into foam cells since the LDL receptor pathway finely tunes the intracellular cholesterol homeostasis? How do patients with hypercholesterolemia, who totally lack the LDL receptor, show severe atherosclerosis? Why are their arteries plenty of foam cell rich lesions? The answer became from the failure to induce cholesterol deposition in macrophages by incubation with LDL *in vitro* even in high concentration [22-26], pointing to a different pathway of influx in these cells. Several studies showed that the LDL receptor is poorly expressed by macrophages and eventually is down-regulated during cholesterol uptake [27-30], and that the LDL receptor is not expressed in human atherosclerotic plaques [30]. In the late 70s, a series of elegant experiments allowed the demonstration that macrophages possess a distinct pathway for cholesterol accumulation. The hypothesis of an alternative pathway of LDL uptake by macrophages [31] postulated that LDL underwent denaturation or chemical modification [24]. It was well known the ability of macrophages of clearing denaturated proteins in vivo [32, 33] and, in fact, macrophages were defined scavenger cells (scavenge, clean away dirt). The first demonstration of massive cholesterol deposition in macrophages was obtained with LDL modified by reacting the protein moyety with organic anhydrides (acetic and maleic), which reacts with amino groups of lysine and changes the protein surface charge [24]. This work has particular relevance in atherosclerosis research because it i) introduced the experimental model for cholesterol accumulation in macrophages; ii) introduced the concept of the scavenger pathway for the LDL accumulation into macrophages; iii) introduced the concept of native LDL; iv) reconciled the LDL paradox.

LDL Modification

In coincidence with the report showing the ability of macrophages to accumulate acetylated LDL, several studies were conducted to identify the mechanistic aspects of the scavenger receptor pathway and of the modification that alters LDL *in vivo*. It was shown that other treatments successfully tranform LDL in a form readily recognized by the scavenger receptor, such as acetoacetylation, maleylation, succinylation, carbamylation, and treatment with glutaraldeyde [15]. It was also shown that the scavenger receptor is able to recognize other ligands such as negatively charged polyanions or maleylated albumin[15].

However, most of these LDL modifications were unlikely to occur *in vivo*. The potential physiological candidate for *in vivo* modification of LDL was discovered by the Fogelman's group, who found that malondiadehyde (MDA), which forms Shiff base bonds with ε-amino groups of lysine of apoB, is an efficient modifier for LDL [34]. MDA is produced during platelet aggregation in equimolar amount with thromoboxane B_2 resulting from the arachidonic acid cascade activation, and the intimate connection between platelet activation and atherosclerosis conferred to MDA the role of *in vivo* plausible reactant for generating m-LDL. An alternative source of MDA in vivo is the lipid peroxidation, classically linked to several disease states including atherosclerosis [35]. Previous data reported as early in 1952 that atherosclerotic lesion contains peroxidized lipids [36]. In addition, it was known for a long time that LDL are extremely prone to oxidation of its lipid content [37, 38], and importantly that autoxidation of LDL in vitro rendered them toxic to cell culture [39]. The presence of lipid oxidation products in atherosclerotic lesions, the susceptibility of LDL to oxidation, the citotoxicity of LDL subjected to oxidation, and the discovery that MDA is an efficient modifier for LDL, together with the impact of the SR pathway in foam cell formation, contributed in the last two decades to boost the research into mechanistic aspects of LDL oxidation resulting in exponential increase of articles about oxidized LDL [40]. Oxidative stress is also able to affect the metabolism of HDL, which appears to function as a counter-transport system for cholesterol from the peripheral tissues to the liver. Among the different potential relation between oxidative stress and HDL metabolism, of particular interest is the recent findings showing that hypochlorous acid/hypochlorite, a potent oxidant generated in vivo by activated phagocytes, alters the physiological properties of HDL by generating a proatherogenic lipoprotein particle [41].

Scavenger Receptor

After the first description of the binding site that mediates the uptake and degradation of modified LDL, the receptor was later called the macrophage scavenger receptor. The binding of ox-LDL to the SR is not mediated only by the derivatized sites of the protein component. It is now known that also oxidized lipid mojeties of the ox-LDL participate in the SR binding [42]. The scavenger receptor belongs to a large family, including at least 6 receptors (see Table 1). The common characteristic shared by almost all the scavenger receptors is the ability to bind with high affinity to a broad range of structurally different ligands. However, it should be noted that each receptor can show functional characteristic and might has a certain specificity for modified LDL. SR-AI/II, which has received the greatest attention, seems to have only a part on the uptake of oxidized LDL. Steinbrecher [43] studied macrophages from wild type and SR-AI/II knockout mice and provided evidence that this receptor accounts for at least 80% of the removal of ac-LDL. Otherwise, only 30% reduction was obtained with ox-LDL indicating that about 70% ox-LDL removal is mediated by another receptor. In agreement, *in vivo* clearance studies in mice showed that the rate of removal of radiolabelled ox-LDL in knockout mice is comparable to the wild type. On the other hand, the same authors [44] showed that SR-AI/II has a potential role in atherogenesis by cross-breading studies in which SRAI/II mice were crossed with apoE-deficient mice, which develope severe atherosclerosis. These authors showed that the SRAI/II/ApoE double knockout mice developed significantly smaller atherosclerostic lesions compared to apoE single-knockout littermates. It should be inferred that these data can not be trasferred directly to humans, as there is considerable variability in modified LDL uptake by the scavenger receptor pathway in various cell types. Rat peritoneal macrophages degrades 4 times faster acetylated LDL compared to mouse peritoneal macrophages [24]. In humans, it is likely that most of the ox-LDL are removed by the CD36 as demonstrated by studies that used macrophages from normal subjects compared to those of subjects lacking the CD36 receptor [45]. These studies indicate that the CD36 receptor could account for 40% of ox-LDL removal. SRBI shows different functionality compared to SRA-I/II. The major function of SRB-I is involved in key pathways of HDL metabolism but recent report showed that SRB-I binds ox-LDL with high affinity and mediates ox-LDL handling to a comparable extent of that of other scavenger receptors [42, 43, 46, 47]. Recently, a novel mechanism of LDL entry in macrophages has been described. Kruth et al. [48] were able to induce macrophage foam cell formation with native LDL by a fluid phase endocytosis mechanism, introducing a potential new pathway different from the scavenger receptor pathway of lipid loading in macrophages. Taking into account that this study used unphysiological manipulation of macrophages with phorbol esters to render them able to accumulate lipids, it remains to be estabilished what is the relevance of this additional mechanism of lipid accumulation in macrophages *in vivo*. Despite more than two decades of research aimed at unraveling mechanims of atherogenesis, the transformation of macrophages in foam cells continues to attract the interest of researchers.

Table 1. Classification of Scavenger Receptors and their Ligands

Receptor	Ligands
Class A	
SRA-I/II/III*	Ox-LDL, acetyl-LDL, polyanions, LPS
MARCO*	Ox-LDL, acetyl-LDL, bacteria
Class B	
SRB-I**	HDL, Ox-LDL
CD36*#	Ox-LDL, apoptotic cells, long chain fatty acids
Others	
CD68*	Ox-LDL, apoptotic cells
SRC$	Ox-LDL, polyanions, acetyl-LDL
SREC^	Ox-LDL, acetyl-LDL
LOX-1^	Ox-LDL, polyanions
FcγRIIB2	Ig G, Ox-LDL?

Cell type: macrophage, ^endothelial cells, #platelets and monocytes, **liver and adrenals, $drosophila macrophage-like cell line (human homologue not yet been reported). MARCO, macrophage receptor with collagenous structure; LOX-1, lectin-like oxidized LDL receptor; SREC, scavenger receptor expressed by endothelial cells; FcγRIIB2, immunoglobulin receptor.

Oxidation of LDL

It is well known that lipid peroxidation occurs via free radicals mediated reactions. The production of oxidized LDL (ox-LDL) was achieved by free radicals generated *in vitro* and *ex vivo* systems using both chemicals and cells. The most used and reproducible chemical system is based on the use of copper ions. Several cell lines, which upon activation generate a flux of free radicals in the extracellular milieau, were shown to be able to oxidize LDL, including endothelial cells [49], smooth muscle cells [50], monocytes [51], and macrophages [52], providing pathophysiological relevance of ox-LDL. Whatever the free radical used, the process of LDL oxidation can be summarized as follow: the free radical attacks high oxidizable lipids, such as polyunsaturated fatty acids that have extractable α-methylenic hydrogens, producing a carbon centered radical that adds oxygen and forms a peroxyl radical [53]. Several reactive oxygen intermediates can be shown *in vitro* to initiate lipid peroxidation, including superoxide, hydrogen peroxide, hydroxyl radical, peroxynitrite, hypochlorous acid, and heme-proteins.

The free radical attack to PUFA is extremely deleterious because it triggers a chain reaction: a single radical is able to recruit~~s~~ several lipid molecules that propagate the reaction far from the initial site of radical flux. As a consequence of lipid peroxidation, a series of oxidation products are formed, including malondialdehyde, responsible for the derivatization of apoB and the consequent recognition of LDL by the scavenger receptor. Other aldehydes produced during LDL oxidation, such as 4-hydroxynonenal, have also higher reactivity compared to MDA. Aldehydes are only an example of the array of products arising from peroxidation of the complex lipid mixture of LDL. The chain reaction can be interrupted if the radical-carrying lipid molecule is intercepted by an hydrogen donor having peculiar chemical-physical characteristics. By virtue of this action, these molecules are currently named chain-breaking antioxidants and the prototypical molecule in biological systems is vitamin E (Figure 2). Lipids in LDL are protected from free radical-mediated oxidation by several lipophilic antioxidants mixed with lipids in the LDL particle, such as α-tocopherol, which is on molar basis the most represented LDL antioxidant.

As a consequence of LDL oxidation either the lipid or the apoprotein components are modified; the apoprotein is derivatized, and the lipid structure is altered by the loss of PUFA, and other lipids, and ~~to~~ by the accumulation of oxidized lipids. Both lipid and protein sites are created on the ox-LDL for binding the SR. In addition to the role in SR receptor recognition, bioactive products arising from oxidation of LDL have further relevance in the context of atherogenesis [54]. Lipid oxidation products induce platelet activation and vasoconstriction [55, 56], are cytotoxic [57, 58] and promote apoptosis [59], inhibit cholesterol de-loading from foam cells [60], induce endothelial dysfunction [61, 62], and contribute to macrophage-mediated matrix breakdown [63]. Free radical-mediated LDL oxidation produces several different types of oxidized lipid, including oxidized phospholipids with PAF-like structure[64], isoprostanes (Figure 3) [65, 66], and oxysterols (Figure 4) [60, 67]. A common characteristic of oxidized lipids is that they can be produced in vitro by oxidizing homogeneous lipid suspension or LDL, and are found in atherosclerotic plaques and plasma by specific and sensitive methods based on mass spectrometry [68, 69]. Isoprostanes are generated by free radical attack to arachidonic acid, and are isomeric to enzymatically formed eicosanoids. Unlike eicosanoids, produced as one/two main products at a given tissue level, an array of isoprostanes isomers may be formed from arachidonic acid, 64 only for $PGF_{2\alpha}$, at sites of oxidative stress. There is considerable interest in the biological activity of isoprostanes and other congeners [55, 56]. In analogy to arachidonic acid, isoprostane-like compounds are generated by other PUFA and found in tissues, including docosahexaenoic acid which is subjected to intensive research in neurodegenaritive disorders [70]. Aside from their functional role in oxidative stress-induced cellular events, isoprostanes have particular relevance as non invasive markers of oxidative stress status *in vivo*. This has been facilitated by their chemical stability and the intense development of high sensitive and specific methods based on mass spectrometry.

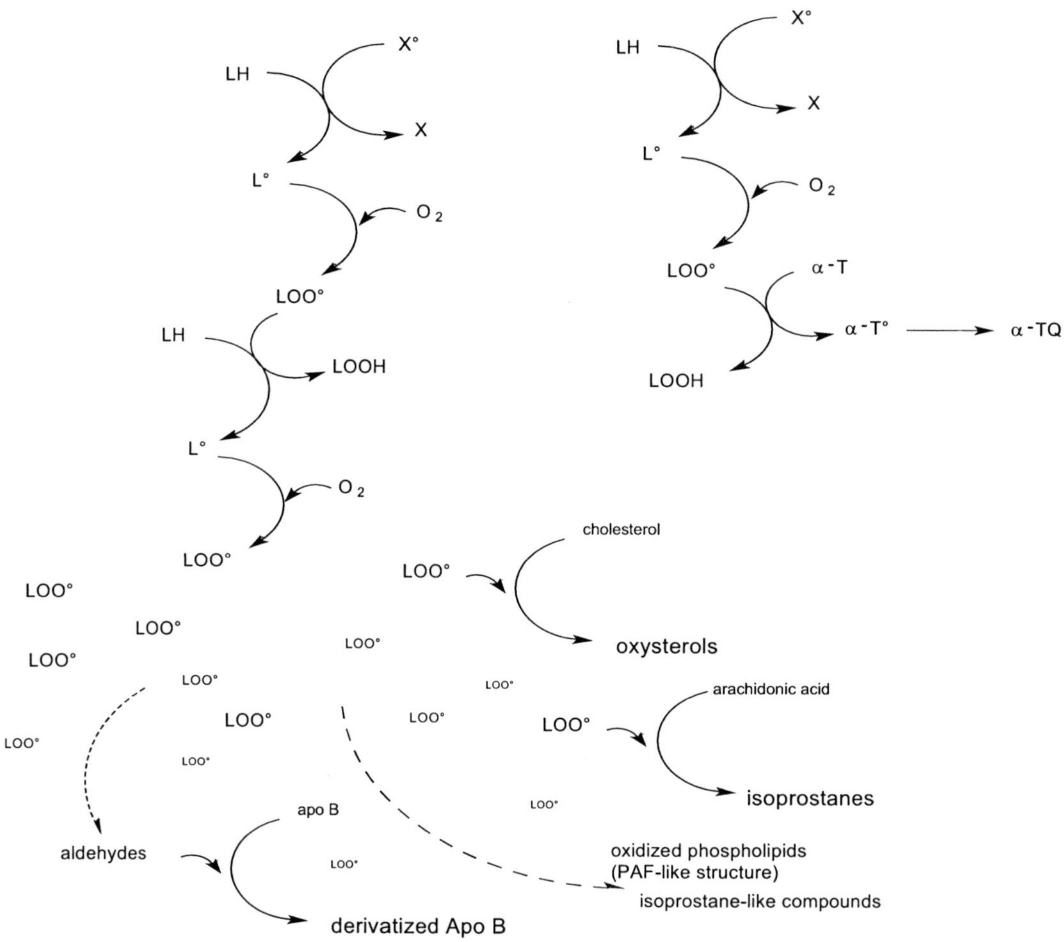

Figure 2. Schematic representation of lipid peroxidation. The initiation phase corresponds to the attack of a free radical (X°) to a polyunsaturated fatty acid (LH), formation of a lipid centered radical (L°) which adds oxygen to form peroxyl radical (LOO°), the main propagating species that expands peroxidation far from the original site. Among end-products of lipid peroxidation oxysterols and isoprostanes are highlighted for their bioactivity. The lipid peroxidation cascade can be interrupted in the presence of chain breaking antioxidants, such as α-tocopherol which traps the radical and transform into a quinone (α-TQ) after the transient appearance of α-tocopheroxyl radical (α-T°). α-T can be regenerated from α-T° at the expenses of thermodinamically suitable reductants like ascorbic acid.

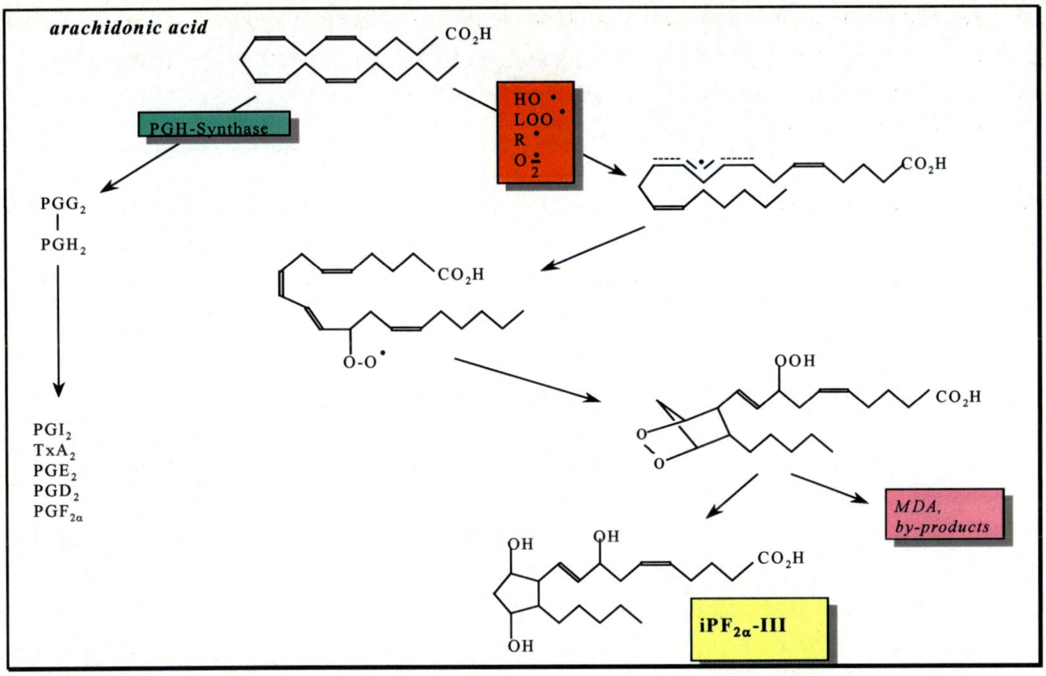

Figure 3. Reaction scheme showing the production of a representative isoprostanes ($iPF_{2\alpha}$-III) by free radical-catalyzed oxidation of arachidonic acid.

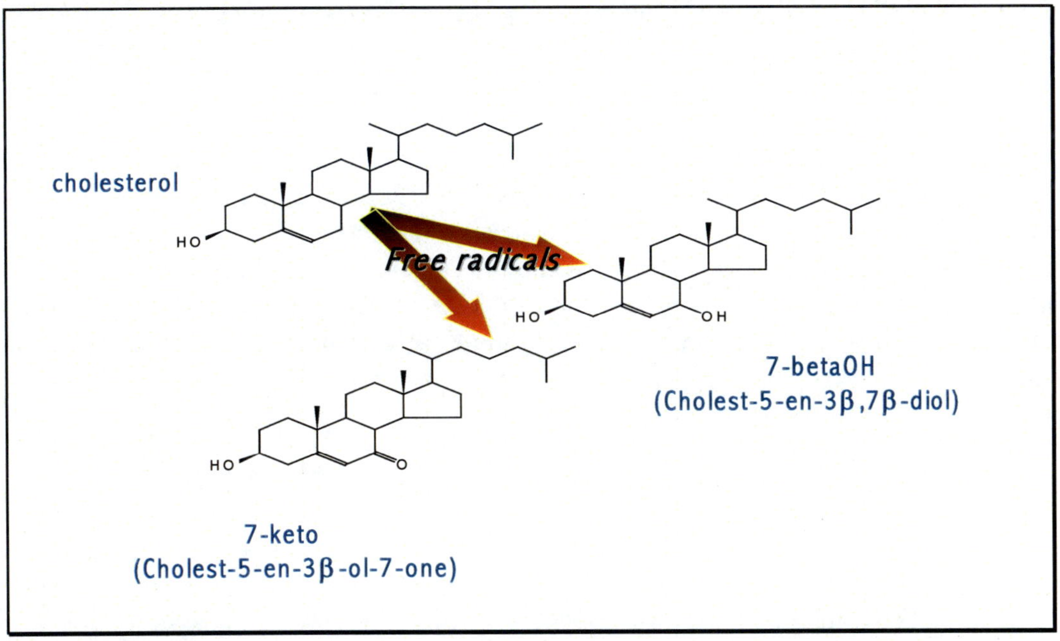

Figure 4. Molecular structure of two free radical-catayzed oxydation products of cholesterol.

Oxysterols are oxygenated derivatives of cholesterol which have been shown to possess many potent biological activities of relevance in atherogenesis [60]. Oxidized derivatives of cholesterol linoleate were reported in 1973 by Harland [71]. Oxysterols can be generated both enzymatically and by free radical attack on the cholesterol backbone. Oxysterols have been ascribed to a number of effects in connection with cholesterol turnover, apoptosis, necrosis, inflammation and immunosuppression [72-75].

Many studies have detected oxysterols in human atherosclerotic plaques, and much of the interest in the role of oxysterols in atherosclerosis can be tracked to the suggestion that atherogenic effects of cholesterol in animal models may have been due to contaminating oxysterols present in the cholesterol supplied [76]. In analogy to isoprostanes, there is considerable interest in oxysterols as non invasive markers of oxidative stress *in vivo*. It is generally considered that 7βhydroxy-cholesterol and 7keto-cholesterol are produced by free radical attack to cholesterol *in vivo* in mammals. Thus, 7βhydroxy-cholesterol and 7keto-cholesterol are currently used as markers of oxidative stress in vivo by sensitive and specific mass spectrometry methods [69, 77].

Both isoprostane and oxysterol research contributed to give unequivocable evidence of the occcurrence of free radical mediated lipid oxidation *in vivo* in the blood stream and at the plaque level. With regard to the latter case, for a long time it has been a great concern about the measurement of lipid oxidation products in atherosclerotic lesions because of the potential artifactual generation of lipid oxidation products during tissue processing. It is to be noted that most of the data on lipid peroxidation in plaque are relatively recent and were mainly based on the TBARS assay. Application of isotope dilution methods allowed the quantitation in carotid plaques of several isoprostanes and oxysterols. In these studies the authors splitted tissue samples with deuterated arachidonic acid and cholesterol and did not obtain artifactual peroxidation of deuterated substrates during tissue processing [57, 68].

Oxidized–LDL is accepted to behave as foreign for the immune system, and autoantibodies against both protein and oxidized lipid epitopes have been characterized and are thought to play a role in immuno-mediated reactions in certain stages of the disease [78-80].

The existence of a link between LDL oxidation and atherogenesis has provided a rationale for testing the effectiveness of antioxidants on the development and progression of atherosclerotic disease. In this context, vitamin E, vitamin C and β-carotene have received the greatest attention, especially with regard to coronary artery disease. These studies are grouped in the following categories: i) experimental atherosclerosis; ii) epidemiologic studies; iii) endothelial dysfunction; iv) imaging studies; v) clinical trials.

Antioxidants and Experimental Atherosclerosis

The oxidant stress hypothesis of atherosclerosis has been corroborated by animal studies that tested the effectiveness of various antioxidant treatments in models of accelerated atherosclerosis. Inhibition of atherosclerotic lesion development was obtained with vitamin E in New Zeland hypercholesterolemic rabbits [81, 82], and in Watanabe Heritable Hyperlipidemic (WHHL) rabbits [83]. A study in Apo-E deficient mice, which develope severe atherosclerosis, showed an early increase in isoprostanes in these animals that was significantly inhibited by vitamin E supplementation, along with a significant decrease in atherosclerotic lesion areas in animals given vitamin E [84]. A confirmation of these findings was provided by the same group in a recent study in which α-tocopherol was shown to reduce atherosclerosis progression by suppressing oxidative stress in the LDL receptor knock out mice, either at early phase of atherogenesis or after the disease is estabilished [85]

The effectiveness of vitamin E supplementation on experimental atherosclerosis was also demonstrated in a rabbit model of restenosis after angioplasty [86]. In analogy with vitamin E, the synthetic antioxidant butylated hydroxytoluene (BHT) inhibited the development of intimal thickening after balloon injury in hypercholesterolemic rabbits [87]. The synthetic antioxidant N,N'-diphenyl-phenylenediamine significantly reduced the extent of atherosclerotic lesions in cholesterol-fed rabbits and in ApoE-deficient mice [88, 89]. Similarly, the antioxidant BO-653 has been shown to inhibit the formation of atherosclerotic lesions in different animal models [90]. Probucol, a cholesterol-lowering drug with antioxidant activity [91], has been shown to inhibit the development of atherosclerotic lesions in WHHL [92] and New Zealand hypercholesterolemic rabbits [93] but enhances the development of atherosclerosis in LDL receptor-null mice [94] and Apo E-deficient mice [95, 96]. However, studies with probucol should be interpreted with caution in the context of an antioxidant mechanism beacause probucol is not an efficient chain breaking antioxidant [97]. Additional findings limit the use of probucol in human: probucol lower HDL levels [98] and prolong QT intervals [99].

The oxidant stress hypothesis of atherosclerosis has recently gained interesting findings by studies concerning premature atherosclerosis reported by Palinski W et al. [100]. In rabbits, maternal hypercholesterolemia is associated with increased fetal and postnatal fatty streaks formation that can be greatly reduced by maternal treatment with cholesterol lowering drugs or antioxidants. These data substantiate the concept that lipid peroxidation plays a major role in the initial phase of the atherosclerotic process.

Antioxidants and Epidemiologic Studies

Data from epidemiologic studies provided evidence of the protective role of vitamin E against coronary heart disease (CHD), either with supplementation or relatively high levels of dietary intake. On the other hand, beta carotene and vitamin C were shown to be less clearly associated with a reduced risk. The MONICA cross-cultural study showed the first strong evidence for the antioxidant hypothesis in coronary heart disease [101]. All essential plasma antioxidants were compared in 16 European populations having different incidences of mortality from ischemic heart disease. Alpha-tocopherol blood levels and α-tocopherol/cholesterol ratio showed a strong inverse correlation to ischemic heart disease mortality. Several prospective cohort studies followed the WHO/MONICA study providing substantial confirmation to this findings. In the US Nurses' Health Study, a significantly reduced risk for myocardial infarction or death from cardiovascular disease was seen in women in the highest versus the lowest quintile of vitamin E intake after adjustment for other known risk factors [102]. In this study, only the use of vitamin E for at least 2 years was associated with significant reduction in the risk of cardiovascular disease. In contrast, neither vitamin C nor beta carotene were found to be associated with a reduced risk for CHD after adjustment for vitamin E intake. The Health Professional Study followed up 39.000 male health professional for 4 years showing that men in the highest quintile of vitamin E intake had a reduced risk for CHD or death [103]. As in the Nurses' Health Study, the reduced risk for cardiovascular events was associated only with the use of vitamin E supplements for at least 2 years. No association between vitamin C intake and reduced risk for CHD was seen. In contrast, beta carotene intake was shown to be associated with reduced cardiovascular risk in partiecipants in the highest quintile, but this benefit was confined only to current smokers. In a subsequent Finnish study, an inverse association between dietary vitamin E and CHD was observed in more than 5.000 men and women followed up for 14 years [104]. In this population, vitamin C intake was associated with a significantly reduced risk for death from CHD only among women. However, this association was not adjusted for vitamin E intake. In the Iowa Women's Health Study, dietary vitamin E was shown to be inversely associated with the risk of death from heart disease in more than 30.000 postmenopausal women [105]. On the other hand, in this study no beneficial effect was seen for vitamin C and vitamin A.

Antioxidants and Endothelial Dysfunction

Endothelial dysfunction is an early manifestation of subclinical atherosclerosis and a risk predictor for coronary heart disease [106]. Endothelial dysfunction is characterized by reduced bioavailability of nitric oxide that can be lost after reacting with other free radicals. Most cardiovascular risk factors, such as hyperlipidemia, hypertension, diabetes, and smoking, were found to be associated either with endothelial dysfunction or increased oxidative stress. Consequently, oxidative stress has been proposed as a common pathogenic mechanism of the deleterious effects of cardiovascular risk factors on endothelial function [106-108]. This hypothesis has been tested in several clinical settings known to be associated to endothelial dysfunction. The assessment of endothelial function was mainly evaluated by measurement of flow-mediated endothelium-dependent vasodilatation (FMD) of the brachial artery during reactive hyperemia [109]. In chronic smokers, the effect of acute and chronic administration of vitamin C and vitamin E on endothelial dysfucntion was tested. Vitamin C was shown to markedly improve endothelium-dependent vasodilatation in chronic smokers in the short term either after oral or parenteral administration [110, 111]. However, the acute improvement of endothelial dysfunction obtained with vitamin C administration was not confirmed in the long-term despite sustained increase of plasma ascorbate concentrations [112]. Oral vitamin E supplementation (from 4 weeks to 4 months), alone or combined with vitamin C, was able to improve endothelial dysfunction in smokers [113, 114]. In contrast with these findings, Neunteufl T. et al. [115] demonstrated that vitamin E supplementation can only attenuate transient impairment of endothelial function after heavy smoking but can not restore chronic endothelial dysfunction after 4-week supplementation.

In most studies supplementation with antioxidant vitamins E and C, alone or in combination, was shown to improve entothelial dysfunction in both hyperlipidemic children and adults [116-118]. The beneficial effect of antioxidants was demonstrated either after intra-arterial infusion or long-term oral supplementation (from 6 weeks to 6 months). In contrast, Stein JH et al. [119] showed that in old patients with hypercholesterolemia endothelial dysfunction did not improve after 1-year administration with antioxidant vitamins and statins.

Intra-arterial infusion of vitamin C was shown to improve endothelial-dependent vasodilatation in patients with type 1 and type 2 diabetes mellitus [120, 121]. Similarly, oral vitamin E supplementation (8-12 weeks) was able to ameliorate endothelial function in both types of diabetes [122, 123]. However, antioxidant therapy with vitamin E plus vitamin C for 6 months was shown to improve endothelial dysfunction in patients with type 1 but not type 2 diabetes [124].

In the context of essential hypertension, the role of antioxidants on endothelial function was mainly studied by testing the effectiveness of intra-arterial infusion of vitamin C on endothelium-dependent vasodilatation either at brachial or coronary level [125-128]. In both models, vitamin C infusion was shown to improve endothelial dysfunction in hypertensive patients. In contrast, acute and chronic treatment with oral vitamin C had not effect on endothelial-dependent vasodilatation [129].

Antioxidants and Arterial Imaging Studies

Carotid intimal-media tickness (IMT), measured by ultrasonography, has been shown to be associated with cardiovascular risk factors and to be predictive of coronary artery disease and stroke, even after traditional risk factors have been taken into consideration [130, 131].

Contrasting findings have been provided by arterial imaging studies with antioxidants (table 2). The ASAP study investigated the effect of vitamin E and vitamin C alone or in combination (272 IU natural vitamin E and 500 mg slow-release vitamin C) on 3- and 6-year progression of carotid atherosclerosis in hypercholesterolemic subjects [132, 133]. Atherosclerosis progression was assessed by ultrasonographic evaluation of the common carotid IMT. After a 3-year follow up the combination of vitamin E and C was able to significantly slow the IMT progression in men but not in women. The effect of vitamin supplementation was more evident in smoking than in non-smoking men. On the other hand, neither vitamin E or vitamin C alone affected IMT progression. The 6-year data provided a confirmation of the above reported findings demonstrating that the combination of vitamin E and C was able to slow down the progression of carotid atherosclerosis in hypercholesterolemic persons. Moreover, the treatment effect was larger in subjects who had lower baseline plasma vitamin C levels and already plaques in the common carotid artery.

The SECURE trial is a sub-study of the HOPE in which the effectiveness of vitamin E supplementation (400 IU/day natural vitamin E) on carotid atherosclerosis progression was evaluated in high-risk patients during a follow-up period of 4.5 years [134]. Vitamin E had a neutral effect on atherosclerosis progression, defined as the annualized progression slope of the carotid IMT. The effect of vitamin E supplementation on carotid IMT progression was also tested by Hodis et al. [135] in a population with LDL cholesterol level $\geq$ 130 mg/dl and no clinical history of cardiovascular disease. Supplementation with 400 IU/day natural vitamin E for a mean follow-up of 3 years had not effect on IMT progression despite reduced circulating oxLDL and reduced LDL oxidative susceptibility.

Table 2. Summary of Imaging Studies with Antioxidants

	Imaging Study	Patients enrolled	Antioxidant Supplements	Daily dosage	Follow-up (years)	Primary endpoint
ASAP [132, 133]	Carotid IMT[§] by ultrasonography	Hypercholesterolemic (n=520)	vitamin E and/or vitamin C	272 IU 500 mg	3 and 6	Combination of vitamin E and C significantly slow the IMT progression in men
SECURE [134]	Carotid IMT[§] by ultrasonography	CHD[*] (n=732)	vitamin E	400 IU	4.5	No effect on the annualized progression slope of carotid IMT
MPV [136]	Coronary angiography	Patients undergoing PTCA (n=317)	probucol and/or multivitamins (vitamin E vitamin C β-carotene)	1000 mg 1400 IU 1000 mg 60.000 IU	0.5	Probucol significantly decreased the rate of restenosis. Multivitamins had not effect on the rate of restenosis
HATS [137]	Coronary angiography	CHD[*] (n=160)	vitamin E vitamin C β-carotene selenium	800 IU 1000 mg 25 mg 100 μg	3	No effect on progression of coronary stenosis
Fang JC et al. [138]	Coronary angiography plus IVUS[‡]	Cardiac-transplant recipients (n=40)	vitamin E vitamin C	800 IU 1000 mg	1	Significant reduction of the progression of coronary atherosclerosis
VEAPS [135]	Carotid IMT[†] by ultrasonography	LDL-c ≥130 mg/dL or 3.37 mmol/L (n=353)	Vitamin E	400 IU	3	Reduction in circulating ox-LDL and LDL oxidizability in vitro, Null effect on IMT

[*] CHD: coronary heart disease.
[†] IMT: intimal-medial thickness.
[‡] IVUS: intravascular ultrasonography.
Note: References are in parentesis in the first column.

The Multivitamins and Probucol trial evaluated the effect of probucol (500 mg twice daily), multivitamins (700 IU synthetic vitamin E, 500 mg vitamin C, and 30,000 IU β-carotene twice daily), or the combination of probucol and multivitamins in reducing the rate of angiographic restenosis after balloon coronary angioplasty [136]. While probucol significantly decreased the rate of restenosis, multivitamins had not significant effect.

The HDL-Atherosclerosis Treatment Study investigated the effect of simvastatin, niacin and antioxidant-vitamin therapy, alone and together, either on coronary stenosis progression or clinical cardiovascular events [137]. The antioxidant therapy included a total daily dosage of 800 IU of vitamin E, 1000 mg of vitamin C, 25 mg of natural beta carotene, and 100 μg of selenium. The primary angiographic end point was the mean change per patient in proximal stenosis after a 3-year follow up. Despite increased plasma vitamin levels and resistance of LDL to *ex vivo* oxidation, in the antioxidant-treated group there was only a trend toward a slowing the progression of coronary stenosis. Thus, the mean percent stenosis increased by 1.8% in the antioxidant-treated group and by 3.9% in the placebo group ($p=0.16$).

Fang JC and collegues [138] recently reported on the effect of vitamin C and E supplementation on progression of transplant-associated atherosclerosis. Fourty cardiac-transplant recipients were randomly assigned to vitamin C plus vitamin E (1000 mg and 800 IU daily, respectively) or placebo for 1 year. The primary end point was the change in average intimal index (plaque area divided by vessel area) measured by intravascular ultrasonography. After 1-year follow up antioxidant vitamins were able to significantly retard the progression of coronary atherosclerosis. Thus, the intimal index increased by 0.8% in the vitamin group and 8% in the placebo group ($p=0.008$).

Concerning imaging studies, several authors use IMT as synonymous of atherosclerosis with consequent confusion in interpretation of antioxidant effects on atherosclerosis. An increased IMT does not fit with the current pathological figure of atherosclerosis, and the biochemical mechanisms that alter the IMT are still unknown. Consequently, caution should be used in interpreting data of antioxidant trials and IMT [139].

Antioxidants and Clinical Trials

Although experimental studies gave encouraging results, the clinical effects of an antioxidant therapy in human atherosclerosis, particularly in coronary heart disease, are still controversial. While observational studies have provided evidence of an inverse relation between coronary artery disease and dietary intake of antioxidants, vitamin E in particular, clinical trials showed contrasting findings (Table 3). The ATBC study analysed the clinical efficacy of 50 mg/day synthetic vitamin E in a population with angiographic evidence of coronary artery disease, and showed no changes in cardiovascular events during a mean follow-up of 5.3 years [140]. The CHAOS study [141] tested the efficacy of vitamin E (400 or 800 IU/day natural vitamin E) in the prevention of cardiovascular death or non-fatal myocardial infraction (MI) in a population with angiographically documented coronary disease. After a mean follow-up of 1.4 years, patients supplemented with vitamin E showed a significant reduction in death from cardiovascular disease and non fatal MI. This effect was due to a very significant reduction in the risk for non fatal MI. In fact, no effect on cardiovascular death alone was seen. In the secondary prevention trial GISSI-Prevenzione [142], the effect of 300 mg/day of synthetic vitamin E was investigated in patients with recent ($\leq$ 3 months) myocardial infarction. The primary combined endpoint was cardiovascular death, non-fatal MI, or stroke during a follow-up of 3.5 years. In contrast to the CHAOS study, GISSI-prevenzione showed that vitamin E did not affect the rate of non fatal myocardial infarction, while showed a non significant trend towards a reduction of cardiovascular death and stroke. The HOPE study is a secondary intervention trial involving 9541 patients who were at high risk for cardiovascular events because of cardiovascular disease or diabetes in addition to one other risk factor [143]. The efficacy of 400 IU/day natural vitamin E on a primary composed outcome (MI, stroke, or death from cardiovascular causes) was tested. In the 4.5 years of follow-up vitamin E did not reduce the incidence of cardiovascular events. The SPACE trial tested the effect of 800 IU/day natural vitamin E supplementation in the prevention of secondary cardiovascular events in haemodialysis patients with history of cardiovascular disease during a follow-up period of 1.4 years [144].

Table 3. Summary of Clinical Trials with Antioxidants

	Prevention	Patients enrolled	Antioxidant Supplements	Daily dosage	Follow-up (years)	Primary endpoint
ATBC [140]	secondary	CHD (n=1424)	vitamin E[S] and/or β-carotene	50 mg 20 mg	5.3	No effect on major coronary events (non-fatal MI and fatal coronary heart disease)
CHAOS [141[	secondary	CHD (n=1035)	vitamin E[N]	400 IU 800 IU	1.4	Significant reduction in the risk of non-fatal MI
GISSI [142]	secondary	CHD (n=2830)	vitamin E[S]	300 mg	3.5	No effect on the combined endpoint (cardiovascular death, non-fatal MI, non-fatal stroke)
HOPE [143]	secondary	CHD (n=4761)	vitamin E[N]	400 IU	4.5	No effect on the combined endpoint (MI, stroke and death from cardiovascular causes)
SPACE [144]	secondary	Haemodialisys (n=196)	vitamin E[N]	800 IU	1.4	Significant reduction in the risk of the combined endpoint (MI, stroke, unstable angina and PVD)
PPP [145]	primary	One or more cardiovascular risk factor* (n=2231)	vitamin E[S]	300 mg	3.6	No effect on the combined endpoint (cardiovascular death, non-fatal MI and non-fatal stroke)
HPS [146]	secondary	CHD, other occlusive arterial disease or diabetes (n=10269)	vitamin E[S] vitamin C β-carotene	600 mg 250 mg 20 mg	5	No effect on mortality from or incidence of any type of vascular disease

* hypertension, hypercholesterolaemia, diabetes, obesity, family history of premature MI.
[S] synthetic.
[N] natural.
CHD: coronary heart disease.
MI: myocardial infarction.
PVD: peripheral vascular disease.
Note: References are in parentesis in the first column.

Vitamin E supplementation significantly reduced the risk of the primary composite endpoint, consisting of fatal and non-fatal myocardial infarction, ischemic stroke, peripheral vascular disease, and unstable angina. The significant reduction in the primary endpoint was largely due to the reduction in total myocardial infarction (70%). In the Primary Prevention Project trial, the investigators assessed the efficacy of vitamin E (300 mg/day from synthetic source) among patients who had one or more cardiovascular risk factors, including hypertension, hypercholesterolaemia, diabetes, or early family history of coronary disease [145]. The primary combined endpoint was the ceumulative rate of cardiovascular death, non-fatal myocardial infarction, and non-fatal stroke during a mean follow-up period of 3.6 years. No effect of vitamin E supplementation on the main combined endpoint or its components was seen. However, the incidence of peripheral artery disease was significantly lower among patients taking vitamin E than controls. The Heart Protection Study assessed the effectiveness of an antioxidant therapy including 600 mg vitamin E, 250 mg vitamin C, and 20 mg beta carotene daily in reducing vascular and non-vascular mortality and morbidity in a large population at high risk because of history of coronary disease, other occlusive arterial disease or diabetes [146]. After a mean follow-up of 5 years, antioxidant vitamin supplementation did not produce any significant reduction in the mortality from, or incidence of any type of vascular disease, cancer, or other major outcome.

Concluding Remarks

In the last two decades a huge number of data have been accumulated in favour of the oxidative stress hypothesis of atherosclerosis. Mechanistic aspects of LDL oxidation at chemical and cellular level, and the associated cellular events represent the most elucidated field of free radical chemistry in biology and medicine. The hypothesis has been validated with the pharmacological tool of antioxidants in animal model of atherosclerosis, and in part in human atherosclerosis. However, the expected benefit of antioxidants in human atherosclerosis did not overlap the beneficial effects obtained in animal models, and the disappointing results of prospective clinical trials have generated debate in the scientific community [139, 147-149]. Many factors could account for the conflicting findings reported by some clinical trials, including the dosage of antioxidant and bioavailability, compliance, changes in dietary habit, and duration of the supplementation period. While it is intuitive that these factors should be controlled in a vitamin supplementation trial, it is far from reality that in some occasions these factors were taken into account. For instance, there is no evidence in the trials that patients were instructed to take vitamin E with meals. Vitamin E is lipophylic and its intestinal absorption is largely favoured if taken with meals [150]. Few studies reported the antioxidant plasma levels that would allow to look at the efficacy of the supplementation. The VEAPS study reported a significant increase in plasma levels of vitamin E (33%) either in the placebo group or in the supplemented group. What about the other trials that did not report the antioxidant plasma levels througout the study? The unexpected increase in plasma vitamin E in the placebo arm of the VEAPS study strongly suggests caution in the interpretation of results of trials that did not report information on plasma antioxidants concentration. On the other hand, some of the above mentioned factors can not be controlled, simply because there is paucity of data about them. In this context, the dosage of antioxidant supplements varied considerably and the dose was selected in the absence of a biochemical rationale. The administration of an antioxidant supplement is expected to increase its plasma/tissue concentration and to suppress oxidative stress *in vivo*. No trial reported information on oxidative stress status and its response to antioxidant supplementation. The antioxidant treatment based on a biochemical rationale has been studied only in experimental atherosclerosis models. In ApoE-deficient mice, that shows elevated urinary levels of isoprostanes, atherosclerosis is retarded only if vitamin E dosage is high enough to suppress urinary isoprostanes [84]. Thus, the antioxidant clinical trials that did not use markers of oxidative stress as entry criteria, introduced a potential selection bias and incorrect sample size calculation. The importance of introducing appropriate markers of oxidative stress to facilitate a correct design of antioxidant trials is supported by a recent study, in which oxysterols and vitamin E were used as a non invasive measure of oxidative stress/antioxidant balance, in patients with advanced carotid atherosclerosis. In these patients, plasma oxysterols are elevated and plasma vitamin E strongly reduced; a similar imabalance is present in atherosclerotic plaques. Vitamin E supplementation suppressed elevated plasma oxysterols and corrected plasma vitamin E. By contrast, the same vitamin E dose did not correct the imbalance in oxidative stress status in carotid plaques [69].

In conclusion, results from supplementation trials that show neutral effect can not be used to disprove a theory based on solid experimenatal evidence but indicate that something is lacking in the translation of the work from the bench to the bedside. The use of realible markers of oxidative stress, measured by sensitive and specific methods, in conjunction with markers of antioxidant status is likely to be necessary to reveal the importance of oxidative stress in human atheroscelrosis, and for a rationale basis for the development of new effective antioxidant strategies based on naturally occurring antioxidants and on new antioxidants with higher potency and favourable pharmacokinetic profile.

References

[1] Stary, H.C., Natural history and histological classification of atherosclerotic lesions: an update. *Arterioscler Thromb Vasc Biol*, 2000. 20(5): p. 1177-8.

[2] Brown, M.S. and J.L. Goldstein, Lipoprotein metabolism in the macrophage: implications for cholesterol deposition in atherosclerosis. *Annu Rev Biochem*, 1983. 52: p. 223-61.

[3] Frostegard, J., et al., Biologically modified LDL increases the adhesive properties of endothelial cells. *Atherosclerosis*, 1991. 90(2-3): p. 119-26.

[4] Frostegard, J., et al., Oxidized low density lipoprotein induces differentiation and adhesion of human monocytes and the monocytic cell line U937. *Proc Natl Acad Sci U S A*, 1990. 87(3): p. 904-8.

[5] Schmidt, K., et al., Activation of soluble guanylate cyclase by nitrovasodilators is inhibited by oxidized low-density lipoprotein. *Biochem Biophys Res Commun*, 1990. 172(2): p. 614-9.

[6] Weis, J.R., et al., Oxidized low-density lipoprotein increases cultured human endothelial cell tissue factor activity and reduces protein C activation. *Faseb J*, 1991. 5(10): p. 2459-65.

[7] Kugiyama, K., et al., Impairment of endothelium-dependent arterial relaxation by lysolecithin in modified low-density lipoproteins. *Nature*, 1990. 344(6262): p. 160-2.

[8] Ross, R., The pathogenesis of atherosclerosis: a perspective for the 1990s. *Nature*, 1993. 362(6423): p. 801-9.

[9] Steinberg, D., Lewis A. Conner Memorial Lecture. Oxidative modification of LDL and atherogenesis. *Circulation*, 1997. 95(4): p. 1062-71.

[10] Stary, H.C., Lipid and macrophage accumulations in arteries of children and the development of atherosclerosis. *Am J Clin Nutr*, 2000. 72(5 Suppl): p. 1297S-1306S.

[11] Napoli, C., et al., Fatty streak formation occurs in human fetal aortas and is greatly enhanced by maternal hypercholesterolemia. Intimal accumulation of low density

lipoprotein and its oxidation precede monocyte recruitment into early atherosclerotic lesions. *J Clin Invest*, 1997. 100(11): p. 2680-90.

[12] Iuliano, L., et al., Preparation and biodistribution of 99m technetium labelled oxidized LDL in man. *Atherosclerosis*, 1996. 126(1): p. 131-41.

[13] Iuliano, L., et al., Radiolabeled native low-density lipoprotein injected into patients with carotid stenosis accumulates in macrophages of atherosclerotic plaque : effect of vitamin E supplementation. *Circulation*, 2000. 101(11): p. 1249-54.

[14] Hevonoja, T., et al., Structure of low density lipoprotein (LDL) particles: basis for understanding molecular changes in modified LDL. *Biochim Biophys Acta*, 2000. 1488(3): p. 189-210.

[15] Esterbauer, H., et al., The role of lipid peroxidation and antioxidants in oxidative modification of LDL. *Free Radic Biol Med*, 1992. 13(4): p. 341-90.

[16] Boren, J., et al., Identification of the low density lipoprotein receptor-binding site in apolipoprotein B100 and the modulation of its binding activity by the carboxyl terminus in familial defective apo-B100. *J Clin Invest*, 1998. 101(5): p. 1084-93.

[17] Brown, M.S. and J.L. Goldstein, A receptor-mediated pathway for cholesterol homeostasis. *Science*, 1986. 232(4746): p. 34-47.

[18] Brown, M.S., S.E. Dana, and J.L. Goldstein, Regulation of 3-hydroxy-3-methylglutaryl coenzyme A reductase activity in cultured human fibroblasts. Comparison of cells from a normal subject and from a patient with homozygous familial hypercholesterolemia. *J Biol Chem*, 1974. 249(3): p. 789-96.

[19] Goldstein, J.L. and M.S. Brown, Molecular medicine. The cholesterol quartet. *Science*, 2001. 292(5520): p. 1310-2.

[20] Blumenthal, R.S., Statins: effective antiatherosclerotic therapy. *Am Heart J*, 2000. 139(4): p. 577-83.

[21] Goldstein, J.L. and M.S. Brown, The low-density lipoprotein pathway and its relation to atherosclerosis. *Annu Rev Biochem*, 1977. 46: p. 897-930.

[22] Wurster, N.B. and D.B. Zilversmit, The role of phagocytosis in the development of atherosclerotic lesions in the rabbit. *Atherosclerosis*, 1971. 14: p. 309-22.

[23] Kar, S. and A.J. Day, Composition and metabolism of lipid in macrophages from normally fed and cholesterol-fed rabbits. *Exp Mol Pathol*, 1978. 28(1): p. 65-75.

[24] Goldstein, J.L., et al., Binding site on macrophages that mediates uptake and degradation of acetylated low density lipoprotein, producing massive cholesterol deposition. *Proc Natl Acad Sci U S A*, 1979. 76(1): p. 333-7.

[25] Basu, S.K., et al., Degradation of low density lipoprotein . dextran sulfate complexes associated with deposition of cholesteryl esters in mouse macrophages. *J Biol Chem*, 1979. 254(15): p. 7141-6.

[26] Weinstein, D.B., T.E. Carew, and D. Steinberg, Uptake and degradation of low density lipoprotein by swine arterial smoot muscle cells with inhibition of cholesterol biosynthesis. *Biochim Biophys Acta*, 1976. 424(3): p. 404-21.

[27] Fogelman, A.M., et al., Factors regulating the activities of the low density lipoprotein receptor and the scavenger receptor on human monocyte-macrophages. *J Lipid Res*, 1981. 22(7): p. 1131-41.

[28] Knight, B.L. and A.K. Soutar, Changes in the metabolism of modified and unmodified low-density lipoproteins during the maturation of cultured blood monocyte-macrophages from normal and homozygous familial hypercholesterolaemic subjects. *Eur J Biochem*, 1982. 125(2): p. 407-13.

[29] Traber, M.G. and H.J. Kayden, Low density lipoprotein receptor activity in human monocyte-derived macrophages and its relation to atheromatous lesions. *Proc Natl Acad Sci U S A*, 1980. 77(9): p. 5466-70.

[30] Yla-Herttuala, S., et al., Gene expression in macrophage-rich human atherosclerotic lesions. 15-lipoxygenase and acetyl low density lipoprotein receptor messenger RNA colocalize with oxidation specific lipid-protein adducts. *J Clin Invest*, 1991. 87(4): p. 1146-52.

[31] Goldstein, J.L. and M.S. Brown, Familial hypercholesterolemia: pathogenesis of a receptor disease. *Johns Hopkins Med J*, 1978. 143(1): p. 8-16.

[32] Mego, J.L. and J.D. McQueen, The Uptake and Degradation of Injected Labeled Proteins by Mouse-Liver Particles. *Biochim Biophys Acta*, 1965. 100: p. 136-43.

[33] Buys, C.H., et al., Rapid uptake by liver sinusoidal cells of serum albumin modified with retention of its compact conformation. *Biochim Biophys Acta*, 1975. 392(1): p. 95-100.

[34] Fogelman, A.M., et al., Malondialdehyde alteration of low density lipoproteins leads to cholesteryl ester accumulation in human monocyte-macrophages. *Proc Natl Acad Sci U S A*, 1980. 77(4): p. 2214-8.

[35] Tappel, A.L., Protection against free radical lipid peroxidation reactions. *Adv Exp Med Biol.*, 1978. 97: p. 111-31.

[36] Glavind, j., et al., Studies on the role of lipoperoxides in human pathology.II. The presence of peroxidized lipids in the atherosclerotic aorta. *Acta Pathol. Microbiol. Scand.*, 1952. 30: p. 1-6.

[37] Schuh, J., G.F. Fairclough, Jr., and R.H. Haschemeyer, Oxygen-mediated heterogeneity of apo-low-density lipoprotein. *Proc Natl Acad Sci U S A*, 1978. 75(7): p. 3173-7.

[38] Ray, B.R., E.O. Davisson, and H.L. Crespi, Experiments on the Degradation of Lipoproteins from Serum. *J. Phys. Chem.*, 1954. 58: p. 841 - 846.

[39] Hessler, J.R., et al., Lipoprotein oxidation and lipoprotein-induced cytotoxicity. *Arteriosclerosis*, 1983. 3(3): p. 215-22.

[40] Hazen, S.L., Oxidation and atherosclerosis. *Free Radic Biol Med*, 2000. 28(12): p. 1683-4.

[41] Marsche, G., et al., Hypochlorite-modified high density lipoprotein, a high affinity ligand to scavenger receptor class B, type I, impairs high density lipoprotein-dependent selective lipid uptake and reverse cholesterol transport. *J Biol Chem*, 2002. 277(35): p. 32172-9.

[42] Gillotte-Taylor, K., et al., Scavenger receptor class B type I as a receptor for oxidized low density lipoprotein. *J Lipid Res*, 2001. 42(9): p. 1474-82.

[43] Dhaliwal, B.S. and U.P. Steinbrecher, Scavenger receptors and oxidized low density lipoproteins. *Clin Chim Acta*, 1999. 286(1-2): p. 191-205.

[44] Suzuki, H., et al., A role for macrophage scavenger receptors in atherosclerosis and susceptibility to infection. *Nature*, 1997. 386(6622): p. 292-6.

[45] Nozaki, S., et al., Reduced uptake of oxidized low density lipoproteins in monocyte-derived macrophages from CD36-deficient subjects. *J Clin Invest*, 1995. 96(4): p. 1859-65.
[46] de Winther, M.P., et al., Macrophage scavenger receptor class A: A multifunctional receptor in atherosclerosis. *Arterioscler Thromb Vasc Biol*, 2000. 20(2): p. 290-7.
[47] Febbraio, M., D.P. Hajjar, and R.L. Silverstein, CD36: a class B scavenger receptor involved in angiogenesis, atherosclerosis, inflammation, and lipid metabolism. *J Clin Invest*, 2001. 108(6): p. 785-91.
[48] Kruth, H.S., et al., Macrophage foam cell formation with native low density lipoprotein. *J Biol Chem*, 2002. 277(37): p. 34573-80.
[49] Henriksen, T., E.M. Mahoney, and D. Steinberg, Enhanced macrophage degradation of low density lipoprotein previously incubated with cultured endothelial cells: recognition by receptors for acetylated low density lipoproteins. *Proc Natl Acad Sci U S A*, 1981. 78(10): p. 6499-503.
[50] Heinecke, J.W., et al., Superoxide-mediated modification of low density lipoprotein by arterial smooth muscle cells. *J Clin Invest*, 1986. 77(3): p. 757-61.
[51] Cathcart, M.K., D.W. Morel, and G.M. Chisolm, 3rd, Monocytes and neutrophils oxidize low density lipoprotein making it cytotoxic. *J Leukoc Biol*, 1985. 38(2): p. 341-50.
[52] Leake, D.S. and S.M. Rankin, The oxidative modification of low-density lipoproteins by macrophages. *Biochem J*, 1990. 270(3): p. 741-8.
[53] Iuliano, L., et al., Protection of low density lipoprotein oxidation by the antioxidant agent IRFI005, a new synthetic hydrophilic vitamin E analogue. *Free Radic Biol Med*, 1999. 26(7-8): p. 858-68.
[54] Marathe, G.K., et al., Bioactive phospholipid oxidation products. *Free Radic Biol Med*, 2000. 28(12): p. 1762-70.
[55] Pratico, D., F(2)-isoprostanes: sensitive and specific non-invasive indices of lipid peroxidation in vivo. *Atherosclerosis*, 1999. 147(1): p. 1-10.
[56] Cracowski, J.L., et al., Vascular biology of the isoprostanes. *J Vasc Res*, 2001. 38(2): p. 93-103.
[57] Chisolm, G.M., et al., 7 beta-hydroperoxycholest-5-en-3 beta-ol, a component of human atherosclerotic lesions, is the primary cytotoxin of oxidized human low density lipoprotein. *Proc Natl Acad Sci U S A*, 1994. 91(24): p. 11452-6.
[58] Cathcart, M.K., et al., Superoxide anion participation in human monocyte-mediated oxidation of low-density lipoprotein and conversion of low-density lipoprotein to a cytotoxin. *J Immunol*, 1989. 142(6): p. 1963-9.
[59] Geng, Y.J. and P. Libby, Evidence for apoptosis in advanced human atheroma. Colocalization with interleukin-1 beta-converting enzyme. *Am J Pathol,* 1995. 147(2): p. 251-66.
[60] Brown, A.J. and W. Jessup, Oxysterols and atherosclerosis. *Atherosclerosis*, 1999. 142(1): p. 1-28.
[61] Rong, J.X., et al., Arterial injury by cholesterol oxidation products causes endothelial dysfunction and arterial wall cholesterol accumulation. *Arterioscler Thromb Vasc Biol*, 1998. 18(12): p. 1885-94.

[62] Rangaswamy, S., et al., Exogenous oxidized low-density lipoprotein injures and alters the barrier function of endothelium in rats in vivo. *Circ Res*, 1997. 80(1): p. 37-44.

[63] Xu, X.P., et al., Oxidized low-density lipoprotein regulates matrix metalloproteinase-9 and its tissue inhibitor in human monocyte-derived macrophages. *Circulation*, 1999. 99(8): p. 993-8.

[64] Subbanagounder, G., A.D. Watson, and J.A. Berliner, Bioactive products of phospholipid oxidation: isolation, identification, measurement and activities. *Free Radic Biol Med*, 2000. 28(12): p. 1751-61.

[65] Lawson, J.A., J. Rokach, and G.A. FitzGerald, Isoprostanes: formation, analysis and use as indices of lipid peroxidation in vivo. *J Biol Chem*, 1999. 274(35): p. 24441-4.

[66] Morrow, J.D., et al., A series of prostaglandin F2-like compounds are produced in vivo in humans by a non-cyclooxygenase, free radical-catalyzed mechanism. *Proc Natl Acad Sci U S A*, 1990. 87(23): p. 9383-7.

[67] Bjorkhem, I. and U. Diczfalusy, Oxysterols: friends, foes, or just fellow passengers? *Arterioscler Thromb Vasc Biol*, 2002. 22(5): p. 734-42.

[68] Pratico, D., et al., Localization of distinct F2-isoprostanes in human atherosclerotic lesions. *J Clin Invest*, 1997. 100(8): p. 2028-34.

[69] Micheletta, F., et al., Vitamin E Supplementation in Patients With Carotid Atherosclerosis. Reversal of Altered Oxidative Stress Status in Plasma But Not in Plaque. *Arterioscler Thromb Vasc Biol*, 2003(in press).

[70] Fam, S.S., et al., Formation of highly reactive A-ring and J-ring isoprostane-like compounds (A4/J4-neuroprostanes) in vivo from docosahexaenoic acid. *J Biol Chem*, 2002. 277(39): p. 36076-84.

[71] Harland, W.A., J.D. Gilbert, and C.J. Brooks, Lipids of human atheroma. 8. Oxidised derivatives of cholesteryl linoleate. *Biochim Biophys Acta*, 1973. 316(3): p. 378-85.

[72] Rong, J.X., et al., Cholesterol oxidation products induce vascular foam cell lesion formation in hypercholesterolemic New Zealand white rabbits. *Arterioscler Thromb Vasc Biol*, 1999. 19(9): p. 2179-88.

[73] Sevanian, A., et al., Characterization of endothelial cell injury by cholesterol oxidation products found in oxidized LDL. *J Lipid Res*, 1995. 36(9): p. 1971-86.

[74] Colles, S.M., K.C. Irwin, and G.M. Chisolm, Roles of multiple oxidized LDL lipids in cellular injury: dominance of 7 beta-hydroperoxycholesterol. *J Lipid Res*, 1996. 37(9): p. 2018-28.

[75] Lizard, G., et al., Characterization and comparison of the mode of cell death, apoptosis versus necrosis, induced by 7beta-hydroxycholesterol and 7-ketocholesterol in the cells of the vascular wall. *Arterioscler Thromb Vasc Biol*, 1999. 19(5): p. 1190-200.

[76] Imai, H., et al., Angiotoxicity and arteriosclerosis due to contaminants of USP-grade cholesterol. *Arch Pathol Lab Med,* 1976. 100(11): p. 565-72.

[77] Iuliano, L., et al., Measurement of oxysterols and alpha-tocopherol in plasma and tissue samples as indices of oxidant stress status. *Anal Biochem*, 2003. 312(2): p. 217-23.

[78] Binder, C.J., et al., Innate and acquired immunity in atherogenesis. *Nat Med,* 2002. 8(11): p. 1218-26.

[79] Horkko, S., et al., Immunological responses to oxidized LDL. *Free Radic Biol Med*, 2000. 28(12): p. 1771-9.

[80] Witztum, J.L. and W. Palinski, Are immunological mechanisms relevant for the development of atherosclerosis? *Clin Immunol*, 1999. 90(2): p. 153-6.

[81] Ozer, N.K., et al., Effect of vitamin E and probucol on dietary cholesterol-induced atherosclerosis in rabbits. *Free Radic Biol Med*, 1998. 24(2): p. 226-33.

[82] Sun, J., et al., beta-Carotene and alpha-tocopherol inhibit the development of atherosclerotic lesions in hypercholesterolemic rabbits. *Int J Vitam Nutr Res*, 1997. 67(3): p. 155-63.

[83] Williams, R.J., et al., Dietary vitamin E and the attenuation of early lesion development in modified Watanabe rabbits. *Atherosclerosis*, 1992. 94(2-3): p. 153-9.

[84] Pratico, D., et al., Vitamin E suppresses isoprostane generation in vivo and reduces atherosclerosis in ApoE-deficient mice. *Nat Med,* 1998. 4(10): p. 1189-92.

[85] Cyrus, T., et al., Vitamin E reduces progression of atherosclerosis in low-density lipoprotein receptor-deficient mice with established vascular lesions. *Circulation*, 2003. 107(4): p. 521-3.

[86] Lafont, A.M., et al., Effect of alpha-tocopherol on restenosis after angioplasty in a model of experimental atherosclerosis. *J Clin Invest*, 1995. 95(3): p. 1018-25.

[87] Freyschuss, A., et al., Antioxidant treatment inhibits the development of intimal thickening after balloon injury of the aorta in hypercholesterolemic rabbits. *J Clin Invest*, 1993. 91(4): p. 1282-8.

[88] Tangirala, R.K., et al., Effect of the antioxidant N,N'-diphenyl 1,4-phenylenediamine (DPPD) on atherosclerosis in apoE-deficient mice. *Arterioscler Thromb Vasc Biol*, 1995. 15(10): p. 1625-30.

[89] Sparrow, C.P., et al., Low density lipoprotein is protected from oxidation and the progression of atherosclerosis is slowed in cholesterol-fed rabbits by the antioxidant N,N'-diphenyl-phenylenediamine. *J Clin Invest*, 1992. 89(6): p. 1885-91.

[90] Cynshi, O., et al., Antiatherogenic effects of the antioxidant BO-653 in three different animal models. *Proc Natl Acad Sci U S A*, 1998. 95(17): p. 10123-8.

[91] Kuzuya, M. and F. Kuzuya, Probucol as an antioxidant and antiatherogenic drug. *Free Radic Biol Med*, 1993. 14(1): p. 67-77.

[92] Steinberg, D., S. Parthasarathy, and T.E. Carew, In vivo inhibition of foam cell development by probucol in Watanabe rabbits. *Am J Cardiol*, 1988. 62(3): p. 6B-12B.

[93] Shaish, A., et al., Beta-carotene inhibits atherosclerosis in hypercholesterolemic rabbits. *J Clin Invest*, 1995. 96(4): p. 2075-82.

[94] Bird, D.A., et al., Effect of probucol on LDL oxidation and atherosclerosis in LDL receptor-deficient mice. *J Lipid Res*, 1998. 39(5): p. 1079-90.

[95] Moghadasian, M.H., et al., Proatherogenic and antiatherogenic effects of probucol and phytosterols in apolipoprotein E-deficient mice: possible mechanisms of action. *Circulation*, 1999. 99(13): p. 1733-9.

[96] Zhang, S.H., et al., Paradoxical enhancement of atherosclerosis by probucol treatment in apolipoprotein E-deficient mice. *J Clin Invest*, 1997. 99(12): p. 2858-66.

[97] Iuliano, L., et al., Fluorescence quenching of dipyridamole associated to peroxyl radical scavenging: a versatile probe to measure the chain breaking antioxidant activity of biomolecules. *Biochim Biophys Acta*, 2000. 1474(2): p. 177-82.

[98] Tikkanen, M.J., E. Helve, and E.A. Nikkila, Treatment of familial and non-familial hypercholesterolaemia: a review of HMG-CoA reductase inhibitors and probucol. *Eur Heart J*, 1987. 8 Suppl E: p. 97-101.

[99] Dujovne, C.A., et al., *Electrocardiographic effects of probucol. A controlled prospective clinical trial.* 1984. 26: p. 735-9.

[100] Palinski, W. and C. Napoli, The fetal origins of atherosclerosis: maternal hypercholesterolemia, and cholesterol-lowering or antioxidant treatment during pregnancy influence in utero programming and postnatal susceptibility to atherogenesis. *Faseb J*, 2002. 16: p. 1348-60.

[101] Gey, K.F., et al., Inverse correlation between plasma vitamin E and mortality from ischemic heart disease in cross-cultural epidemiology. *Am J Clin Nutr*, 1991. 53(1 Suppl): p. 326S-334S.

[102] Stampfer, M.J., et al., Vitamin E consumption and the risk of coronary disease in women. *N Engl J Med,* 1993. 328(20): p. 1444-9.

[103] Rimm, E.B., et al., Vitamin E consumption and the risk of coronary heart disease in men. *N Engl J Med*, 1993. 328(20): p. 1450-6.

[104] Knekt, P., et al., Antioxidant vitamin intake and coronary mortality in a longitudinal population study. *Am J Epidemiol*, 1994. 139(12): p. 1180-9.

[105] Kushi, L.H., et al., Dietary antioxidant vitamins and death from coronary heart disease in postmenopausal women. *N Engl J Med.*, 1996. 334: p. 1156-62.

[106] Bonetti, P.O., L.O. Lerman, and A. Lerman, Endothelial dysfunction: a marker of atherosclerotic risk. *Arterioscler Thromb Vasc Biol*, 2003. 23(2): p. 168-75.

[107] Cai, H. and D.G. Harrison, Endothelial dysfunction in cardiovascular diseases: the role of oxidant stress. *Circ Res*, 2000. 87(10): p. 840-4.

[108] Tomasian, D., J.F. Keaney, and J.A. Vita, Antioxidants and the bioactivity of endothelium-derived nitric oxide. *Cardiovasc Res*, 2000. 47(3): p. 426-35.

[109] Corretti, M.C., et al., Guidelines for the ultrasound assessment of endothelial-dependent flow-mediated vasodilation of the brachial artery: a report of the International Brachial Artery Reactivity Task Force. *J Am Coll Cardiol*, 2002. 39(2): p. 257-65.

[110] Heitzer, T., H. Just, and T. Munzel, Antioxidant vitamin C improves endothelial dysfunction in chronic smokers. *Circulation*, 1996. 94(1): p. 6-9.

[111] Hirai, N., et al., Insulin resistance and endothelial dysfunction in smokers: effects of vitamin C. *Am J Physiol Heart Circ Physiol*, 2000. 279(3): p. H1172-8.

[112] Raitakari, O.T., et al., Oral vitamin C and endothelial function in smokers: short-term improvement, but no sustained beneficial effect. *J Am Coll Cardiol,* 2000. 35(6): p. 1616-21.

[113] Antoniades, C., et al., Effects of antioxidant vitamins C and E on endothelial function and thrombosis/fibrinolysis system in smokers. *Thromb Haemost*, 2003. 89(6): p. 990-5.

[114] Heitzer, T., et al., Effect of vitamin E on endothelial vasodilator function in patients with hypercholesterolemia, chronic smoking or both. *J Am Coll Cardiol*, 1999. 33(2): p. 499-505.

[115] Neunteufl, T., et al., Effects of vitamin E on chronic and acute endothelial dysfunction in smokers. *J Am Coll Cardiol,* 2000. 35(2): p. 277-83.

[116] Engler, M.M., et al., Antioxidant vitamins C and E improve endothelial function in children with hyperlipidemia: Endothelial Assessment of Risk from Lipids in Youth (EARLY) Trial. *Circulation*, 2003. 108(9): p. 1059-63.

[117] Mietus-Snyder, M. and M.J. Malloy, Endothelial dysfunction occurs in children with two genetic hyperlipidemias: improvement with antioxidant vitamin therapy. *J Pediatr,* 1998. 133(1): p. 35-40.

[118] Ting, H.H., et al., Vitamin C improves endothelium-dependent vasodilation in forearm resistance vessels of humans with hypercholesterolemia. *Circulation*, 1997. 95(12): p. 2617-22.

[119] Stein, J.H., et al., The effects of lipid-lowering and antioxidant vitamin therapies on flow-mediated vasodilation of the brachial artery in older adults with hypercholesterolemia. *J Am Coll Cardiol,* 2001. 38(7): p. 1806-13.

[120] Timimi, F.K., et al., Vitamin C improves endothelium-dependent vasodilation in patients with insulin-dependent diabetes mellitus. *J Am Coll Cardiol,* 1998. 31(3): p. 552-7.

[121] Ting, H.H., et al., Vitamin C improves endothelium-dependent vasodilation in patients with non-insulin-dependent diabetes mellitus. *J Clin Invest*, 1996. 97(1): p. 22-8.

[122] Skyrme-Jones, R.A., et al., Vitamin E supplementation improves endothelial function in type I diabetes mellitus: a randomized, placebo-controlled study. *J Am Coll Cardiol,* 2000. 36(1): p. 94-102.

[123] Paolisso, G., et al., Chronic vitamin E administration improves brachial reactivity and increases intracellular magnesium concentration in type II diabetic patients. *J Clin Endocrinol Metab*, 2000. 85(1): p. 109-15.

[124] Beckman, J.A., et al., Oral Antioxidant Therapy Improves Endothelial Function in Type 1 but not Type 2 Diabetes Mellitus. *Am J Physiol Heart Circ Physio*l, 2003([Epub ahead of print]).

[125] Solzbach, U., et al., Vitamin C improves endothelial dysfunction of epicardial coronary arteries in hypertensive patients. *Circulation*, 1997. 96(5): p. 1513-9.

[126] Taddei, S., et al., Vitamin C improves endothelium-dependent vasodilation by restoring nitric oxide activity in essential hypertension. *Circulation*, 1998. 97(22): p. 2222-9.

[127] Natali, A., et al., Effect of vitamin C on forearm blood flow and glucose metabolism in essential hypertension. *Arterioscler Thromb Vasc Biol*, 2000. 20(11): p. 2401-6.

[128] Sherman, D.L., et al., Pharmacological concentrations of ascorbic acid are required for the beneficial effect on endothelial vasomotor function in hypertension. *Hypertension*, 2000. 35(4): p. 936-41.

[129] Duffy, S.J., et al., Effect of ascorbic acid treatment on conduit vessel endothelial dysfunction in patients with hypertension. *Am J Physiol Heart Circ Physiol,* 2001. 280(2): p. H528-34.

[130] O'Leary, D.H., et al., Carotid-artery intima and media thickness as a risk factor for myocardial infarction and stroke in older adults. Cardiovascular Health Study Collaborative Research Group. *N Engl J Med*, 1999. 340(1): p. 14-22.

[131] Mukherjee, D. and J.S. Yadav, Carotid artery intimal-medial thickness: indicator of atherosclerotic burden and response to risk factor modification. *Am Heart J*, 2002. 144(5): p. 753-9.

[132] Salonen, R.M., et al., Six-year effect of combined vitamin C and E supplementation on atherosclerotic progression: the Antioxidant Supplementation in Atherosclerosis Prevention (ASAP) Study. *Circulation*, 2003. 107(7): p. 947-53.

[133] Salonen, J.T., et al., Antioxidant Supplementation in Atherosclerosis Prevention (ASAP) study: a randomized trial of the effect of vitamins E and C on 3-year progression of carotid atherosclerosis. *J Intern Med,* 2000. 248(5): p. 377-86.

[134] Lonn, E., et al., Effects of ramipril and vitamin E on atherosclerosis: the study to evaluate carotid ultrasound changes in patients treated with ramipril and vitamin E (SECURE). *Circulation*, 2001. 103(7): p. 919-25.

[135] Hodis, H.N., et al., Alpha-tocopherol supplementation in healthy individuals reduces low-density lipoprotein oxidation but not atherosclerosis: the Vitamin E Atherosclerosis Prevention Study (VEAPS). *Circulation*, 2002. 106(12): p. 1453-9.

[136] Tardif, J.C., et al., Probucol and multivitamins in the prevention of restenosis after coronary angioplasty. Multivitamins and Probucol Study Group. *N Engl J Med*, 1997. 337(6): p. 365-72.

[137] Brown, B.G., et al., Simvastatin and niacin, antioxidant vitamins, or the combination for the prevention of coronary disease. *N Engl J Med*, 2001. 345(22): p. 1583-92.

[138] Fang, J.C., et al., Effect of vitamins C and E on progression of transplant-associated arteriosclerosis: a randomised trial. *Lancet*, 2002. 359(9312): p. 1108-13.

[139] Iuliano, L., et al., Effect of ramipril and vitamin E on atherosclerosis. *Circulation*, 2002. 105(2): p. E5-6.

[140] Rapola, J.M., et al., Randomised trial of alpha-tocopherol and beta-carotene supplements on incidence of major coronary events in men with previous myocardial infarction. *Lancet*, 1997. 349(9067): p. 1715-20.

[141] Stephens, N.G., et al., Randomised controlled trial of vitamin E in patients with coronary disease: Cambridge Heart Antioxidant Study (CHAOS). *Lancet*, 1996. 347(9004): p. 781-6.

[142] GISSI-prevenzione.Investigators., Dietary supplementation with n-3 polyunsaturated fatty acids and vitamin E after myocardial infarction: results of the GISSI-Prevenzione trial. Gruppo Italiano per lo Studio della Sopravvivenza nell'Infarto miocardico. *Lancet*, 1999. 354(9177): p. 447-55.

[143] Yusuf, S., et al., Vitamin E supplementation and cardiovascular events in high-risk patients. The Heart Outcomes Prevention Evaluation Study Investigators. *N Engl J Med,* 2000. 342(3): p. 154-60.

[144] Boaz, M., et al., Secondary prevention with antioxidants of cardiovascular disease in endstage renal disease (SPACE): randomised placebo-controlled trial. *Lancet*, 2000. 356(9237): p. 1213-8.

[145] de Gaetano, G., Low-dose aspirin and vitamin E in people at cardiovascular risk: a randomised trial in general practice. Collaborative Group of the Primary Prevention Project. *Lancet*, 2001. 357(9250): p. 89-95.

[146] MRC/BHF Heart Protection Study of antioxidant vitamin supplementation in 20,536 high-risk individuals: a randomised placebo-controlled trial. *Lancet*, 2002. 360(9326): p. 23-33.
[147] Violi, F., F. Micheletta, and L. Luliano, Antioxidant strategy for cardiovascular disease. *Lancet*, 2001. 357(9269): p. 1704; author reply 1706.
[148] Iuliano, L., Antioxidant therapy in vascular disease. *Haematologica*, 2001. 86(11 Suppl 2): p. 22-4.
[149] Jialal, I. and S. Devaraj, Antioxidants and atherosclerosis: don't throw out the baby with the bath water. *Circulation*, 2003. 107(7): p. 926-8.
[150] Micheletta F, Natoli S, Misuraca M, Sbarigia E, Diczfalusy U, Iuliano L. Vitamin E supplementation in patients with carotid atherosclerosis: reversal of altered oxidative stress status in plasma but not in plaque. *Arterioscler Thromb Vasc Biol.* 2004;24(1):136-40.

In: Focus on Atherosclerosis Research
Editor: Leon V.Clark, pp. 215-251

ISBN 1-59454-044-6

Chapter X

Vascular Oxidant Stress and Endothelial Dysfunction in Hyperhomocysteinemia

Norbert Weiss[*]
Medizinische Poliklinik – Innenstadt, Klinikum der Universität München
München, Germany

Abstract

Mildly elevated plasma homocysteine levels are an independent risk factor for atherothrombotic vascular disease in the coronary, cerebrovascular, and peripheral arterial circulation. Endothelial dysfunction contributes to the vascular disorders linked to hyperhomocysteinemia. Dysfunctional endothelium becomes manifested by impaired endothelium-dependent regulation of vascular tone and blood flow, by increased recruitment and adhesion of circulating inflammatory cells to the endothelium, and by a loss of endothelial cell antithrombotic function. There is increasing evidence that implicates oxidative stress as being contributory to homocysteine's deleterious effects on the vasculature. Elevated levels of homocysteine may lead to increased generation of superoxide by a biochemical mechanism involving nitric oxide synthase, and, to less extent, by an increase in the chemical oxidation of homocysteine and other aminothiols in the circulation. The resultant increase in superoxide levels is further amplified by homocysteine-dependent alterations in the function of cellular antioxidant enzymes such as cellular glutathione peroxidase or extracellular superoxide dismutase. The increased vascular oxidant stress finally results in decreased bioavailability of the endothelium-derived antiatherogenic signaling molecular nitric oxide via oxidative inactivation. This plays a central role in the molecular mechanisms underlying the effects of homocysteine on endothelial function. Scavenging of superoxide anion by either superoxide dismutase

* Address correspondence to: Priv.-Doz. Dr. med. Norbert Weiss, Medizinische Poliklinik – Innenstadt, Klinikum der Universität München, Pettenkoferstrasse 8a, D-80336 München, Germany, phone: +49-89-5160-3564, fax: +49-89-5160-3559, email: Norbert.Weiss@med.uni-muenchen.de

or 4,5-dihydroxybenzene 1,3-disulfonate (Tiron) reverses endothelial dysfunction in hyperhomocysteinemic animal models and in isolated aortic rings incubated with homocysteine. Similarly, homocysteine-induced endothelial dysfunction is also reversed by increasing the concentrations of the endogenous antioxidant glutathione or by overexpressing cellular glutathione peroxidase in animal models of mild hyperhomocysteinemia. Taken together, these findings strongly suggest that the adverse vascular effects of homocysteine are at least partly mediated by oxidative inactivation of nitric oxide. Clinical studies have shown that lowering of plasma homocysteine levels by supplementation of folic acid reverses homocysteine-induced endothelial dysfunction. Whether or not this translates into a reduction of clinical events associated with hyperhomocysteinemia is currently under investigation.

Abbreviations

CBS:	cysthathionine β-synthase.
$CBS^{(-/+)}$:	heterozygous cystathionine β-synthase knock out mice.
$CBS^{(+/+)}$:	wildtype littermates of heterozygous $CBS^{(-/+)}$ mice.
CCR2:	chemokine receptor 2.
DCF:	2',7'-dichlorofluoresceine.
eNOS:	endothelial nitric oxide synthase.
GPx-1:	cellular isoform of glutathione peroxidase.
H_2O:	water.
H_2O_2:	hydrogen peroxide.
ICAM-1:	intracellular adhesion molecule 1.
IL-8:	interleukin 8.
L-NAME:	L-nitroarginine methyl ester.
MCP-1:	macrophage chemoattraction protein.
NF-κB:	nuclear factor κB.
OH^-:	hydroxide anion.
$^{\bullet}OH$:	hydroxyl radical.
OTC:	L-2-oxo-4-thiazolidine carboxylate.
RAGE:	receptor for advanced glycation endproducts.
TF:	tissue factor.
Tiron:	4,5-dihydroxybenzene 1,3 disulfonate.
VCAM-1:	vascular cell adhesion molecule 1.

Introduction

Elevated plasma levels of homocysteine (hyperhomocysteinemia) have been established as a risk factor for atherothrombotic vascular disease during the last two decades. The pathobiological mechanisms by which homocysteine promotes atherothrombosis are not completely understood as yet. One well-supported but not universally accepted hypothesis suggests that hyperhomocysteinemia leads to increased oxidant stress in the vasculature. This effect may be central to its promotion of vascular pathology, partly by decreasing the bioavailability of the endothelial antiatherogenic signaling molecule nitric oxide. The resulting endothelial dysfunction contributes to decreased vasodilator capacity, activation of circulating leukocytes and platelets, activation of prothrombotic and inhibition of fibrinolytic mechanisms, and stimulation of vascular smooth muscle cell proliferation. All these effects participate in the initiation and progression of atherosclerotic lesions and thrombus formation [1]. The purpose of this article is to review recent advances in understanding the mechanisms by which elevated levels of homocysteine induce vascular oxidant stress, the impact of such stress on endothelial function in experimental models of hyperhomocysteinemia and in the clinical setting, and to discuss potential clinical consequences and therapeutic approaches.

Mildly Elevated Plasma Levels of Homocysteine are an Independent Risk Factor for Atherothrombotic Vascular Disease

As early as 1969, clinical studies in children with rare inborn errors of homocysteine metabolism, that lead to markedly elevated plasma homocysteine levels up to 30 times the normal range, suggested that severe hyperhomocysteinemia is associated with the development of premature atherosclerosis and thromboembolism, besides other clinical abnormalities [2, 3]. An international questionnaire survey that compiled data from 629 patients with severe hyperhomocysteinemia due to cystathionine β -synthase deficiency showed that untreated patients suffer one thrombembolic event per 25 patient-years. [4]. Treatment of hyperhomocysteinemia in these patients with high doses of vitamin B_6 in vitamin B_6 responsive patients or with a methionine-restricted diet, supplemented with cystine and betaine in vitamin B_6 non-responders leads to a significant reduction in thromboembolic events. This has been shown in a multicenter observational study that assessed the effectiveness of long-term homocysteine-lowering treatment in reducing vascular risk in 158 patients during 2,822 patient-years of treatment. There were 17 vascular events in the treated patients compared to 112 vascular events that would have been expected according the natural history of the disease (relative risk of 0.09 [95% CI 0.036 to 0.228; $P <$ 0.0001] [5]. These data argue for a causality of hyperhomocysteinemia in increasing the cardiovascular risk in these patients.

Based on these observations it was proposed, first by McCully, that even mildly elevated plasma homocysteine levels, as frequently found in the general population, may play a role in the etiology of vascular diseases.

There is no consensus on what levels constitute hyperhomocysteinemia at the present time. Recent data show that an increased risk for atherosclerotic vascular diseases seems to be associated with fasting homocysteine levels above 12 µmol/L (see below) [6]. In slight modification of the initial classification for several types of hyperhomocysteinemia in humans in relation to fasting total plasma homocysteine concentrations, mild or – synonymously used – moderate hyperhomocysteinemia refers to plasma levels between 12 and 30 µmol/L, intermediate hyperhomocysteinemia to levels between 31 and 100 µmol/L, and severe hyperhomocysteinemia to concentrations > 100 µmol/L, respectively [6, 7].

During the last 20 years, a great number of retrospective case-control studies and prospective nested case-control studies nearly uniformally established the link between mild hyperhomocysteinemia and atherothrombotic vascular diseases in the general population, as clinically manifested by an increased prevalence and incidence of coronary artery disease, cerebrovascular disease, and peripheral vascular disease. Mild hyperhomocysteinemia appears to be an independent risk factor for arterial vascular disease as this relationship persists after statistical adjustment for conventional risk factors. A meta-analysis by Boushey and colleagues [8] using data from 27 studies published before 1995 indicated that a 5 μmol/L increase in plasma homocysteine above median levels of 10 μmol/L is associated with a significant and graded increase in the risk of coronary artery disease (odds ratio 1.6 [95 % confidence intervall: 1.4 to 1.7]), cerebrovascular disease (odds ratio 1.5 [1.3 to 1.9]), and peripheral vascular disease (odds ratio 6.8 [2.9 to 15.8]). Comparing elevated homocysteine levels to other established vascular risk factors, the authors calculated that a 5 μmol/L increase in plasma homocysteine levels is equivalent to a 0.5 mmol/L (20 mg/dL) increase in plasma cholesterol levels in increasing the risk for myocardial infarction. From data of a recent case-control study conducted in nine European centers it was estimated that the cardiovascular risk associated with elevated homocysteine levels detected both under basal conditions (> 12 μmol/L) and after an oral methionine challenge (net increase > 27 μmol/L over basal levels or absolute increase > 38 μmol/L) is comparable to the risk associated with hyperlipidemia or smoking, but somewhat lower than that of hypertension [6]. From these studies it has been estimated that ten percent of the population's coronary artery disease risk appears attributable to plasma homocysteine levels [8].

Prospective cohort studies, which are generally accepted as providing the most rigorous evidence, have, however, yielded some inconclusive results in linking homocysteine to vascular disease. Most of the prospective studies have provided evidence for mild hyperhomocysteinemia as a major risk factor for atherothrombotic vascular disease after adjustment for conventional risk factors [9-21], although some studies have not [22-27]. These conflicting results might be partly explained by the different ethnic background and lifestyle of the specific study's participants and by the sample size. Lifestyle issues appear to be especially important as a source of potential bias owing to multivitamin use by study subjects. For example, the Atherosclerosis Risk in Community Trial, the largest prospective trial with a negative outcome, did not provide detailed information about vitamin supplementation as a potential confounding variable [24-26].

Nevertheless, the reasons for the variable results are unknown. Several arguments, however, weigh in favor of the positive studies: (a) nearly all of the retrospective case-control studies show a clear-cut association between hyperhomocysteinemia and atherothrombosis; (b) the magnitude of cardiovascular risk correlates with plasma homocysteine levels in many studies [15, 19, 20, 28, 29]; (c) several meta-analyses of the retrospective case-control and of the prospective population based studies showed similar, consistent results favoring hyperhomocysteinemia as a vascular risk factor [8, 20, 30-33]; (d) several studies documented an association between mortality in patients with preexisting coronary or peripheral vascular disease and plasma homocysteine levels independent of tradional risk factors [15, 34-39], or with cerebrovascular events in patients with significant stenosis of the carotid artery [40], suggesting that homocysteine may promote acute thrombotic events leading to cardiac death or stroke; (e) plasma homocysteine levels are related to the extent of atherosclerosis in coronary and peripheral arteries [33, 41-44] and to greater carotid artery intimal-medial wall thickness in cross-sectional studies of subjects with subclinical vascular disease [45-47]; and (f) mildly elevated plasma homocysteine levels induce endothelial vascular dysfunction, a key event in the pathogenesis of atherothrombosis, and lowering of plasma homocysteine levels may have a favorable effect on vascular function. The debate about hyperhomocysteinemia as a cardiovascular risk factor will continue, as long as results of intervention studies aimed at reducing cardiovascular events by homocysteine lowering treatment, are still pending. Several of these studies are performed at the moment.

Endothelial Dysfunction as an Initial Step in Atherosclerosis

The pathobiological mechanisms that lead to the atherogenic propensity associated with elevated plasma levels of homocysteine are not completely understood as yet. Experimental evidence suggests that a key event in the vascular pathobiology of hyperhomocysteinemia is the development of functional endothelial dysfunction and structural endothelial injury.

The healthy endothelium plays a central role in cardiovascular control [48]. By its complex endocrine and paracrine functions the endothelium acts to maintain vascular homeostasis through regulation of vascular tone, through multiple interactions with cells in the vessel wall and the lumen, like vascular smooth muscle cells, circulating platelets, monocytes and leukocytes, and through control of blood fluidity and coagulation [49]. Functionally intact endothelium exerts potent antiatherothrombotic effects by maintaing a low arterial tone at rest, by inhibiting leukocyte-endothelial interactions, by attenuating platelet activation, by inhibiting smooth muscle cell proliferation, by activation of fibrinolysis, and by inhibition of the coagulation cascade.

In contrast, endothelial dysfunction plays a crucial role in the pathogenesis and progression of atherothrombotic vascular disease, both as a target and a mediator of the disease process [50, 51]. Endothelial dysfunction, which may be characterized as an imbalance between vasorelaxing and –contracting substances, between procoagulant and anticoagulant factors, between proinflammatory and antiinflammatory mediators, and between inhibitors and promotors of vascular cell growth, is an early and persistent vascular abnormality in the evolution of atherothrombotic disease (Figure 1).

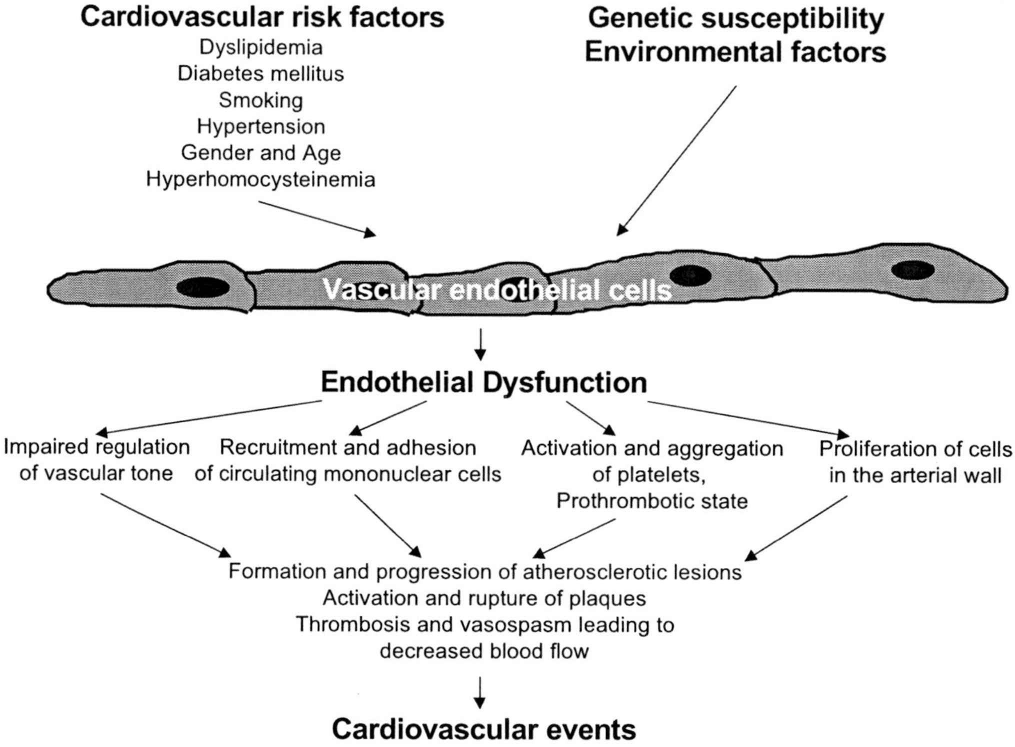

Figure 1. Endothelial function plays a key role in the delovpment of atherosclerosis and cardiovascular disease events. Cardiovascular risk factors together with intrinsic genetic susceptibility and environmental factors contribute to the development of endothelial dysfunction. This affects vascular homeostasis and promotes the formation and progression of atherosclerotic lesions, activation and rupture of atherosclerotic plaques, and thrombosis and vasospasm resulting in cardiovascular disease events.

The impairment of endothelium-dependent regulation of vascular tone can be clinically determined as one manifestation of endothelial dysfunction by non-invasive and invasive techniques [52, 53]. The presence of atherosclerotic lesions in the coronary arteries is associated with endothelial dysfunction in this vascular bed [54-57]. Endothelial dysfunction can also be detected in vessels distant from the clinically affected vascular bed, indicating that endothelial dysfunction seems to be a general abnormality in patients with atherosclerotic vascular disease [58-61]. In patients with atherosclerotic risk factors, impaired endothelial vasodilator function is notable early in the process of atherogenesis, even prior to the development of frank atherosclerotic lesions. Its extent is related to the intensity and duration of risk factor exposure, and treatment of risk factors results in reversal or attenuation of endothelial dysfunction [50, 51, 62]. These observations suggest that the endothelium is both a target and a mediator of atherothrombosis. Endothelial vasodilator function may therefore serve as an index integrating the overall stress imposed by vascular risk factors. Indeed, clinical studies have demonstrated that endothelial vasodilator dysfunction is a predictive marker for the future development of coronary artery disease and cardiovascular events in patients free of overt cardiovascular disease [63], and is associated with a worse prognosis in the setting of existing coronary artery disease predicting coronary disease progression and cardiovascular event rate [64-66]. These findings underscore the central role of the endothelium in vascular function and its regulation.

Impairment of endothelium-dependent vasomotor responses as one indicator of endothelial dysfunction is indicative of a reduction in the bioavailability of the endothelium-derived signaling molecule nitric oxide [67, 68]. Nitric oxide is released from endothelial cells in response to increased shear stress [69] and certain biochemical stimuli [70], like acetylcholine or bradykinin. Endothelium-derived nitric oxide may function as an endogenous antiatherogenic molecule as it is able to mediate all antiatherothrombotic functions of the endothelium mentioned above [71]. In contrast, a reduction in the bioavailability of nitric oxide will lead to the loss of an important component in the endogenous defense against vascular injury, inflammation, and thrombosis, all key events involved throughout the course of atherosclerosis, and thereby play a key role in the development of atherosclerotic vascular disease.

Endothelial Dysfunction is a Key Event in the Vascular Pathobiology of Hyperhomocysteinemia

Impairment of Endothelium-Dependent Vasodilation in Hyperhomocysteinemia

Animal models of mild hyperhomocysteinemia, induced by vitamin-deficient and methionine-enriched diets in cynomolgus monkeys [72] or in rats [73, 74], by heterozygous disruption of the cystathionine β-synthase gene in mice ($CBS^{(-/+)}$ mice) [75-77], or by the combination of genetic and dietary approaches [78], uniformly showed an impairment of endothelium-dependent vasoreactivity and regulation of blood flow. The impaired endothelium-dependent vasoreactivity was evident in conductance vessels (i.e. aortic ring preparations) as well as in resistance vessels (i.e. mesenteric arterioles). In contrast, endothelium-independent vasodilation was maintained in these animal models (Figure 2). Numerous studies in humans confirmed these findings. Healthy humans free of overt cardiovascular disease, with either acutely elevated plasma homocysteine levels following an oral methionine challenge [79-89] or with chronic, mild hyperhomocysteinemia [90-92] showed impaired endothelium-dependent vasodilator function, but preserved endothelium-independent vasodilator responses. This could also be observed in the macrovasculature as well in the microvasculature [93].

Homocysteine, and not methionine or vitamin deficiencies, appears to be responsible for these adverse vascular effects. This is indicated by the facts that mice with a genetic defect in homocysteine transsulfuration have mild hyperhomocysteinemia of 1.5 to 1.8-fold the level in wild type mice, but have normal plasma methionine levels and were provided a chow that yields normal B vitamin levels [75-77]. Thus, a mild increase in homocysteine alone is sufficient to induce endothelial dysfunction. Futhermore, impaired endothelial function in humans can be induced by an oral methionine as well as by an oral homocysteine challenge; the latter does not lead to increased plasma levels of methionine [87, 88]. Further analyses of different homocysteine species and their time-dependent concentrations in plasma after oral homocysteine or methionine challenges revealed that peak reduction in flow-mediated dilation coincided with maximal concentrations of reduced homocysteine, but not with free oxidized homocystine, protein-bound oxidized homocystine, or related species [87], suggesting that reduced homocysteine is responsible for vascular dysfunction *in-vivo*.

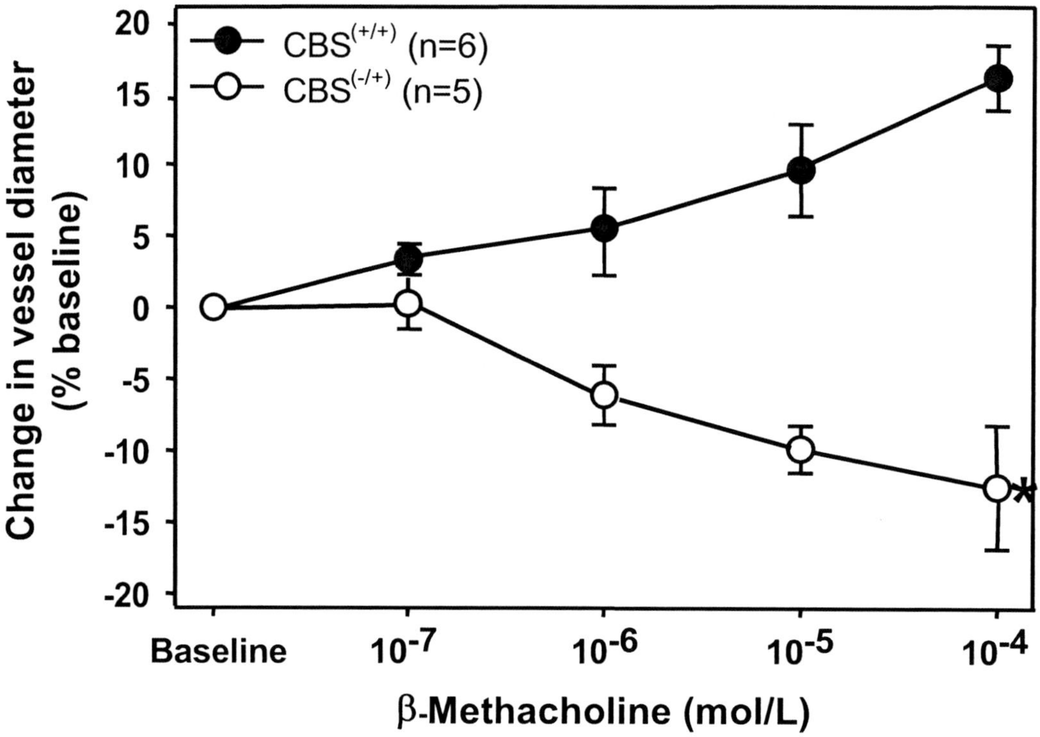

Figure 2. Mesenteric microvascular response to superfusion with β-methacholine in mildly hyperhomocysteinemic heterozygous cystathionine β-synthase knockout mice ($CBS^{(-/+)}$) and their wild type littermates ($CBS^{(+/+)}$). Superfusion of the mesentery with the endothelium-dependent vasodilator β-methacholine, which stimulates endothelial nitric oxide production, results in dose-dependent vasodilation in wild type mice. In contrast, mesenteric arterioles from $CBS^{(-/+)}$ mice showed paradoxical. Superfusion of the mesentery with the endothelium-independent vasodilator sodium nitroprusside lead to dose-dependent vasodilation in both groups of animals (not shown). This reaction in $CBS^{(-/+)}$ mice is indicative of endothelial dysfunction due to a decrease in bioavailable nitric oxide as cholinergic stimulation of vascular smooth muscle cells with β-methacholine will lead to vasoconstriction of mesenteric arterioles when unopposed by endothelium-derived nitric oxide (Ref. 76, 77).

The observation that homocysteine impairs endothelium-dependent vasodilator function induced by shear stress, acetylcholine or bradykinin but does not impair endothelium-independent vasodilation induced by sodium nitroprusside or nitroglycerin indicates that the effects of homocysteine are presumably due to a decrease in bioavailable endogenous or endothelium-derived nitric oxide. This conclusion is supported by the observation that the accumulation of cyclic guanosine monophosphate, the second messenger of nitric oxide-mediated vasodilation, was decreased in aortas from mildly hyperhomocysteinemic $CBS^{(-/+)}$ mice compared to their wild-type littermates [75], and that the plasma levels of the nitric oxide-derived endproducts nitrite and nitrate were significantly lower in hyperhomocysteinemic subjects than in healthy controls [92]. In accordance with these *in-*

vivo findings, homocysteine has been shown to decrease the production and/or bioactivity of nitric oxide and of S-nitrosothiols by cultured endothelial cells [76, 94, 95] and isolated arteries *ex-vivo* [96].

Homocysteine-Induced Endothelial Dysfunction Augments the Recruitment and Adhesion of Circulating Mononuclear Cells to the Endothelium

In addition to an impairment of endothelium-dependent vascular reactivity, homocysteine-induced endothelial dysfunction may be manifest as an activation of endothelial cells leading to increased expression of chemokines and adhesion molecules resulting in increased recruitment of circulating inflammatory blood cells and their increased adhesion at the endothelium. All these events are known to play an important role in the pathogenesis of early atherosclerotic lesions and their progression [97]. Incubation of cultured endothelial cells with homocysteine has been shown to increase the expression of the chemoattractant proteins macrophage chemoattraction protein-(MCP)-1 and interleukin-(IL)-8 leading to increased chemotaxis of human peripheral blood monocytes and U937 cells [98], and to increase the expression of adhesion molecules on endothelial cells, like intercellular adhesion molecule-(ICAM)-1, that mediate monocyte adhesion to endothelial cells [99]. Homocysteine has also been shown to stimulate the expression of the MCP-1 receptor (CCR2) in human monocytes, an effect that is possibly mediated by oxygen-derived free radicals [100]. Mildly hyperhomocysteinemic $CBS^{(-/+)}$ mice showed increased plasma levels of soluble P-selectin and increased expression of P-selectin on aortic tissue compared to wild-type mice; these effects were reversible with antioxidant therapy [77]. In hyperlipidemic apolipoprotein E knockout mice that had been made hyperhomocysteinemic by a methionine-enriched diet depleted in B-vitamins, hyperhomocysteinemia was associated with increased expression of vascular cell adhesion molecule-(VCAM)-1, tissue factor (TF), receptor for advanced glycation endproducts (RAGE) and matrix metalloproteinase-(MMP)-9 in aortic sections [101]. These findings correlated with increased atherosclerotic lesion area and complexity, pointing out that hyperhomocysteinemia may promote atherosclerotic lesion progression via an enhanced vascular inflammatory response. The mechanisms underlying these observations are not yet clear. Reduced bioavailability of nitric oxide promoting P-selectin expression [77] and/or activation of the redox sensitive transcription factor NF-κB regulating ICAM-1, VCAM-1, TF and RAGE expression [99, 101] may be involved in vascular inflammatory response to hyperhomocysteinemia.

Inhibition of Endothelial Cell Thromboresistance by Homocysteine

Experimental work from several groups suggests that homocysteine may alter endothelial cell function to promote thrombosis. Homocysteine has been shown to impair the ability of endothelial cells to inhibit platelet aggregation *in-vitro*, likely by reducing nitric oxide bioavailability [102], or by decreasing the biosynthesis of prostacyclin [103, 104], although the latter data are still somewhat controversial [105]. In addition, exposure of endothelial cells to homocysteine increases tissue factor expression [106] and the activation of factor V to factor Va [107], thus favoring thrombin generation through initiation and propagation of the coagulation cascade. These procoagulant effects of homocysteine may be potentiated by inhibition of endogenous anticoagulants and inhibition of fibrinolysis by homocysteine. Homocysteine has been shown to inhibit thrombomodulin expression and activity [108, 109], and thrombomodulin-dependent activation of the anticoagulant protein C by thrombin [110]. In addition, homocysteine may inhibit inactivation of factor Va by activated protein C through homocysteinylation of the cofactor [111], and may reduce the binding of antithrombin III to the endothelial surface through inhibition of heparan sulfate expression [112]. Inhibition of endothelial-dependent fibrinolysis by homocysteine is suggested by the finding of impaired binding and activity of tissue-type plasminogen activator on the endothelial cell surface [113]. This effect is mediated by derivatization of the tail domain of annexin II by homocysteine at a cysteine residue, which inhibits tissue-type plasminogen activator binding to the endothelial cell surface [114]. All these observations suggest that hyperhomocysteinemia promotes thrombosis. However, most of the *in-vitro* studies mentioned above were performed with homocysteine concentrations above pathophysiologically relevant levels, or were reproduced with cysteine, and, thus, could be related to the nonspecific reactivity of the sulfhydryl group of homocysteine [102, 106, 112].

Support of the hypothesis that mild hyperhomocysteinemia induces a prothrombotic state *in-vivo* is derived from animal studies and clinical observations. The view that platelet hyperreactivity occurs in hyperhomocysteinemic states is supported by the finding that dietary folate deficiency in rats induces platelet hyperreactivity and increases macrophage tissue factor activity, effects that correlate with increases in homocysteine levels [115]. Plasma homocysteine concentrations induced by methionine challenge in rats correlated with platelet aggregation and increased synthesis of the proaggregant thromboxane A_2 [116]. Increased thromboxane A_2 synthesis has also been documented in homocystinuric patients [117, 118]. In addition, plasma homocysteine concentrations were found to correlate with plasma fibrinogen levels in mildly hyperhomocysteinemic patients with coronary heart disease [44], and increased levels of von Willebrand factor were found in mildly hyperhomocysteinemic patients with premature arterial disease [119] and in healthy individuals after methionine challenge [120]. Together with a reduction of the antiplatelet effects of nitric oxide by homocysteine [102], these further supports the view that increased platelet activation and aggregation may be involved in the development of thrombosis in hyperhomocysteinemia.

Promotion of Atherosclerosis Lesion Size by Homocysteine

Recent animal studies suggest that homocysteine not only leads to functional vascular changes but, in addition, may promote atherosclerotic lesion development and progression. Apolipoprotein E-deficient mice were fed diets that raised plasma homocysteine levels up to 36 to 47 μmol/L [101, 121]. Hyperlipidemic apolipoprotein E knockout mice spontaneously develop complex atherosclerotic lesions, but when fed a hyperhomocysteinemic diet for 8 weeks or 3 months, respectively, they develop atherosclerotic lesions in the aorta that were of greater size and complexity than those seen in apolipoprotein E knockout mice fed normal chows. Double knockout mice with targeted deletions of the genes for apolipoprotein E and cystathionine β-synthase demonstrated greater atherosclerotic lesion size in the aortic arch compared to apolipoprotein E knockout mice that increased with cystathionine β-synthase gene deletion in a dose-dependent manner. In hyperhomocysteinemic mice, the increases of plasma homocysteine levels were significantly correlated with increases in the atherosclerotic lesion area in the aortic arch [122]. These reports demonstrate that hyperhomocysteinemia due to dietary manipulation or due to disruption of the cystathionine β-synthase promotes the development of atherosclerosis in a susceptible animal model.

In conclusions these data indicate that elevated plasma levels of homocysteine interfere with several functions of the endothelium that regulate vascular homeostasis and protect the vasculature from the development of atherosclerotic lesions and thrombosis (Figure 3).

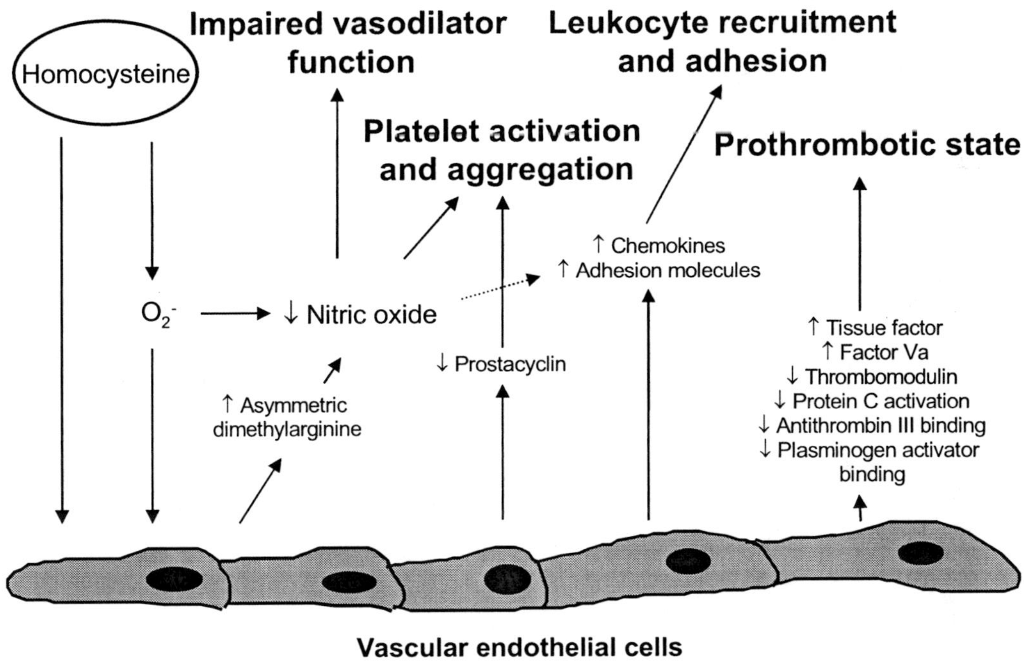

Figure 3. Postulated mechanisms and effects of homocysteine-induced endothelial dysfunction. Exposure of endothelial cells to homocysteine leads to a decrease in the bioavailability of nitric oxide resulting in impaired endothelium-dependent vasodilator function. This may be caused by increased inactivation of nitric oxide by reactive oxygen species and/or increased levels of the nitric oxide synthase inhibitor asymmetric dimethylarginine. Decreased nitric oxide bioavailability together with a decreased endothelial synthesis of prostacyclin promotes platelet activation and aggregation. Homocysteine induces the expression of several chemokines and adhesion molecules leading to increased recruitment and adhesionor circulating inflammatory cell. This may be caused by a decrease in nictric oxide bioavailability and/or increased vascular oxidant stress leading to an activation of the redox-sensitive transcription factor NF-κB that stimulates several inflammatory pathways. Furthermore, homocysteine may promote thrombus formation by activating endothelium-dependent prothrombotic mechanisms, like the expression of tissue factor or the activity of factor Va, and by inhibiting antithrombotic and fibrinolytic mechanisms.

Hyperhomocysteinemia is Associated with Increased Superoxide Output

Experiments in hyperhomocysteinemic animals and incubation of cultured endothelial cells with homocysteine indicate that elevated levels of homocysteine lead to increased formation of reactive oxygen species. Aortic tissue from mildly hyperhomocysteinemic heterozygous cystathionine β-synthase deficient ($CBS^{-/+}$) mice produces more superoxide anion compared to that from wild type mice [75], as measured by the lucigenin chemiluminescence assay or the cytochrome c assay. The same effect could be demonstrated in cultured endothelial cells incubated with homocysteine [95, 123].

Superoxide anion reacts with nitric oxide in a diffusion-limited reaction to form peroxynitrite [124], which does not induce vasorelaxation [125], and itself is highly reactive. The potential physiological relevance of this reaction is demonstrated by immunohistochemial and immunoblot data showing an increase in protein tyrosine nitration, a marker of peroxynitrite-dependent protein oxidation, in vascular tissues from mildly hyperhomocysteinemic mice [75], and in lysates from aortic rings cultured in homocysteine-containing media [126].

The role for hydrogen peroxide, which is produced from superoxide via superoxide dismutase, in homocysteine-induced endothelial toxicity *in-vitro* is somewhat more nebulous. Early studies demonstrated that catalase inhibits the homocysteine-induced lysis of endothelial cells in the presence of transition metals or ceruloplasmin [127, 128], but this lytic effect of homocysteine was only observed at concentrations far exceeding those observed in the most severe cases of hyperhomocysteinemia. More recently, we and others [95, 129, 130] have observed a homocysteine dose-dependent increase in the fluorescence of an hydrogen peroxide-sensitive fluorescent probe, 2',7'-dichlorofluoresceine (DCF), in cultured endothelial cells, but the interpretation of these results is tempered by the observation that DCF also fluoresces when it reacts with peroxynitrite and reactive oxygen species other than hydrogen peroxide.

Mechanisms of Increased Vascular Oxidant Stress in Hyperhomocysteinemia

Elevation of the Oxidation Rate of Homocysteine in Hyperhomocysteinemia

Homocysteine is an intermediate in the methionine cycle, the function of which is to generate one-carbon methyl groups in the form of S-adenosylmethionine for transmethylation reactions [131]. The essential amino acid methionine is the precursor of homocysteine. In mammals, homocysteine can be diverted from the methionine cycle into the transsulfuration pathway to generate the non-essential amino acid cysteine, or it can be remethylated to methionine. Chemically, methionine contains a sulfide sulfur (R-S-R'), whereas homocysteine and cysteine are sulfhydryl compounds (R-SH). Compounds containing a free sulfhydryl group are known as "thiols". Other biologically relevant low-molecular-weight thiols are glutathione, coenzyme A, and dihydrolipoic acid. Under aerobic conditions (i.e., in the presence of molecular oxygen as an electron acceptor) and at physiological pH, thiols such as homocysteine, oxidize to form disulfides, according to the general reaction 2 RSH + $O_2 \Leftrightarrow$ RSSR + H_2O_2. In plasma, this reaction can be catalyzed by transition metals such as copper and cobalt (the former present in the circulation associated with albumin and ceruloplasmin, the latter with cobalamin). Homocysteine has classically been thought to autooxidize readily via this mechanism to form homocystine, oxidize other thiols such as cysteine and glutathione to form mixed disulfides, or oxidize free cysteine residues on proteins and peptides to form mixed disulfides (Figure 4). This notion is supported by the observation that 1% or less of total homocysteine species in the plasma of healthy humans under physiological conditions consists of free homocysteine, whereas 5-15% consists of homocystine, another 5-15% are mixed disulfides (cysteine-homocysteine or glutathione-homocysteine), and more than 70% is protein-bound [132]. Moreover, a decreased ratio of reduced to total aminothiols in plasma has been demonstrated in experimental hyperhomocysteinemia after a methionine challenge [132-134] and in hyperhomocysteinemic patients [135-138]. More recent evidence, however, suggests that oxidized forms of homocysteine in plasma arise primarily via disulfide exchange, with only a small fraction resulting from direct homocysteine oxidation [139]. Hence, the contribution of direct homocysteine oxidation to homocysteine-induced vascular oxidative stress may not be as important as enzymatic mechanisms.

Involvement of Nitric Oxide Synthase in the Generation of Reactive Oxygen Species

Recent *in-vitro* experiments suggest that homocysteine-induced oxidant stress is stereospecific for the naturally-occuring L-isoform, indicating a biochemical rather than chemical basis for the effect [140]. In these experiments, increased lipid peroxidation was observed in endothelial cells incubated with L-homocysteine, but not with D-homocysteine, as measured by total isoprostane $F_{2\alpha}$-III and the levels of thiobarbituric acid reactive substances (TBARs) in the supernatant of endothelial cell cultures. Stereospecificity was also observed for DCF fluorescence *in-situ*, indicating a parallel effect on overall reactive oxygen species content. Enhanced lipid peroxidation and DCF-fluorescence was not observed with either cysteine or glutathione, and was not due to extracellular thiol oxidation, as it could be fully replicated with the oxidized disulfide form of homocysteine, L-homocystine.

Mechanistically, this pro-oxidant effect seems to be dependent upon both endothelial nitric oxide synthase and superoxide anion. Pharmacological inhibition of endothelial nitric oxide synthase with L-nitroarginine methyl ester (L-NAME) completely blocked the homocysteine-dependent stimulation of reactive oxygen species formation as measured by DCF-fluorescence. Furthermore, L-NAME completely abrogated both the homocysteine-dependent increases in isoprostane $F_{2\alpha}$-III formation and DCF-fluorescence. Similarly, transfection of endothelial cells with superoxide dismutase cDNA blocked the effect of homocysteine on endothelial cell lipid peroxidation, whereas addition of extracellular superoxide dismutase had no effect. Importantly, neither the addition of extracellular catalase nor loading cells with catalase had any effect on lipid peroxidation, indicating that hydrogen peroxide is not involved in this process.

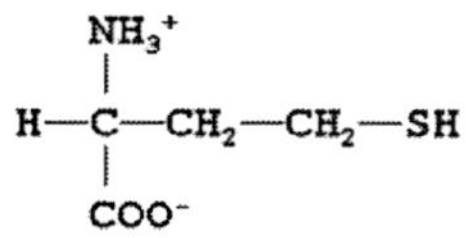

Homocysteine
MW 135.19

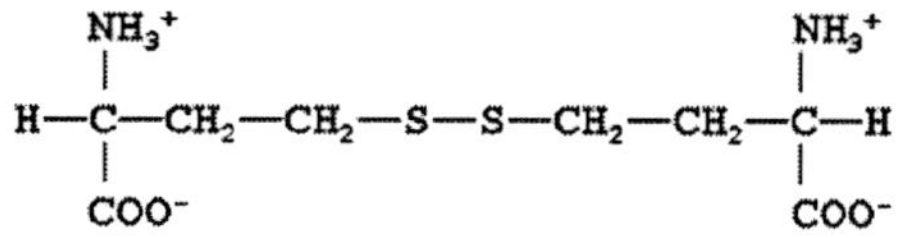

Homocystine
MW 268.36

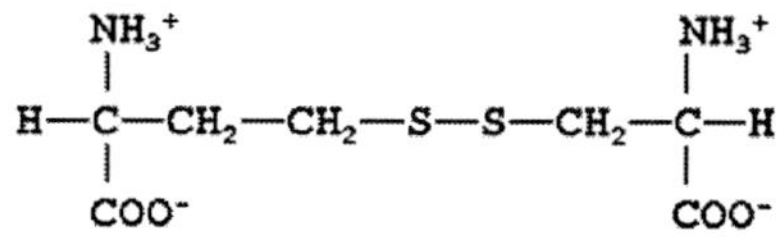

Homocysteine-cysteine mixed disulfide
MW 254.33

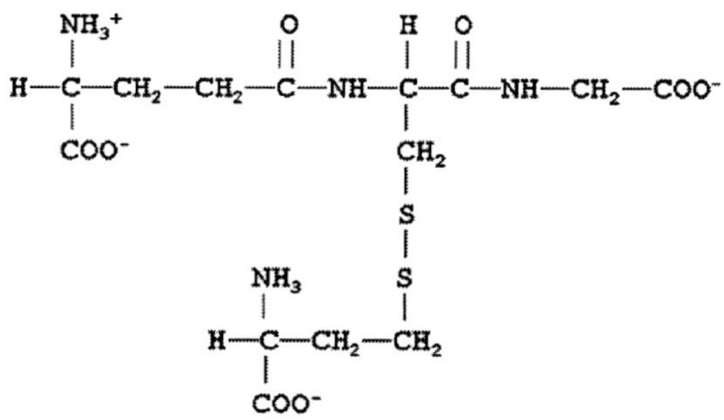

Homocysteine-glutathione mixed disulfide
MW 439.49

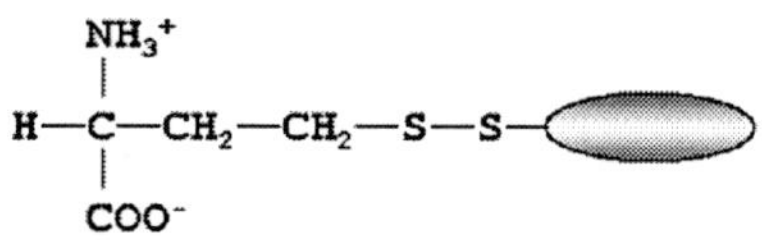

Protein-bound homocysteine

Figure 4. Chemistry of homocysteine and different homocysteine oxidation products.

The most straightforward explanation for these results would be that homocysteine induces increased formation of peroxynitrite. Unlike superoxide, peroxynitrite is both able to initiate lipid peroxidation [141] and to react with DCF [142]. Peroxynitrite has a short half-life, which makes its detection difficult in biological systems [143], but may react with cellular tyrosine residues to form nitrosated endproducts. As mentioned earlier, immunostaining for one such endproduct, 3-nitrotyrosine, was positive in aortic tissue from mildly hyperhomocysteinemic $CBS^{(-/+)}$ mice compared with wild type mice [75], as was an immunoblot analysis for 3-nitrotyrosine of protein lysates from aortic rings cultured in homocysteine-containing media [126].

Although the precise mechanism by which homocysteine may induce increased peroxynitrite formation remains to be elucidated, two potential mechanisms include endothelial nitric oxide synthase "uncoupling" [124, 144], in which endothelial nitric oxide synthase is the source of superoxide (and peroxynitrite); and increased superoxide production from other enzymatic sources, in which some of the excess superoxide reacts with nitric oxide to form peroxynitrite (Figure 5).

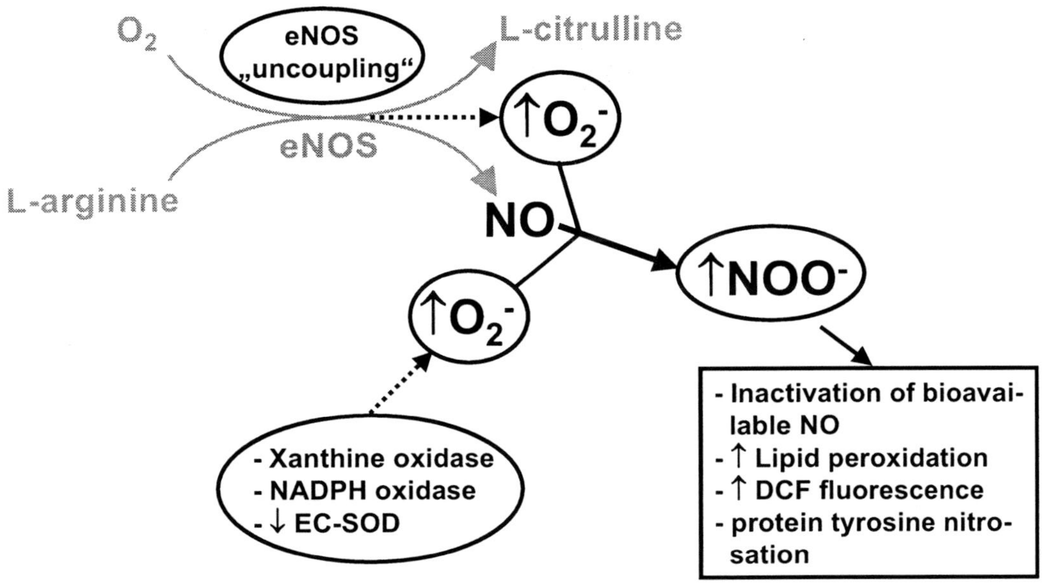

Figure 5. Hypothetical mechanism of nitric oxide synthase-dependent generation of reactive oxygen species in hyperhomocysteinema. L-homocysteine and L-homocystine, but the not the respective D-isoforms, increase memebrane lipid peroxidation and DCF-fluorescence in cultured endothelial cells. This effect is dependent of endothelial nitric oxide synthase (eNOS) and superoxide anion (O_2^-), as L-NAME and transfection of endothelial cells with superoxide dismutase cDNA, respectively, block the effect of homocysteine. Increased O_2^- may result from "uncoupling" of eNOS and/or from other enzymatic sources. Some of the excess O_2^-reacts with nitric oxide (NO) to form peroxynitrite ($ONOO^-$). This reaction leads to biological inactivation of NO. Furthermore, $ONNO^-$ initiates lipid peroxidation and increases DCF-fluorescence.

Inhibition of Cellular Antioxidant Enzymes by Homocysteine

The cellular defense system against reactive oxygen species includes several antioxidant enzymes and non-enzymatic antioxidants, such as α-tocopherol, ascorbic acid, β-carotene, and glutathione. Homocysteine has been shown in particular to disrupt the normal function of two important cellular antioxidant enzymes, the cellular isoform of glutathione peroxidase (GPx-1) and superoxide dismutase.

Homocysteine, but not other low-molecular-weight thiols, decreases both the expression and specific activity of GPx-1 as shown *in-vitro* and *in-vivo* [76, 77, 94, 145-147]. This key enzyme for the cellular defense against oxidant stress uses glutathione to reduce hydrogen peroxide and lipid peroxides to their respective alcohols [148], and may also act as a peroxynitrite reductase [149]. Transition metal ions such as iron and copper catalyze the breakdown of hydrogen peroxide to form hydroxyl radical ($^{\bullet}OH$), which is highly reactive and causes lipid peroxidation, among its numerous effects; and hydroxide anion (OH^-), which promotes alkaline tissue damage. This process is offset in part by catalase and GPx-1-dependent reduction of hydrogen peroxide (H_2O_2) to water (H_2O). Elevated levels of lipid peroxides are accompanied by an increase in peroxyl radicals, which can inactivate nitric oxide through the formation of lipid peroxynitrites [150, 151]. Thus a deficiency of GPx-1 may lead to a decrease in bioavailable nitric oxide via at least two mechanisms, an increase in reactive oxygen species and an increase in lipid hydroperoxides. In support of this hypothesis we have shown previously that mice deficient in GPx-1 have endothelial dysfunction owing to decreased bioavailable endothelium-derived nitric oxide; increased oxidative stress and increased lipid hydroperoxide generation as measured by increased tissue isoprostane $F_{2\alpha}$-III and hepatic phospholipid hydroperoxide levels; and increased nitrosative stress, as indicated by increased immunostaining for 3-nitrotyrosine in aortic tissue of GPx-1 knockout mice compared to wild type mice [152].

Extracellular superoxide dismutase is a secreted glycoprotein with an affinity for heparin-like glycosaminoglycans. It is present in the circulation in equilibrium between the glycosaminoglycans on the endothelial surface and the plasma phase, and acts as the principal enzymatic scavenger of superoxide in the extracellular space [153, 154]. The plasma levels of extracellular superoxide dismutase correlate positively with plasma homocysteine levels in homocystinuric patients [155] and in patients with mild hyperhomocysteinemia [156]. This effect is caused by decreased binding of extracellular superoxide dismutase to endothelial cell surfaces by alterations in the endothelial heparan sulfate proteoglycan by homocysteine [157]. This effect may result in a loss of the ability to protect the endothelial surface from oxidative stress, although this hypothesis has not yet been definitively proven.

In summary, experimental data suggest that elevated levels of homocysteine lead to increased vascular generation of reactive oxygen species, primarily by a mechanism involving nitric oxide synthase, and secondarily due to increased homocysteine oxidation. These effects are further amplified by alterations in the function of important cellular antioxidant enzymes. The resultant vascular oxidant stress may lead to vascular dysfunction (Figure 6).

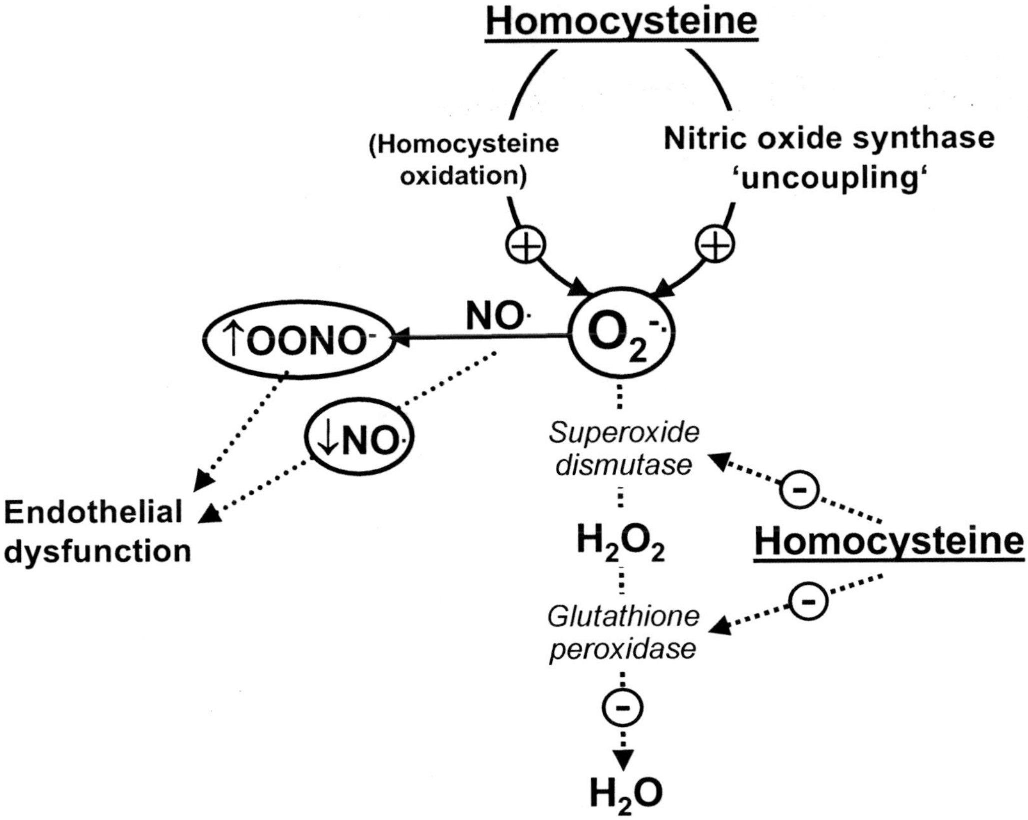

Figure 6. Postulated mechanisms of homocysteine-induced vascular oxidant stress. Homocysteine leads to increased generation of reactive oxygen species like superoxide anion (O_2^-), by a nitric oxide synthase dependent mechanism, and to less extent due to increased homocysteine oxidation. Inhibition of antioxidant enzymes, like superoide dismutase and glutathione peroxidase further amplifies these effects. Superoxide anion reacts with nitric oxide to form peroxynitrite ($ONOO^-$) thereby loosing its antiatherothrombotic activity.

Impact of Increased Vascular Oxidant Stress on Homocysteine-Induced Endothelial Dysfunction and Reduced Nitric Oxide Bioavailability

As outline above, hyperhomocysteinemia leads to endothelial dysfunction due to a decrease in the bioavailability of nitric oxide. Decreased nitric oxide bioavailability may either reflect an absolute deficit of nitric oxide due to decreased biosynthesis, impaired availability of bioactive nitric oxide or enhanced nitric oxide inactivation [71, 158-162].

Decreased nitric oxide biosynthesis may caused by decreased expression and activity of endothelial nitric oxide synthase, by alterations of membrane receptors in the arterial wall that interact with physiological or biochemical stimuli to induce nitric oxide synthase, by decreased availability or affinitiy of L-arginine, the substrate for nitric oxide synthase, or by decreased availability of cofactors for nitric oxide synthesis, especially tetrahydrobiopterin.

Decreased nitric oxide bioavailability in hyperhomocysteinemia is not due to a decrease in the expression of endothelial nitric oxide synthase. Incubation of cultured endothelial cells with homocysteine does not affect mRNA or protein levels of endothelial nitric oxide synthase [94, 95]. Furthermore, immunostaining of aortic tissue with an anti-endothelial nitric oxide synthase antibody showed similar staining of the endovascular surface in mildly hyperhomocysteinemic $CBS^{(-/+)}$ mice compared with $CBS^{(+/+)}$ [75]. Hyperhomocysteinemia, however, leads to increased levels of asymmetric dimethylarginine an endogenous inhibitor of nitric oxide synthase, which may impair nitric oxide biosynthesis. Dimethylarginines are likely formed from the degradation of proteins containing methylated arginine residues [163, 164]. Methylated arginines are either excreted in the urine or metabolized to citrulline by the enzyme dimethylarginine dimethylaminohydrolase. Cynomolgus monkeys fed a methionine-rich, folic acid-depleted diet free of choline, and healthy humans after an oral methionine challenge showed increased plasma levels of asymmetric dimethylarginine that correlated with endothelial dysfunction [86, 165]. In addition, when endothelial cells or nonvascular cells were exposed to homocysteine or to its precursor methionine, concentration of asymmetric dimethylarginine in the cell culture medium increased in a dose- and time-dependent fashion, associated with reduced nitric oxide synthesis by endothelial cells. The homocysteine-induced increase in asymmetric dimethylarginine by cells was associated with a reduction in the activity of dimethylarginine dimethylaminohydrolase, the enzyme that degrades asymmetric dimethylarginine. This inhibitory effect of homocysteine appears to be either caused by direct binding and functional interference of the reactive sulfhydryl group of homocysteine with cysteine groups of dimethylarginine dimethylaminohydrolase, as it can be prevented by a thiol antioxidant that protects sulfhydryl groups, and by the reducing agent dithiothreitol [166], or by increased oxidant stress under conditions of elevated homocysteine, as it has been previously shown that oxidative stress impairs the activity of dimethylarginine dimethylaminohydrolase and increases dimethylarginine elaboration by endothelial cells [167].

Impaired bioavailability and bioactivity of nitric oxide may further be due to enhanced degradation by increased free radicals and/or oxidation-sensitive mechanisms, by impaired nitric oxide diffusion from the endothelium to vascular smooth muscle cells followed by decreasd sensitivity to its vasodilator action, and by impaired interaction of nitric oxide with guanylate cyclase resulting in decreased cyclic GMP production. An increase in reactive oxygen species leading to inactivation of nitric oxide seems to play a major role in mediating the adverse effects of homocysteine on endothelial function.

As outlined above, elevated levels of homocysteine are associated with an increase in reactive oxygen species that correlates with endothelial dysfunction *in-vitro* and *in-vivo*. Conversely, scavenging of reactive oxygen species has been shown to restore nitric oxide bioavailability and endothelial function in hyperhomocysteinemia.

In-vitro studies have shown that a scavenger of peroxynitrite and a cell-permeant superoxide dismutase mimetic, or overexpression of the cellular antioxidant enzyme intracellular glutathione peroxidase reversed the homocysteine induced suppression of NO production by endothelial cells [76, 95]. Furthermore, scavenging of superoxide anion has been shown to prevent homocysteine-induced upregulation of inflammatory pathways in endothelial cells leading to increased adhesion molecule expression and monocyte adhesion

[99] and to abolish the stimulatory effect of homocysteine on CCR2 expression as well as the binding activity of the receptor on monocytes [100].

Enhanced formation of reactive oxygen species also seems to be a critical determinant of homocysteine-induced endothelial dysfunction *in-vivo*. Scavenging of superoxide anion by either superoxide dismutase or 4,5-dihydroxybenzene 1,3 disulfonate (Tiron) has been shown to reverse endothelial dysfunction in hyperhomocysteinemic animal models [75, 168] and in isolated aortic rings incubated with homocysteine [123].

The intracellular redox buffer system consisting of reduced glutathione and glutathione disulfide together with glutathione reductase and glutathione peroxidase plays a central role in the cellular defense against oxidant stress and has an important role in maintaining endothelial function. This point is underscored by the findings that nitric oxide synthesis is impaired in glutathione-depleted human endothelial cells and, conversely, bolstering cellular glutathione content with glutathione monoethyl ester results in enhanced nitric oxide production [169]. Furthermore, mice deficient in cellular glutathione peroxidase have endothelial dysfunction. This is due to increased vascular oxidative stress with resulting reduction in the bioavailability of nitric oxide [152]. The role of the cellular redox state in homocysteine-induced endothelial dysfunction is emphasized by the observation that treatment of hyperhomocysteinemic CBS$^{(-/+)}$ mice with the intracellular cysteine donor L-2-oxo-4-thiazolidine carboxylate (OTC) restores endothelial function [77]. This agent increases intracellular levels of reduced glutathione and total thiols and thereby shifts the cellular redox state to a more reduced environment. OTC treated mildly hyperhomocysteinimic CBS$^{(-/+)}$ mice had normal vasodilatory capacity of mesenteric arterioles during superfusion with the endothelium-dependent vasodilator methacholine or bradykinin and less P-selectin expression on the aortic endothelial surface and on adherent platelets. In contrast, untreated CBS$^{(-/+)}$ mice demonstrated impaired endothelium-dependent vasodilation and increased aortic P-selectin expression compared to wildtype mice. The specific role of cellular glutathione peroxidase in homocysteine-induced endothelial dysfunction is demonstrated by the observations that overexpression of cellular glutathione peroxidase in hyperhomocysteinemic CBS$^{(-/+)}$ mice restores the normal endothelium-dependent vasodilator response. Overexpression of cellular glutathione peroxidase in cultured endothelial cells attenuates the homocysteine-induced decrease in nitric oxide release from these cells [76]. These findings demonstrate that increasing the cellular antioxidant capacity by either increasing the concentration of reduced glutathione or overexpression of cellular glutathione peroxidase compensates for the adverse effects of homocysteine on endothelial function, and suggests that the adverse vascular effects of homocysteine are at least partly mediated by oxidative inactivation of nitric oxide.

This view is further supported by studies in humans that demonstrated that pretreatment with the antioxidant vitamins C or E prevents endothelial and platelet dysfunction associated with hyperhomocysteinemia induced by oral methionine challenge [81, 83, 89, 170].

Hyperhomocysteinemia-induced oxidant stress not only leads to biological inactivation of nitric oxide but may also lead to lipid peroxidation at the cell surface and direct oxidation of low-density lipoprotein [171-173], and may indirectly contribute to oxidative modification of low-density lipoprotein mediated by vascular cells *in-vitro* [171-174]. All these events are known to be important in the initial steps of atherothrombosis. In accordance with these *in-*

vitro findings, animals fed a methionine-rich diet leading to hyperhomocysteinemia showed an increase in plasma and aortic thiobarbituric acid reactive substances that was related to the development of atherosclerotic lesions [129, 175, 176]. Mildly hyperhomocysteinemic $CBS^{(-/+)}$ mice showed increased plasma levels of the isoprostane $F_{2\alpha}$-III, an endproduct of the non-enzymatic peroxidation of arachidonic acid [75]. An association of elevated plasma homocysteine levels with enhanced *in-vivo* lipid and protein oxidation in humans is suggested by the correlation of plasma homocysteine with plasma F_2-isoprostanes [177], and by increased levels of malondialdehyde, conjugated dienes, and oxidized protein in hyperhomocysteinemic patients [178].

Potential Clinical Consequences and Therapeutic Options

Plasma homocysteine levels are strongly and inversely correlated with plasma levels of folate and to a lesser extent with vitamin B_6 and vitamin B_{12} levels [179, 180]. Conversely, numerous clinical studies, including 12 randomized and placebo controlled intervention studies, have uniformly shown that supplementation with these vitamins effectively lowers plasma homocysteine levels. A recently performed meta-analysis of the published controlled clinical trials indicated [181] that folic acid supplementation reduces homocysteine levels by 25 % (95 % CI: 23 to 28 %) with similar effects in a daily dose range of 0.5 to 5 mg. An additional reduction in plasma homocysteine is produced by vitamin B_{12} (mean dose 0.5 mg), whereas vitamin B_6 in a mean dose of 16.5 mg did not have any significant effect. A comparison of different strategies to increase the folic acid supply to about 600 μg per day showed that the daily consumption of folic acid-fortified breakfast cereals or the intake of folic acid tablets is more effective than the consumption of folic acid-rich natural food in lowering plasma homocysteine levels [182]. The reason for this observation is that despite adequate dietary instruction of participants, daily folic acid intake increased to a smaller extent in subjects consuming natural food compared to subjects using folic acid supplements. Unwanted side effects of vitamin supplementation using physiological doses (folic acid up to 10 mg per day, vitamin B_6 up to 100 mg per day) have not been observed thus far; however, folic acid supplementation in subjects who are vitamin B_{12} deficient may unmask the adverse effects of this deficiency state. Although vitamin B_{12} has only a small additional homocysteine-lowering effect, it may be added primarily to avoid the theoretical risk of neuropathy of folic acid therapy in vitamin B_{12}-deficient patients, even those with intrinsic factor deficiency or malabsorption states [183-185].

Although lowering of plasma homocysteine levels by vitamin supplementation has been shown to be efficient and safe, conclusive data from randomized and placebo controlled intervention trials, proving a beneficial effect of vitamin supplementation on morbidity and mortality from atherothrombotic vascular disease, are not yet available. Several prospective intervention trials are currently ongoing in Europe, North America, and Australia that are designed to investigate the effect of vitamin supplementation on clinical endpoints of cardiovascular, cerebrovascular, peripheral arterial occlusive and venous thromboembolic diseases in primary or secondary prevention. More than 60,000 patients will be included in these trials; which are likely to have important therapeutic impact.

There is some uncontrolled evidence from observations in patients with severe hyperhomocysteinemia due to cystathionine β-synthase deficiency that homocysteine-lowering therapies may prevent vascular complications. A study on the natural history of untreated severe hyperhomocysteinemia in 629 patients disclosed that patients beyond the age of ten years suffer one vascular event per 25 patient-years [4]. A multicenter observational study documented the effects of long-term homocysteine-lowering treatment for severe hyperhomocysteinemia due to cystathionine β-synthase deficency on vascular outcome. A total of 158 patients with 2822 patient-years of treatment were observed. Treatment with vitamin B_6 alone in vitamin B_6-responsive patients, or in combination with betaine in vitamin B_6-non-responsive patients, resulted in a marked reduction of plasma homocysteine from pretreatment levels that, however, usually remained moderately elevated. There were 17 vascular events in 12 treated patients. Without treatment, 112 vascular events would have been expected, which represents a relative risk for vascular events in treated patients of 0.09 (95 % CI 0.036 to 0.228; $P < 0.0001$) [5]. These findings show that treatment regimens designed to lower plasma homocysteine significantly reduce cardiovascular risk in severely hyperhomocysteinemic patients due to cystathionine β-synthase deficiency despite imperfect biochemical control.

Some encouraging short-term clinical studies investigated the effects of vitamin supplementation on surrogate markers of vascular disease in patients with mild hyperhomocysteinemia. Plasma levels of von Willebrand factor and soluble thrombomodulin are established biochemical markers of endothelial injury or activation [186, 187] and are increased in patients with hyperhomocysteinemia [119]. Treatment of young patients with peripheral arterial occlusive disease and elevated plasma homocysteine levels with folic acid (5 mg per day) and pyridoxine (250 mg per day) in an uncontrolled, longitudinal study decreased plasma homocysteine levels and significantly reduced the plasma levels of von Willebrand factor and thrombomodulin [188]. An uncontrolled, longitudinal study on the progression of carotid plaque area in patients with no atherosclerotic risk factors except homocysteine levels greater than 14 μmol/L demonstrated that the atherosclerotic plaque area increased in patients with hyperhomocysteinemia at a rate of 0.31 ± 0.39 cm^2/year, but remained stable after initiation of vitamin supplementation (rate of progression -0.05 ± 0.25 cm^2/year; $P = 0.02$) [189]. In a randomized and placebo-controlled trial among 158 healthy siblings of patients with premature atherothrombotic disease, participants were treated with folic acid and vitamin B_6 for 2 years. Approximately two third of the participants had elevated plasma homocysteine levels at baseline that significantly decreased with treatment. Vitamin treatment was associated with a decreased rate of abnormal exercise stress tests (odds ratio 0.40 [0.17 to 0.93]; $P = 0.035$) [190].

The effect of oral folic acid supplementation in hyperhomocysteinemic subjects on endothelial function as assessed by flow-mediated endothelium-dependent dilation of the brachial arterial using high resolution ultrasound has been investigated in several double-blind, randomized, placebo-controlled crossover trials. Folic acid supplementation (5 mg/day for six weeks) compared to placebo lowered total plasma homocysteine levels (8.7 ± 2.5 vs. 12.1 ± 3.6 μmol/L; $P = 0.003$) and improved arterial endothelial function (increase in diameter + 0.08 ± 0.05 mm vs. 0.04 ± 0.04 mm; P = 0.015) in healthy adults without conventional atherosclerotic risk factors but mildly elevated homocysteine levels at baseline (> 13 μmol/L) [191]. These findings were reproduced by others in similarly designed studies [92, 192] and also in subjects with acute hyperhomocysteinemia following an oral methionine challenge [193], or in patients with existing coronary artery disease and elevated plasma homocysteine levels [194-197]. This effect may be a consequence of an increase in bioavailable nitric oxide after folic acid supplementation as indicated by improved endothelium-dependent vasodilation but not of endothelium-independent vasodilation in all cited studies and by an increase in the plasma level of nitric oxide-derived endproducts by folic acid supplementation [92]. The improvement of flow-mediated dilation correlates closely with the reduction in free plasma homocysteine [195], independent of changes in total or protein-bound homocysteine, folate, and vitamin B_{12} [195, 196], supporting the view that reduced homocysteine is the species that is deleterious to the endothelium. Some studies that investigated the effect of vitamin supplementation on endothelial dysfunction in hyperhomocysteinemic subjects, however, could not document a positive effect [198-200], which can partly be explained by insufficient lowering of plasma homocysteine levels by folic acid in patients with predialysis renal failure (post-treatment homocysteine levels: 15.1 [14.1 to 16.2] μmol/L) [198], use of lower doses of folic acid compared to positive studies (400 μg supplements) [199], or inadequate sample size [201].

In addition, recent data suggest that folic acid may improve endothelial function and nitric oxide synthesis by mechanisms independent from its homocysteine-lowering effect [202]. Recent evidence suggests that 5-methyltetrahydrofolate, the reduced form of dietary folic acid, may bind to a pteridine binding domain in nitric oxide synthases that is similar to the folate-binding site of dihydrofolate reductase. This may facilitate tetrahydrobiopterin's radical-mediated reductive activation of the ferrous heme-O_2 moiety of the enzyme, which is involved in the catalytic formation of nitric oxide, and may thereby increase nitric oxide production [203].

In conclusion, the studies reviewed here provide evidence that lowering plasma homocysteine levels by B-vitamin supplementation improves endothelial dysfunction as indicated by a reduction of elevated biochemical markers of endothelial injury or activation, and by an improvement of endothelium-dependent vasodilator function. These encouraging findings support the need for large, randomized, controlled trials aimed at investigating the effect of vitamin supplementation on clinical endpoints of atherothrombotic vascular disease.

The results of the first published intervention trials, however, gave some conflicting results. A prospective, double-blind, randomized trial investigated the effect of lowering plasma homocysteine levels with a combination of folic acid (1 mg), vitamin B_{12} (400 μg), and pyridoxine (10 mg) or placebo on restenosis after successful coronary angioplasty in 205 patients. After six months of follow-up, the minimal luminal diameter was significantly greater in the vitamin-treated group (1.72 ± 0.76 vs. 1.45 ± 0.88 mm; P = 0.02), and was accompanied by a significantly lower rate of restenosis and need for revascularization of the target lesion compared with placebo-treated patients [204]. This effect could also be observed in the subgroup of patients with percutaneous intervention of small coronary arteries (reference diameter 2.9 mm), which tend to have a worse outcome than lesions in wider arteries [205]. Whether or not this favorable effect of homocysteine-lowering on restenosis translates in a reduction of major cardiac events or mortality is not known yet.

In the first published secondary intervention study with folic acid in patients with stable coronary artery disease on statin treatment that looked at hard clinical outcomes, nearly 600 patients were randomized to either open-label folic acid 0.5 mg per day or standard care [206]. Folic acid treatment significantly decreased plasma homocysteine levels by 18 % but did not reduce the primary end point of all cause mortality and a composite of vascular events. As baseline plasma homocysteine levels were a significant risk factor in a multifactorial survival model with adjustments for clinical factors, this study implies that elevated plasma homocysteine may be a risk factor but not a modifiable risk factor for an adverse outcome in these kind of patients.

Before any firm conclusions can be drawn on the use of homocysteine lowering strategies for the prevention of clinical consequences of atherothrombotic diseases, the results of more trials have to become available.

In addition, plasma homocysteine levels cannot be normalized by vitamin supplementation in all cases. For example, hyperhomocysteinemia persists even with folic acid treatment (1-5 mg/day) in many patients with chronic renal failure [207]. Alternative therapeutic options directed at reducing the cardiovascular risk associated with hyperhomocysteinemia are warranted in these subgroups of patients. Antioxidant therapy seems to be one such strategy. In this context it is worth mentioning that, although the results of trials of antioxidants for reducing cardiovascular morbidity and mortality in the general population were disappointing and did not show a benefit [208], there are two clinical trials in patients with chronic renal failure that showed a significant benefit of α-tocopherol or the thiol antioxidant N-acetylcysteine on cardiovascular morbidity and mortality [209, 210]. Whether or not this relates to an effect of antioxidants on improving homocysteine-induced endothelial dysfunction in patients with chronic renal failure remains to be determined. Nevertheless the experimental data reviewed here and the results of these two clinical trials encourage further evaluation of the effect of antioxidant strategies in selected groups of patients at risk for cardiovascular disease, as in patients with hyperhomocysteinemia who do not respond to folic acid supplementation.

References

[1] Weiss N, Keller C, Hoffmann U, Losclazo J. Endothelial dysfunction and atherothrombosis in mild hyperhomocysteinemia. *Vasc Med* 2002;7(3):227-39.

[2] McCully KS. Vascular pathology of homocysteinemia: implications for the pathogenesis of arteriosclerosis. *Am J Pathol* 1969;56(1):111-28.

[3] Mudd SH, Levy HL, Skovby F. Disorders of Transsulfuration. In: Scriver CR, Beaudet AL, Sly WS, Valle D, editors. *The metabolic and molecular bases of inherited diseases*. New York: McGraw-Hill Company; 1995. p. 1279-327.

[4] Mudd SH, Skoby F, Levy HL, Pettigrew LKD, Wilcken B, Pyeritz R.E., et al. The natural history of homocystinuria due to cystathionine beta-synthase deficiency. *Am J Hum Genet* 1985;37(1):1-31.

[5] Yap S, Boers GH, Wilcken B, Wilcken DE, Brenton DP, Lee PJ, et al. Vascular outcome in patients with homocystinuria due to cystathionine beta-synthase deficiency treated chronically: a multicenter observational study. *Arterioscler Thromb Vasc Biol* 2001;21(12):2080-5.

[6] Graham IM, Daly LE, Refsum HM, Robinson K, Brattstrom LE, Ueland PM, et al. Plasma homocysteine as a risk factor for vascular disease. The European Concerted Action Project. *JAMA* 1997;277(22):1775-81.

[7] Kang SS, Wong PWK, Malinow MR. Hyperhomocyst(e)inemia as a risk factor for occlusive vascular disease. *Annu Rev Nutr* 1992;12:279-98.

[8] Boushey CJ, Beresford SAA, Omenn GS, Motulsky AG. A quantitative assessement of plasma homocysteine as a risk factor for vascular disease. *JAMA* 1995;274(13):1049-57.

[9] Arnesen E, Refsum H, Bonaa KH, Ueland PM, Forde OH, Nordrehaug JE. Serum total homocysteine and coronary heart disease. *Int J Epidemiol* 1995;24(4):704-9.

[10] A'Brook R, Tavendale R, Tunstall-Pedoe H. Homocysteine and coronary risk in the general population: analysis from the Scottish Heart Health Study and Scottish MONICA surveys. *Eur Heart J* 1998;19 (Suppl):8 (Abstr).

[11] Bostom AG, Silbershatz H, Rosenberg IH, Selhub J, D'Agostino RB, Wolf PA, et al. Nonfasting plasma total homocysteine levels and all-cause and cardiovascular disease mortality in elderly Framingham men and women. *Arch Intern Med* 1999;159(10):1077-80.

[12] Bots ML, Launer LJ, Lindemans J, Hoes AW, Hofman A, Witteman JC, et al. Homocysteine and short-term risk of myocardial infarction and stroke in the elderly: the Rotterdam Study. *Arch Intern Med* 1999;159(1):38-44.

[13] Kark JD, Selhub J, Adler B, Gofin J, Abramson JH, Friedman G, et al. Nonfasting plasma total homocysteine level and mortality in middle-aged and elderly men and women in Jerusalem. *Ann Intern Med* 1999;131(5):321-30.

[14] Moustapha A, Naso A, Nahlawi M, Gupta A, Arheart KL, Jacobsen DW, et al. Prospective study of hyperhomocysteinemia as an adverse cardiovascular risk factor in end-stage renal disease. *Circulation* 1998;97(2):138-41.

[15] Nygard O, Nordrehaug JE, Refsum H, Ueland PM, Farstad M, Vollset SE. Plasma homocysteine levels and mortality in patients with coronary artery disease. *N Engl J Med* 1997;337(4):230-36.

[16] Perry IJ, Refsum H, Morris RW, Ebrahim SB, Ueland PM, Shaper AG. Prospective study of serum total homocysteine concentration and risk of stroke in middle-aged British men. *Lancet* 1995;346(8987):1395-8.

[17] Petri M, Roubenoff R, Dallal GE, Nadeau MR, Selhub J, Rosenberg IH. Plasma homocysteine as a risk factor for atherothrombotic events in systemic lupus erythematosus. *Lancet* 1996;348(9035):1120-24.

[18] Ridker PM, Manson JE, Buring JE, Shih J, Matias M, Hennekens CH. Homocysteine and risk of cardiovascular disease among postmenopausal women. *JAMA* 1999;281(19):1817-21.

[19] Stampfer MJ, Malinow MR, Willett WC, Newcomer LM, Upson B, Ullmann D, et al. A prospective study of plasma homocyst(e)ine and risk of myocardial infarction in US physicians. *JAMA* 1992;268(7):877-81.

[20] Wald NJ, Watt HC, Law MR, Weir DG, McPartlin J, Scott JM. Homocysteine and ischemic heart disease: results of a prospective study with implications regarding prevention. *Arch Intern Med* 1998;158(8):862-7.

[21] Whincup PH, Refsum H, Perry IJ, Morris R, Walker M, Lennon L, et al. Serum total homocysteine and coronary heart disease: prospective study in middle aged men. *Heart* 1999;82(4):448-54.

[22] Alfthan G, Pekkanen J, Jauhiainen M, Pitkaniemi J, Karvonen M, Tuomilehto J, et al. Relation of serum homocysteine and lipoprotein(a) concentrations to atherosclerotic disease in a prospective Finnish population based study. *Atherosclerosis* 1994;106(1):9-19.

[23] Evans RW, Shaten BJ, Hempel JD, Cutler JA, Kuller LH. Homocyst(e)ine and risk of cardiovascular disease in the Multiple Risk Factor Intervention Trial. *Arterioscler Thromb Vasc Biol* 1997;17(10):1947-53.

[24] Folsom AR, Nieto FJ, McGovern PG, Tsai MY, Malinow MR, Eckfeldt JH, et al. Prospective study of coronary heart disease incidence in relation to fasting total homocysteine, related genetic polymorphisms, and B vitamins: the Atherosclerosis Risk in Communities (ARIC) study. *Circulation* 1998;98(3):204-10.

[25] Kuller LH, Evans RW. Homocysteine, vitamins, and cardiovascular disease. *Circulation* 1998;98(3):196-9.

[26] Ubbink JB, Fehily AM, Pickering J, Elwood PC, Vermaak WJ. Homocysteine and ischaemic heart disease in the Caerphilly cohort. *Atherosclerosis* 1998;140(2):349-56.

[27] Verhoef P, Hennekens CH, Malinow MR, Kok FJ, Willett WC, Stampfer MJ. A prospective study of plasma homocyst(e)ine and risk of ischemic stroke. *Stroke* 1994;25(10):1924-30.

[28] Hopkins PN, Wu LL, Wu J, Hunt SC, James BC, Vincent GM, et al. Higher plasma homocyst(e)ine and increased susceptibility to adverse effects of low folate in early familial coronary artery disease. *Arterioscler Thromb Vasc Biol* 1995;15(9):1314-20.

[29] Pancharuniti N, Lewis CA, Sauberlich HE, Perkins LL, Go RC, Alvarez JO, et al. Plasma homocyst(e)ine, folate, and vitamin B-12 concentrations and risk for early-onset coronary artery disease. *Am J Clin Nutr* 1994;59(4):940-8.

[30] Danesh J, Lewington S. Plasma homocysteine and coronary heart disease: systematic review of published epidemiological studies. *J Cardiovasc Risk* 1998;5(4):229-32.

[31] Moller J, Nielsen GM, Tvedegaard KC, Andersen NT, Jorgensen PE. A meta-analysis of cerebrovascular disease and hyperhomocysteinaemia. *Scand J Clin Lab Invest* 2000;60(6):491-9.

[32] Ueland PM, Refsum H, Beresford SA, Vollset SE. The controversy over homocysteine and cardiovascular risk. *Am J Clin Nutr* 2000;72(2):324-32.

[33] Schnyder G, Pin R, Roffi M, Flammer Y, Hess OM. Association of plasma homocysteine with the number of major coronary arteries severely narrowed. *Am J Cardiol* 2001;88(9):1027-30.

[34] Alfthan G, Aro A, Gey KF. Plasma homocysteine and cardiovascular disease mortality. *Lancet* 1997;349(9049):397.

[35] Anderson JL, Muhlestein JB, Horne BD, Carlquist JF, Bair TL, Madsen TE, et al. Plasma homocysteine predicts mortality independently of traditional risk factors and C-reactive protein in patients with angiographically defined coronary artery disease. *Circulation* 2000;102(11):1227-32.

[36] Christen WG, Ajani UA, Glynn RJ, Hennekens CH. Blood levels of homocysteine and increased risks of cardiovascular disease: causal or casual? *Arch Intern Med* 2000;160(4):422-34.

[37] Hoogeveen EK, Kostense PJ, Jakobs C, Dekker JM, Nijpels G, Heine RJ, et al. Hyperhomocysteinemia increases risk of death, especially in type 2 diabetes : 5-year follow-up of the Hoorn Study. *Circulation* 2000;101(13):1506-11.

[38] Stehouwer CD, Gall MA, Hougaard P, Jakobs C, Parving HH. Plasma homocysteine concentration predicts mortality in non-insulin-dependent diabetic patients with and without albuminuria. *Kidney Int* 1999;55(1):308-14.

[39] Taylor LM, Jr., Moneta GL, Sexton GJ, Schuff RA, Porter JM. Prospective blinded study of the relationship between plasma homocysteine and progression of symptomatic peripheral arterial disease. *J Vasc Surg* 1999;29(1):8-19.

[40] Streifler JY, Rosenberg N, Chetrit A, Eskaraev R, Sela BA, Dardik R, et al. Cerebrovascular events in patients with significant stenosis of the carotid artery are associated with hyperhomocysteinemia and platelet antigen-1 (Leu33Pro) polymorphism. *Stroke* 2001;32(12):2753-8.

[41] Chao CL, Tsai HH, Lee CM, Hsu SM, Kao JT, Chien KL, et al. The graded effect of hyperhomocysteinemia on the severity and extent of coronary atherosclerosis. *Atherosclerosis* 1999;147(2):379-86.

[42] van den Berg M, Stehouwer CD, Bierdrager E, Rauwerda JA. Plasma homocysteine and severity of atherosclerosis in young patients with lower-limb atherosclerotic disease. *Arterioscler Thromb Vasc Biol* 1996;16(1):165-71.

[43] Verhoef P, Kok FJ, Kruyssen DA, Schouten EG, Witteman JC, Grobbee DE, et al. Plasma total homocysteine, B vitamins, and risk of coronary atherosclerosis. *Arterioscler Thromb Vasc Biol* 1997;17(5):989-95.

[44] von Eckardstein A, Malinow MR, Upson B, Heinrich J, Schulte H, Schonfeld R, et al. Effects of age, lipoproteins, and hemostatic parameters on the role of homocyst(e)inemia as a cardiovascular risk factor in men. *Arterioscler Thromb* 1994;14(3):460-4.

[45] Malinow MR, Nieto FJ, Szklo M, Chambless LE, Bond G. Carotid artery intimal-medial wall thickening and plasma homocyst(e)ine in asymptomatic adults. The Atherosclerosis Risk in Communities Study. *Circulation* 1993;87(4):1107-13.

[46] Selhub J, Jacques PF, Bostom AG, D'Agostino RB, Wilson PW, Belanger AJ, et al. Association between plasma homocysteine concentrations and extracranial carotid-artery stenosis. *N Engl J Med* 1995;332(5):286-91.

[47] Willinek WA, Ludwig M, Lennarz M, Holler T, Stumpe KO. High-normal serum homocysteine concentrations are associated with an increased risk of early

atherosclerotic carotid artery wall lesions in healthy subjects. *J Hypertens* 2000;18(4):425-30.

[48] Vane JR, Anggard EE, Botting RM. Regulatory functions of the vascular endothelium. *N Engl J Med* 1990;323(1):27-36.

[49] Worthley SG, Osende JI, Helft G, Badimon JJ, Fuster V. Coronary artery disease: pathogenesis and acute coronary syndromes. *Mt Sinai J Med* 2001;68(3):167-81.

[50] Vita JA, Treasure CB, Nabel EG, McLenachan JM, Fish RD, Yeung AC, et al. Coronary vasomotor response to acetylcholine relates to risk factors for coronary artery disease. *Circulation* 1990;81(2):491-7.

[51] Nishimura RA, Lerman A, Chesebro JH, Ilstrup DM, Hodge DO, Higano ST, et al. Epicardial vasomotor responses to acetylcholine are not predicted by coronary atherosclerosis as assessed by intracoronary ultrasound. *J Am Coll Cardiol* 1995;26(1):41-9.

[52] Luscher TF, Noll G. Endothelial function as an end-point in interventional trials: concepts, methods and current data. *J Hypertens Suppl* 1996;14(2):S111-9; discussion S19-21.

[53] Edmundowicz D. Noninvasive Studies of Coronary and Peripheral Arterial Bloodflow. *Curr Atheroscler Rep* 2002;4(5):381-5.

[54] Ludmer PL, Selwyn AP, Shook TL, Wayne RR, Mudge GH, Alexander RW, et al. Paradoxical vasoconstriction induced by acetylcholine in atherosclerotic coronary arteries. *N Engl J Med* 1986;315(17):1046-51.

[55] Cox DA, Vita JA, Treasure CB, Fish RD, Alexander RW, Ganz P, et al. Atherosclerosis impairs flow-mediated dilation of coronary arteries in humans. *Circulation* 1989;80(3):458-65.

[56] Zeiher AM, Drexler H, Wollschlager H, Just H. Modulation of coronary vasomotor tone in humans. Progressive endothelial dysfunction with different early stages of coronary atherosclerosis. *Circulation* 1991;83(2):391-401.

[57] Quyyumi AA, Dakak N, Andrews NP, Husain S, Arora S, Gilligan DM, et al. Nitric oxide activity in the human coronary circulation. Impact of risk factors for coronary atherosclerosis. *J Clin Invest* 1995;95(4):1747-55.

[58] Yataco AR, Corretti MC, Gardner AW, Womack CJ, Katzel LI. Endothelial reactivity and cardiac risk factors in older patients with peripheral arterial disease. *Am J Cardiol* 1999;83(5):754-8.

[59] Angerer P, Negut C, Stork S, von Schacky C. Endothelial function of the popliteal artery in patients with coronary artery disease. *Atherosclerosis* 2001;155(1):187-93.

[60] Kuvin JT, Patel AR, Sliney KA, Pandian NG, Rand WM, Udelson JE, et al. Peripheral vascular endothelial function testing as a noninvasive indicator of coronary artery disease. *J Am Coll Cardiol* 2001;38(7):1843-9.

[61] Chan SY, Mancini GB, Kuramoto L, Schulzer M, Frohlich J, Ignaszewski A. The prognostic importance of endothelial dysfunction and carotid atheroma burden in patients with coronary artery disease. *J Am Coll Cardiol* 2003;42(6):1037-43.

[62] McLenachan JM, Williams JK, Fish RD, Ganz P, Selwyn AP. Loss of flow-mediated endothelium-dependent dilation occurs early in the development of atherosclerosis. *Circulation* 1991;84(3):1273-8.

[63] Schindler TH, Hornig B, Buser PT, Olschewski M, Magosaki N, Pfisterer M, et al. Prognostic value of abnormal vasoreactivity of epicardial coronary arteries to sympathetic stimulation in patients with normal coronary angiograms. *Arterioscler Thromb Vasc Biol* 2003;23(3):495-501.

[64] Schachinger V, Britten MB, Zeiher AM. Prognostic impact of coronary vasodilator dysfunction on adverse long- term outcome of coronary heart disease. *Circulation* 2000;101(16):1899-906.

[65] Suwaidi JA, Hamasaki S, Higano ST, Nishimura RA, Holmes DR, Jr., Lerman A. Long-term follow-up of patients with mild coronary artery disease and endothelial dysfunction. *Circulation* 2000;101(9):948-54.

[66] Heitzer T, Schlinzig T, Krohn K, Meinertz T, Munzel T. Endothelial dysfunction, oxidative stress, and risk of cardiovascular events in patients with coronary artery disease. *Circulation* 2001;104(22):2673-8.

[67] Guerra R, Jr., Brotherton AF, Goodwin PJ, Clark CR, Armstrong ML, Harrison DG. Mechanisms of abnormal endothelium-dependent vascular relaxation in atherosclerosis: implications for altered autocrine and paracrine functions of EDRF. *Blood Vessels* 1989;26(5):300-14.

[68] Luscher TF, Richard V, Tschudi M, Yang ZH, Boulanger C. Endothelial control of vascular tone in large and small coronary arteries. *J Am Coll Cardiol* 1990;15(3):519-27.

[69] Cooke JP, Stamler J, Andon N, Davies PF, McKinley G, Loscalzo J. Flow stimulates endothelial cells to release a nitrovasodilator that is potentiated by reduced thiol. *Am J Physiol* 1990;259(3 Pt 2):H804-12.

[70] Ignarro LJ, Harbison RG, Wood KS, Kadowitz PJ. Activation of purified soluble guanylate cyclase by endothelium-derived relaxing factor from intrapulmonary artery and vein: stimulation by acetylcholine, bradykinin and arachidonic acid. *J Pharmacol Exp Ther* 1986;237(3):893-900.

[71] Napoli C, Ignarro LJ. Nitric oxide and atherosclerosis. *Nitric Oxide* 2001;5(2):88-97.

[72] Lentz SR, Sobey CG, Piegors DJ, Bhopatkar MY, Faraci FM, Malinow MR, et al. Vascular dysfunction in monkeys with diet-induced hyperhomocyst(e)inemia. *J Clin Invest* 1996;98(1):24-9.

[73] Ungvari Z, Pacher P, Rischak K, Szollar L, Koller A. Dysfunction of nitric oxide mediation in isolated rat arterioles with methionine diet-induced hyperhomocysteinemia. *Arterioscler Thromb Vasc Biol* 1999;19(8):1899-904.

[74] Bagi Z, Ungvari Z, Szollar L, Koller A. Flow-induced constriction in arterioles of hyperhomocysteinemic rats is due to impaired nitric oxide and enhanced thromboxane A(2) mediation. *Arterioscler Thromb Vasc Biol* 2001;21(2):233-7.

[75] Eberhardt RT, Forgione MA, Cap A, Leopold JA, Rudd MA, Trolliet M, et al. Endothelial dysfunction in a murine model of mild hyperhomocyst(e)inemia. *J Clin Invest* 2000;106(4):483-91.

[76] Weiss N, Zhang YY, Heydrick S, Bierl C, Loscalzo J. Overexpression of cellular glutathione peroxidase rescues homocyst(e)ine-induced endothelial dysfunction. *Proc Natl Acad Sci U S A* 2001;98(22):12503-8.

[77] Weiss N, Heydrick S, Zhang YY, Bierl C, Cap A, Loscalzo J. Cellular redox state and endothelial dysfunction in mildly hyperhomocysteinemic cystathionine beta-synthase-deficient mice. *Arterioscler Thromb Vasc Biol* 2002;22(1):34-41.

[78] Lentz SR, Erger RA, Dayal S, Maeda N, Malinow MR, Heistad DD, et al. Folate dependence of hyperhomocysteinemia and vascular dysfunction in cystathionine beta-synthase-deficient mice. *Am J Physiol Heart Circ Physiol* 2000;279(3):H970-5.

[79] Bellamy MF, McDowell IF, Ramsey MW, Brownlee M, Bones C, Newcombe RG, et al. Hyperhomocysteinemia after an oral methionine load acutely impairs endothelial function in healthy adults. *Circulation* 1998;98(18):1848-52.

[80] Hanratty CG, McAuley DF, McGurk C, Young IS, Johnston GD. Homocysteine and endothelial vascular function. *Lancet* 1998;351(9111):1288-9.

[81] Chambers JC, McGregor A, Jean-Marie J, Obeid OA, Kooner JS. Demonstration of rapid onset vascular endothelial dysfunction after hyperhomocysteinemia: an effect reversible with vitamin C therapy. *Circulation* 1999;99(9):1156-60.

[82] Chambers JC, Obeid OA, Kooner JS. Physiological increments in plasma homocysteine induce vascular endothelial dysfunction in normal human subjects. *Arterioscler Thromb Vasc Biol* 1999;19(12):2922-7.

[83] Kanani PM, Sinkey CA, Browning RL, Allaman M, Knapp HR, Haynes WG. Role of oxidant stress in endothelial dysfunction produced by experimental hyperhomocyst(e)inemia in humans. *Circulation* 1999;100(11):1161-8.

[84] Lambert J, van den Berg M, Steyn M, Rauwerda JA, Donker AJ, Stehouwer CD. Familial hyperhomocysteinaemia and endothelium-dependent vasodilatation and arterial distensibility of large arteries. *Cardiovasc Res* 1999;42(3):743-51.

[85] Chao CL, Kuo TL, Lee YT. Effects of methionine-induced hyperhomocysteinemia on endothelium-dependent vasodilation and oxidative status in healthy adults. *Circulation* 2000;101(5):485-90.

[86] Boger RH, Lentz SR, Bode-Boger SM, Knapp HR, Haynes WG. Elevation of asymmetrical dimethylarginine may mediate endothelial dysfunction during experimental hyperhomocyst(e)inaemia in humans. *Clin Sci (Lond)* 2001;100(2):161-7.

[87] Chambers JC, Ueland PM, Wright M, Dore CJ, Refsum H, Kooner JS. Investigation of relationship between reduced, oxidized, and protein-bound homocysteine and vascular endothelial function in healthy human subjects. *Circ Res* 2001;89(2):187-92.

[88] Hanratty CG, McGrath LT, McAuley DF, Young IS, Johnston GD. The effects of oral methionine and homocysteine on endothelial function. *Heart* 2001;85(3):326-30.

[89] Raghuveer G, Sinkey CA, Chenard C, Stumbo P, Haynes WG. Effect of vitamin E on resistance vessel endothelial dysfunction induced by methionine. *Am J Cardiol* 2001;88(3):285-90.

[90] Woo KS, Chook P, Lolin YI, Cheung AS, Chan LT, Sun YY, et al. Hyperhomocyst(e)inemia is a risk factor for arterial endothelial dysfunction in humans. *Circulation* 1997;96(8):2542-4.

[91] Tawakol A, Omland T, Gerhard M, Wu JT, Creager MA. Hyperhomocyst(e)inemia is associated with impaired endothelium- dependent vasodilation in humans. *Circulation* 1997;95(5):1119-21.

[92] Holven KB, Holm T, Aukrust P, Christensen B, Kjekshus J, Andreassen AK, et al. Effect of folic acid treatment on endothelium-dependent vasodilation and nitric oxide-derived end products in hyperhomocysteinemic subjects. *Am J Med* 2001;110(7):536-42.

[93] Tawakol A, Forgione MA, Stuehlinger M, Alpert NM, Cooke JP, Loscalzo J, et al. Homocysteine impairs coronary microvascular dilator function in humans. *J Am Coll Cardiol* 2002;40(6):1051-58.

[94] Upchurch GR, Jr., Welch GN, Fabian AJ, Freedman JE, Johnson JL, Keaney JF, Jr., et al. Homocyst(e)ine decreases bioavailable nitric oxide by a mechanism involving glutathione peroxidase. *J Biol Chem* 1997;272(27):17012-7.

[95] Zhang X, Li H, Jin H, Ebin Z, Brodsky S, Goligorsky MS. Effects of homocysteine on endothelial nitric oxide production. *Am J Physiol Renal Physiol* 2000;279(4):F671-8.

[96] Li N, Yi FX, Rute E, Zhang DX, Slocum GR, Zou AP. Effects of homocysteine on intracellular nitric oxide and superoxide levels in the renal arterial endothelium. *Am J Physiol Heart Circ Physiol* 2002;283(3):H1237-43.

[97] Ross R. Atherosclerosis - an inflammatory disease. *N Engl J Med* 1999;340(2):115-26.

[98] Poddar R, Sivasubramanian N, DiBello PM, Robinson K, Jacobsen DW. Homocysteine induces expression and secretion of monocyte chemoattractant protein-1 and interleukin-8 in human aortic endothelial cells: implications for vascular disease. *Circulation* 2001;103(22):2717-23.

[99] Postea O, Kroetz F, Hillebrand S, Banas B, Zahler S, Keller C, et al. Sterospecific and redox-sensitve increase in monocyte adhesion to endothelial cells by homocysteine. *J Inherit Metab Dis* 2003;26(S1):20.

[100] Wang G, O K. Homocysteine stimulates the expression of monocyte chemoattractant protein-1 receptor (CCR2) in human monocytes: possible involvement of oxygen free radicals. *Biochem J* 2001;357(Pt 1):233-40.

[101] Hofmann MA, Lalla E, Lu Y, Ryu Gleason M, Wolf BM, Tanji N, et al. Hyperhomocysteinemia enhances vascular inflammation and accelerates atherosclerosis in a murine model. *J Clin Invest* 2001;107(6):675-83.

[102] Stamler JS, Osborne JA, Jaraki O, Rabbani LE, Mullins M, Singel D, et al. Adverse vascular effects of homocysteine are modulated by endothelium- derived relaxing factor and related oxides of nitrogen. *J Clin Invest* 1993;91(1):308-18.

[103] Graeber JE, Slott JH, Ulane RE, Schulman JD, Stuart MJ. Effect of homocysteine and homocystine on platelet and vascular arachidonic acid metabolism. *Pediatr Res* 1982;16(6):490-3.

[104] Panganamala RV, Karpen CW, Merola AJ. Peroxide mediated effects of homocysteine on arterial prostacyclin synthesis. *Prostaglandins Leukot Med* 1986;22(3):349-56.

[105] Wang J, Dudman NP, Wilcken DE. Effects of homocysteine and related compounds on prostacyclin production by cultured human vascular endothelial cells. *Thromb Haemost* 1993;70(6):1047-52.

[106] Fryer RH, Wilson BD, Gubler DB, Fitzgerald LA, Rodgers GM. Homocysteine, a risk factor for premature vascular disease and thrombosis, induces tissue factor activity in endothelial cells. *Arterioscler Thromb* 1993;13(9):1327-33.

[107] Rodgers GM, Kane WH. Activation of endogenous factor V by a homocysteine-induced vascular endothelial cell activator. *J Clin Invest* 1986;77(6):1909-16.

[108] Lentz SR, Sadler JE. Inhibition of thrombomodulin surface expression and protein C activation by the thrombogenic agent homocysteine. *J Clin Invest* 1991;88(6):1906-14.

[109] Lentz SR, Malinow MR, Piegors DJ, Bhopatkar-Teredesai M, Faraci FM, Heistad DD. Consequences of hyperhomocyst(e)inemia on vascular function in atherosclerotic monkeys. *Arterioscler Thromb Vasc Biol* 1997;17(11):2930-4.

[110] Rodgers GM, Conn MT. Homocysteine, an atherogenic stimulus, reduces protein C activation by arterial and venous endothelial cells. *Blood* 1990;75(4):895-901.

[111] Undas A, Williams EB, Butenas S, Orfeo T, Mann KG. Homocysteine inhibits inactivation of factor Va by activated protein C. *J Biol Chem* 2001;276(6):4389-97.

[112] Nishinaga M, Ozawa T, Shimada K. Homocysteine, a thrombogenic agent, suppresses anticoagulant heparan sulfate expression in cultured porcine aortic endothelial cells. *J Clin Invest* 1993;92(3):1381-6.

[113] Hajjar KA. Homocysteine-induced modulation of tissue plasminogen activator binding to its endothelial cell membrane receptor. *J Clin Invest* 1993;91(6):2873-9.

[114] Hajjar KA, Mauri L, Jacovina AT, Zhong F, Mirza UA, Padovan JC, et al. Tissue plasminogen activator binding to the annexin II tail domain. Direct modulation by homocysteine. *J Biol Chem* 1998;273(16):9987-93.

[115] Durand P, Prost M, Blache D. Pro-thrombotic effects of a folic acid deficient diet in rat platelets and macrophages related to elevated homocysteine and decreased n-3 polyunsaturated fatty acids. *Atherosclerosis* 1996;121(2):231-43.

[116] Durand P, Lussier-Cacan S, Blance D. Acute methionine load-induced hyperhomocysteinemia enhances platelet aggregation, thromboxane biosynthesis, and macrophage-derived tissue factor activity in rats. *FASEB J* 1997;11(13):1157-68.

[117] Di Minno G, Davi G, Margaglione M, Cirillo F, Grandone E, Ciabattoni G, et al. Abnormally high thromboxane biosynthesis in homozygous homocystinuria. Evidence for platelet involvement and probucol-sensitive mechanism. *J Clin Invest* 1993;92(3):1400-6.

[118] Davi G, Di Minno G, Coppola A, Andria G, Cerbone AM, Madonna P, et al. Oxidative stress and platelet activation in homozygous homocystinuria. *Circulation* 2001;104(10):1124-8.

[119] Freyburger G, Labrouche S, Sassoust G, Rouanet F, Javorschi S, Parrot F. Mild hyperhomocysteinemia and hemostatic factors in patients with arterial vascular diseases. *Thromb Haemost* 1997;77(3):466-71.

[120] Constans J, Blann AD, Resplandy F, Parrot F, Seigneur M, Renard M, et al. Endothelial dysfunction during acute methionine load in hyperhomocysteinaemic patients. *Atherosclerosis* 1999;147(2):411-3.

[121] Zhou J, Moller J, Danielsen CC, Bentzon J, Ravn HB, Austin RC, et al. Dietary supplementation with methionine and homocysteine promotes early atherosclerosis but not plaque rupture in ApoE-deficient mice. *Arterioscler Thromb Vasc Biol* 2001;21(9):1470-6.

[122] Wang H, Jiang X, Yang F, Gaubatz JW, Ma L, Magera MJ, et al. Hyperhomocysteinemia accelerates atherosclerosis in cystathionine beta-synthase and

apolipoprotein E double knock-out mice with and without dietary perturbation. *Blood* 2003;101(10):3901-7.

[123] Lang D, Kredan MB, Moat SJ, Hussain SA, Powell CA, Bellamy MF, et al. Homocysteine-induced inhibition of endothelium-dependent relaxation in rabbit aorta: role for superoxide anions. *Arterioscler Thromb Vasc Biol* 2000;20(2):422-7.

[124] Pritchard KA, Jr., Groszek L, Smalley DM, Sessa WC, Wu M, Villalon P, et al. Native low-density lipoprotein increases endothelial cell nitric oxide synthase generation of superoxide anion. *Circ Res* 1995;77(3):510-8.

[125] Gryglewski RJ, Palmer RM, Moncada S. Superoxide anion is involved in the breakdown of endothelium-derived vascular relaxing factor. *Nature* 1986;320:454-6.

[126] Mujumdar VS, Aru GM, Tyagi SC. Induction of oxidative stress by homocyst(e)ine impairs endothelial function. *J Cell Biochem* 2001;82(3):491-500.

[127] Wall RT, Harlan JM, Harker LA, Striker GE. Homocysteine-induced endothelial cell injury in-vitro: a model for the study of vascular injury. *Thromb Res* 1980;18(1-2):113-21.

[128] Starkebaum G, Harlan JM. Endothelial cell injury due to copper-catalyzed hydrogen peroxide generation from homocysteine. *J Clin Invest* 1986;77(4):1370-6.

[129] Toborek M, Hennig B. Dietary methionine imbalance, endothelial cell dysfunction and atherosclerosis. *Nutr Res* 1996;16:1251-66.

[130] Chern CL, Huang RF, Chen YH, Cheng JT, Liu TZ. Folate deficiency-induced oxidative stress and apoptosis are mediated via homocysteine-dependent overproduction of hydrogen peroxide and enhanced activation of NF-kappaB in human Hep G2 cells. *Biomed Pharmacother* 2001;55(8):434-42.

[131] Mato JM, Avila MA, Corrales FJ. Biosynthesis of S-adenosylmethionine. In: Carmel R, Jacobsen DW, editors. *Homocysteine in health and disease*. Cambridge: Cambrige University Press; 2001. p. 47-62.

[132] Mansoor MA, Svardal AM, Schneede J, Ueland PM. Dynamic relation between reduced, oxidized, and protein-bound homocysteine and other thiol components in plasma during methionine loading in healthy men. *Clin Chem* 1992;38(7):1316-21.

[133] Guttormsen AB, Mansoor AM, Fiskerstrand T, Ueland PM, Refsum H. Kinetics of plasma homocysteine in healthy subjects after peroral homocysteine loading. *Clin Chem* 1993;39(7):1390-7.

[134] Mansoor MA, Guttormsen AB, Fiskerstrand T, Refsum H, Ueland PM, Svardal AM. Redox status and protein binding of plasma aminothiols during the transient hyperhomocysteinemia that follows homocysteine administration. *Clin Chem* 1993;39(6):980-5.

[135] Andersson A, Lindgren A, Hultberg B. Effect of thiol oxidation and thiol export from erythrocytes on determination of redox status of homocysteine and other thiols in plasma from healthy subjects and patients with cerebral infarction. *Clin Chem* 1995;41(3):361-6.

[136] Mansoor MA, Ueland PM, Aarsland A, Svardal AM. Redox status and protein binding of plasma homocysteine and other aminothiols in patients with homocystinuria. *Metabolism* 1993;42(11):1481-5.

[137] Mansoor MA, Ueland PM, Svardal AM. Redox status and protein binding of plasma homocysteine and other aminothiols in patients with hyperhomocysteinemia due to cobalamin deficiency. *Am J Clin Nutr* 1994;59(3):631-5.

[138] Mansoor MA, Bergmark C, Svardal AM, Lonning PE, Ueland PM. Redox status and protein binding of plasma homocysteine and other aminothiols in patients with early-onset peripheral vascular disease. Homocysteine and peripheral vascular disease. *Arterioscler Thromb Vasc Biol* 1995;15(2):232-40.

[139] Sengupta S, Wehbe C, Majors AK, Ketterer ME, DiBello PM, Jacobsen DW. Relative roles of albumin and ceruloplasmin in the formation of homocystine, homocysteine-cysteine-mixed disulfide, and cystine in circulation. *J. Biol. Chem.* 2001;276(50):46896-904.

[140] Weiss N, Heydrick SJ, Postea O, Keller C, Keaney JFJ, Loscalzo J. Influence of hyperhomocysteinemia on the cellular redox state - Impact on homocysteine-induced endothelial dysfunction. *Clin Chem Lab Med* 2003;41(11):1455-61..

[141] Radi R, Beckman JS, Bush KM, Freeman BA. Peroxynitrite-induced membrane lipid peroxidation: the cytotoxic potential of superoxide and nitric oxide. *Arch Biochem Biophys* 1991;288(2):481-7.

[142] Possel H, Noack H, Augustin W, Keilhoff G, Wolf G. 2,7-Dihydrodichlorofluorescein diacetate as a fluorescent marker for peroxynitrite formation. *FEBS Lett* 1997;416(2):175-8.

[143] Rubbo H, Freeman BA. Nitric oxide regulation of lipid oxidation reactions: formation and analysis of nitrogen-containing oxidized lipid derivatives. *Methods Enzymol* 1996;269:385-94.

[144] Vasquez-Vivar J, Kalyanaraman B, Martasek P. The role of tetrahydrobiopterin in superoxide generation from eNOS: enzymology and physiological implications. *Free Radic Res* 2003;37(2):121-7.

[145] Outinen PA, Sood SK, Pfeifer SI, Pamidi S, Podor TJ, Li J, et al. Homocysteine-induced endoplasmic reticulum stress and growth arrest leads to specific changes in gene expression in human vascular endothelial cells. *Blood* 1999;94(3):959-67.

[146] Weiss N, Zhang Y, Loscalzo J. Homocyst(e)ine impairs cellular glutathione peroxidase expression and promotes endothelial dysfunction in an animal model of hyperhomocyst(e)inemia. *Circulation* 2000;102:II-238 (Abstr).

[147] Huang R-FS, Hsu Y-C, Lin H-L, Yang FL. Folate depletion and elevated plasma homocysteine promote oxidative stress in rat livers. *J Nutr* 2001;131:33-8.

[148] Flohé L. The selenoprotein glutathione peroxidase. In: Dolphin D, Poulson R, Avramovic O, editors. *Glutathione: chemical, biochemical and medical aspects:* John Wiley & Sons, Inc.; 1989. p. 644-731.

[149] Sies H, Sharov VS, Klotz LO, Briviba K. Glutathione peroxidase protects against peroxynitrite-mediated oxidations. A new function for selenoproteins as peroxynitrite reductase. *J Biol Chem* 1997;272(44):27812-7.

[150] O'Donnell VB, Freeman BA. Interactions between nitric oxide and lipid oxidation pathways: implications for vascular disease. *Circ Res* 2001;88(1):12-21.

[151] Rubbo H, Radi R, Trujillo M, Telleri R, Kalyanaraman B, Barnes S, et al. Nitric oxide regulation of superoxide and peroxynitrite-dependent lipid peroxidation. Formation of

novel nitrogen-containing oxidized lipid derivatives. *J Biol Chem* 1994; 269(42):26066-75.

[152] Forgione MA, Weiss N, Heydrick S, Cap A, Klings ES, Bierl C, et al. Cellular glutathione peroxidase deficiency and endothelial dysfunction. *Am J Physiol Heart Circ Physiol* 2002;282(4):H1255-61.

[153] Karlsson K, Marklund SL. Heparin-induced release of extracellular superoxide dismutase to human blood plasma. *Biochem J* 1987;242(1):55-9.

[154] Adachi T, Yamada H, Futenma A, Kato K, Hirano K. Heparin-induced release of extracellular-superoxide dismutase form (V) to plasma. *J Biochem (Tokyo)* 1995;117(3):586-90.

[155] Wilcken DE, Wang XL, Adachi T, Hara H, Duarte N, Green K, et al. Relationship between homocysteine and superoxide dismutase in homocystinuria: possible relevance to cardiovascular risk. *Arterioscler Thromb Vasc Biol* 2000;20(5):1199-202.

[156] Wang XL, Duarte N, Cai H, Adachi T, Sim AS, Cranney G, et al. Relationship between total plasma homocysteine, polymorphisms of homocysteine metabolism related enzymes, risk factors and coronary artery disease in the Australian hospital-based population. *Atherosclerosis* 1999;146(1):133-40.

[157] Yamamoto M, Hara H, Adachi T. Effects of homocysteine on the binding of extracellular-superoxide dismutase to the endothelial cell surface. *FEBS Lett* 2000;486(2):159-62.

[158] Anderson TJ, Gerhard MD, Meredith IT, Charbonneau F, Delagrange D, Creager MA, et al. Systemic nature of endothelial dysfunction in atherosclerosis. *Am J Cardiol* 1995;75(6):71B-74B.

[159] Drexler H. Nitric oxide and coronary endothelial dysfunction in humans. *Cardiovasc Res* 1999;43(3):572-9.

[160] Ignarro LJ, Cirino G, Casini A, Napoli C. Nitric oxide as a signaling molecule in the vascular system: an overview. *J Cardiovasc Pharmacol* 1999;34(6):879-86.

[161] Loscalzo J, Vita J. *Nitric oxide and the cardiovascular system.* Totawa, NJ: Humana Press; 2000.

[162] Govers R, Rabelink TJ. Cellular regulation of endothelial nitric oxide synthase. *Am J Physiol Renal Physiol* 2001;280(2):F193-206.

[163] McDermott JR. Studies on the catabolism of Ng-methylarginine, Ng, Ng-dimethylarginine and Ng, Ng-dimethylarginine in the rabbit. *Biochem J* 1976;154(1):179-84.

[164] Vallance P, Leone A, Calver A, Collier J, Moncada S. Endogenous dimethylarginine as an inhibitor of nitric oxide synthesis. *J Cardiovasc Pharmacol* 1992;20(Suppl 12):S60-2.

[165] Boger RH, Bode-Boger SM, Sydow K, Heistad DD, Lentz SR. Plasma concentration of asymmetric dimethylarginine, an endogenous inhibitor of nitric oxide synthase, is elevated in monkeys with hyperhomocyst(e)inemia or hypercholesterolemia. *Arterioscler Thromb Vasc Biol* 2000;20(6):1557-64.

[166] Stuhlinger MC, Tsao PS, Her JH, Kimoto M, Balint RF, Cooke JP. Homocysteine impairs the nitric oxide synthase pathway: role of asymmetric dimethylarginine. *Circulation* 2001;104(21):2569-75.

[167] Ito A, Tsao PS, Adimoolam S, Kimoto M, Ogawa T, Cooke JP. Novel mechanism for endothelial dysfunction: dysregulation of dimethylarginine dimethylaminohydrolase. *Circulation* 1999;99(24):3092-5.

[168] Zhang F, Slungaard A, Vercellotti GM, Iadecola C. Superoxide-dependent cerebrovascular effects of homocysteine. *Am J Physiol* 1998;274(6 Pt 2):R1704-11.

[169] Ghigo D, Alessio P, Foco A, Bussolino F, Costamagna C, Heller R, et al. Nitric oxide synthesis is impaired in glutathione-depleted human umbilical vein endothelial cells. *Am J Physiol* 1993;265(3 Pt 1):C728-32.

[170] Nappo F, De Rosa N, Marfella R, De Lucia D, Ingrosso D, Perna AF, et al. Impairment of endothelial functions by acute hyperhomocysteinemia and reversal by antioxidant vitamins. *JAMA* 1999;281(22):2113-8.

[171] Heinecke JW, Rosen H, Suzuki LA, Chait A. The role of sulfur-containing amino acids in superoxide production and modification of low density lipoprotein by arterial smooth muscle cells. *J Biol Chem* 1987;262(21):10098-103.

[172] Hirano K, Ogihara T, Miki M, Yasuda H, Tamai H, Kawamura N, et al. Homocysteine induces iron-catalyzed lipid peroxidation of low-density lipoprotein that is prevented by alpha-tocopherol. *Free Radic Res* 1994;21(5):267-76.

[173] Wood JL, Graham A. Structural requirements for oxidation of low-density lipoprotein by thiols. *FEBS Lett* 1995;366(1):75-80.

[174] Sparrow CP, Olszewski J. Cellular oxidation of low density lipoprotein is caused by thiol production in media containing transition metal ions. *J Lipid Res* 1993;34(7):1219-28.

[175] Toborek M, Kopieczna-Grzebieniak E, Drozdz M, Wieczorek M. Increased lipid peroxidation as a mechanism of methionine-induced atherosclerosis in rabbits. *Atherosclerosis* 1995;115(2):217-24.

[176] Matthias D, Becker CH, Riezler R, Kindling PH. Homocysteine induced arteriosclerosis-like alterations of the aorta in normotensive and hypertensive rats following application of high doses of methionine. *Atherosclerosis* 1996;122(2):201-16.

[177] Voutilainen S, Morrow JD, Roberts LJ, 2nd, Alfthan G, Alho H, Nyyssonen K, et al. Enhanced in-vivo lipid peroxidation at elevated plasma total homocysteine levels. *Arterioscler Thromb Vasc Biol* 1999;19(5):1263-6.

[178] Ventura P, Panini R, Verlato C, Scarpetta G, Salvioli G. Peroxidation indices and total antioxidant capacity in plasma during hyperhomocysteinemia induced by methionine oral loading. *Metabolism* 2000;49(2):225-8.

[179] Selhub J, Jacques PF, Wilson PWF, Rush D, Rosenberg IH. Vitamin status and intake as primary determinants of homocysteinemia in an elderly population. *JAMA* 1993;270(22):2693-8.

[180] Ubbink JB. The role of vitamins in the pathogenesis and treatment of hyperhomocyst(e)inaemia. *J Inherit Metab Dis* 1997;20(2):316-25.

[181] Clarke R, Armitage J. Vitamin supplements and cardiovascular risk: review of the randomized trials of homocysteine-lowering vitamin supplements. *Semin Thromb Hemost* 2000;26(3):341-48.

[182] Riddell LJ, Chisholm A, Williams S, Mann JI. Dietary strategies for lowering homocysteine concentrations. *Am J Clin Nutr* 2000;71(6):1448-54.

[183] Lindenbaum J, Healton EB, Savage DG, Brust JC, Garrett TJ, Podell ER, et al. Neuropsychiatric disorders caused by cobalamin deficiency in the absence of anemia or macrocytosis. *N Engl J Med* 1988;318(26):1720-8.

[184] Stabler SP, Allen RH, Savage DG, Lindenbaum J. Clinical spectrum and diagnosis of cobalamin deficiency. *Blood* 1990;76(5):871-81.

[185] Savage DG, Lindenbaum J. Folate-cobalamin interactions. In: Bailey LB, editor. *Folate in Health and Disease*. New York: Marcel Dekker Inc; 1995. p. 237-85.

[186] Boneu B, Abbal M, Plante J, Bierme R. Letter: Factor-VIII complex and endothelial damage. *Lancet* 1975;1(7922):1430.

[187] Seigneur M, Dufourcq P, Conri C, Constans J, Mercie P, Pruvost A, et al. Levels of plasma thrombomodulin are increased in atheromatous arterial disease. *Thromb Res* 1993;71(6):423-31.

[188] Van den Berg M, Boers GH, Franken DG, Blom HJ, Van Kamp GJ, Jakobs C, et al. Hyperhomocysteinaemia and endothelial dysfunction in young patients with peripheral arterial occlusive disease. *Eur J Clin Invest* 1995;25(3):176-81.

[189] Peterson JC, Spence JD. Vitamins and progression of atherosclerosis in hyper-homocyst(e)inaemia. *Lancet* 1998;351(9098):263.

[190] Vermeulen EG, Stehouwer CD, Twisk JW, van den Berg M, de Jong SC, Mackaay AJ, et al. Effect of homocysteine-lowering treatment with folic acid plus vitamin B6 on progression of subclinical atherosclerosis: a randomised, placebo- controlled trial. *Lancet* 2000;355(9203):517-22.

[191] Bellamy MF, McDowell IF, Ramsey MW, Brownlee M, Newcombe RG, Lewis MJ. Oral folate enhances endothelial function in hyperhomocysteinaemic subjects. *Eur J Clin Invest* 1999;29(8):659-62.

[192] Woo KS, Chook P, Lolin YI, Sanderson JE, Metreweli C, Celermajer DS. Folic acid improves arterial endothelial function in adults with hyperhomocystinemia. *J Am Coll Cardiol* 1999;34(7):2002-6.

[193] Chao CL, Chien KL, Lee YT. Effect of short-term vitamin (folic acid, vitamins B6 and B12) administration on endothelial dysfunction induced by post-methionine load hyperhomocysteinemia. *Am J Cardiol* 1999;84(11):1359-61.

[194] Title LM, Cummings PM, Giddens K, Genest JJ, Jr., Nassar BA. Effect of folic acid and antioxidant vitamins on endothelial dysfunction in patients with coronary artery disease. *J Am Coll Cardiol* 2000;36(3):758-65.

[195] Chambers JC, Ueland PM, Obeid OA, Wrigley J, Refsum H, Kooner JS. Improved vascular endothelial function after oral B vitamins: An effect mediated through reduced concentrations of free plasma homocysteine. *Circulation* 2000;102(20):2479-83.

[196] Doshi SN, McDowell IF, Moat SJ, Lang D, Newcombe RG, Kredan MB, et al. Folate improves endothelial function in coronary artery disease: an effect mediated by reduction of intracellular superoxide? *Arterioscler Thromb Vasc Biol* 2001;21(7):1196-202.

[197] Doshi SN, McDowell IF, Moat SJ, Payne N, Durrant HJ, Lewis MJ, et al. Folic acid improves endothelial function in coronary artery disease via mechanisms largely independent of homocysteine lowering. *Circulation* 2002;105(1):22-6.
[198] Thambyrajah J, Landray MJ, McGlynn FJ, Jones HJ, Wheeler DC, Townend JN. Does folic acid decrease plasma homocysteine and improve endothelial function in patients with predialysis renal failure? *Circulation* 2000;102(8):871-5.
[199] Pullin CH, Ashfield-Watt PA, Burr ML, Clark ZE, Lewis MJ, Moat SJ, et al. Optimization of dietary folate or low-dose folic acid supplements lower homocysteine but do not enhance endothelial function in healthy adults, irrespective of the methylenetetrahydrofolate reductase (C677T) genotype. *J Am Coll Cardiol* 2001;38(7):1799-805.
[200] van Dijk RA, Rauwerda JA, Steyn M, Twisk JW, Stehouwer CD. Long-term homocysteine-lowering treatment with folic acid plus pyridoxine is associated with decreased blood pressure but not with improved brachial artery endothelium-dependent vasodilation or carotid artery stiffness: a 2-year, randomized, placebo-controlled trial. *Arterioscler Thromb Vasc Biol* 2001;21(12):2072-9.
[201] Thambyrajah J, Landray MJ, Jones HJ, McGlynn FJ, Wheeler DC, Townend JN. A randomized double-blind placebo-controlled trial of the effect of homocysteine-lowering therapy with folic acid on endothelial function in patients with coronary artery disease. *J Am Coll Cardiol* 2001;37(7):1858-63.
[202] Gori T, Burstein JM, Ahmed S, Miner SE, Al-Hesayen A, Kelly S, et al. Folic acid prevents nitroglycerin-induced nitric oxide synthase dysfunction and nitrate tolerance: a human in-vivo study. *Circulation* 2001;104(10):1119-23.
[203] Loscalzo J. Folate and nitrate-induced endothelial dysfunction: a simple treatment for a complex pathobiology. *Circulation* 2001;104(10):1086-8.
[204] Schnyder G, Roffi M, Pin R, Flammer Y, Lange H, Eberli FR, et al. Decreased rate of coronary restenosis after lowering of plasma homocysteine levels. *N Engl J Med* 2001;345(22):1593-600.
[205] Schnyder G, Roffi M, Flammer Y, Pin R, Eberli FR, Meier B, et al. Effect of homocysteine-lowering therapy on restenosis after percutaneous coronary intervention for narrowings in small coronary arteries. *Am J Cardiol* 2003;91(10):1265-9.
[206] Liem A, Reynierse-Buitenwerf GH, Zwinderman AH, Jukema JW, van Veldhuisen DJ. Secondary prevention with folic acid: effects on clinical outcomes. *J Am Coll Cardiol* 2003;41(12):2105-13.
[207] van Guldener C, Stehouwer CD. Homocysteine-lowering treatment: an overview. *Expert Opin Pharmacother* 2001;2(9):1449-60.
[208] Clarke R, Armitage J. Antioxidant vitamins and risk of cardiovascular disease. Review of large-scale randomised trials. *Cardiovasc Drugs Ther* 2002;16(5):411-5.
[209] Boaz M, Smetana S, Weinstein T, Matas Z, Gafter U, Iaina A, et al. Secondary prevention with antioxidants of cardiovascular disease in endstage renal disease (SPACE): randomised placebo-controlled trial. *Lancet* 2000;356(9237):1213-8.
[210] Tepel M, van der Giet M, Statz M, Jankowski J, Zidek W. The antioxidant acetylcysteine reduces cardiovascular events in patients with end-stage renal failure: a randomized, controlled trial. *Circulation* 2003;107(7):992-5.

Index

#

β

A

B

C

D

E

F

G

H

I

L

M

N

O

P

R

S

T

V

W